The Cyclic Peptide Toxins of *Amanita* and Other Poisonous Mushrooms

Amanita phalloides. (Drawing: Tamara Clark; tamaraclark.com)

Jonathan Walton

The Cyclic Peptide Toxins of *Amanita* and Other Poisonous Mushrooms

Springer

Jonathan Walton
United States Department of Energy Plant Research Lab
and Department of Plant Biology
Michigan State University
East Lansing, MI, USA

ISBN 978-3-030-08299-4 ISBN 978-3-319-76822-9 (eBook)
https://doi.org/10.1007/978-3-319-76822-9

Softcover re-print of the Hardcover 1st edition 2018

Printed on acid-free paper

This Springer imprint is published by the registered company Springer International Publishing AG part of Springer Nature.
The registered company address is: Gewerbestrasse 11, 6330 Cham, Switzerland

Preface

Poisonous mushrooms have fascinated and horrified people for thousands of years. Mushrooms in the large and charismatic genus *Amanita* in particular have featured in countless historical anecdotes involving emperors, kings, and popes, and occur throughout popular culture from children's stories to murder mysteries. Scientists have not been immune to the fascination of poisonous mushrooms, as attested by a professional literature of over 10,000 papers covering every aspect of the responsible chemical principles. Three Nobel laureates have worked on the chemistry and biochemistry of the most deadly of the *Amanita* toxins – Heinrich Wieland (Nobel Prize 1927), his son-in-law Feodor Lynen (1964), and Roger Kornberg (2006). Lynen's 1937 PhD dissertation on the isolation of phalloidin ("Über die Giftstoffe des Knollenblätterpilzes") is available in many libraries. The peptide toxins of poisonous mushrooms continue to provide fertile ground for discoveries in toxicology, synthetic organic chemistry, enzymology, and fungal genomics.

It is just one of the many remarkable traits of the *Amanita* cyclic peptides that the major toxins, amatoxins and phallotoxins, should have completely different cellular targets even though they are made by the same organisms through the same biosynthetic pathway and are chemically closely related. The amatoxins are potent and specific inhibitors of one of the most important enzymes in all eukaryotes, RNA polymerase II, whereas the phallotoxins bind with high affinity to actin, which comprises the essential cytoskeleton of all eukaryotes and is central to the action of muscles. Recently, it has been found that poisonous *Amanita* mushrooms make dozens of additional and previously unknown cyclic peptides. Perhaps some of them will also have equally dramatic and diverse biological activities.

This book owes a large debt to Theodore Wieland's monograph on the same topic, *Peptides of Poisonous Amanita Mushrooms* (Springer 1986). Nonetheless, the current book is substantially different in content, primarily due to the immense progress that has been made on our understanding of all aspects of poisonous mushrooms in the past 30 years. Most of our new knowledge has been made possible by extraordinary technical progress in the fields of chemistry and biology. The refinement and subsequent widespread adoption of HPLC, NMR spectroscopy, mass spectrometry (MS), and combinations thereof, especially LC/MS, have made toxin identification and quantitation easier, more sensitive, and more accurate. Advances in protein structure through X-ray crystallography have enabled a detailed understanding of the interaction of α-amanitin with its target, RNA polymerase II, and of the enzymes involved in toxin biosynthesis. The exponential growth of the field of molecular genetics has impacted *Amanita* toxin research in numerous ways, including discovery of the toxin biosynthetic genes and new insights into the evolutionary relationships of poisonous mushrooms and other fungi. All this new knowledge has enabled new biotechnological and medical applications unforeseen by Professor Wieland.

In addition to the desirability of updating its predecessor, a major impetus for writing this book was the author's persistent struggle to become, and stay, educated about the multifarious/nefarious aspects of the *Amanita* toxins. Among both laypeople and scientists, the same questions tend to arise over and over: Why do the mushrooms make them? How are the toxins taken up by cells and organs? Which chemical functional groups are critical for activity? Why are they so toxic? Do edible mushrooms make the toxins? How are they biosynthesized? How did toxin biosynthesis evolve multiple times in unrelated mushroom lineages? Answering these questions to my own satisfaction required an immersion in a large and often contradictory literature, much of it dating back half a century or more. Every time I had educated myself on one of these questions, it seemed logical to write down the relevant citations and, when possible, the answers, before they were forgotten. Thus, the essence of this book took shape.

In this day and age, no one can expect to be expert in all the fields touched upon by the *Amanita* cyclic peptides, and the author can claim to be an expert in only one or two relevant subdisciplines. It is impossible to cover so much scientific territory – chemistry, mycology, biochemistry, ecology, evolution, etc. – without making numerous errors and risking offense to honored colleagues with misrepresentations (or, worse, neglect) of their inestimable contributions. In this interactive age, it seems reasonable to extend an invitation to readers to communicate their comments and corrections (email: walton@msu.edu), toward the eventual establishment of a discussion forum that could benefit the entire *Amanita* toxin community. In the meantime, please accept the author's apologies for the inevitable inaccuracies and oversights.

East Lansing, MI, USA Jonathan Walton

Acknowledgments

Since I knew little about poisonous mushrooms until relatively late in my career as a plant physiologist/fungal molecular biologist, I owe a huge debt to the many people who generously shared their knowledge and passion. In particular, I am grateful to the members of my laboratory who took our understanding of the *Amanita* cyclic peptides into uncharted territory, especially Heather Hallen-Adams (currently in the Food Science and Technology Department, University of Nebraska-Lincoln), Hong Luo (Kunming Institute of Botany, Chinese Academy of Sciences), John Scott-Craig, Richard Ransom (Funite, LLC, and University of Michigan), and R. Michael Sgambelluri.

I would like to thank Rod Tulloss, who runs the world's best website on all things *Amanita* (www.amanitaceae.org), and Britt Bunyard, the publisher and Editor-in-Chief of *Fungi* magazine (www.fungimag.com). Both not only supplied specimens but also condescended to edify this neophyte about the nuances of *Amanita* taxonomy; together we someday hope to resolve "cryptic speciation" in *A. bisporigera*, in my mind a question of major importance for understanding the evolutionary origins and depth of cyclic peptide diversity. I thank Todd Mitchell, MD, for his pioneering efforts to bring the amanitin antidote, silybin (Legalon® SIL), to the United States, and for informative discussions on the medical aspects of peptide toxin poisonings at Gayle's Bakery in Capitola. My Chinese colleagues Zhuliang Yang (Kunming Institute of Botany, Chinese Academy of Sciences), Zuohong Chen (Hunan Normal University), and Tai-hui Li (Guangdong Institute of Microbiology, Guangzhou) have not only made many outstanding contributions to *Amanita* biology and toxicology but also generously shared their knowledge and hospitality.

Debbie Viess and David Rust (Bay Area Mycological Society), organizers of the annual Pt. Reyes Fungus Fair, shared their leftover specimens with me. Catherine (Cat) Adams (University of California, Berkeley) took me on an excursion to collect *A. phalloides* in Marin County. Else Vellinga and Tom Bruns (UC Berkeley), Mary Berbee and Brandon Landry (University of British Columbia), Britney Ramsey (Oregon), Joshua Birkebak (Assured Bio, Oakridge, TN), Sara Epis and Davide Sassera (Universitá degli Studi di Milano), Randy Garrett (Pacific Collegiate School, Santa Cruz), Anne Pringle (University of Wisconsin, Madison), and Adolf and Oluna Ceska (Victoria, British Columbia) shared their observations and provided specimens of *Lepiota* and *Galerina.* Heinz Faulstich (University of Heidelberg), Todd Mitchell (Dominican Hospital, Santa Cruz), and Heather Hallen-Adams read parts of the manuscript and made many valuable suggestions. Needless to say, all the remaining errors and omissions are my responsibility alone.

The U.S. Department of Energy, Office of Basic Energy Sciences, and the U.S. National Institutes of Health funded my mushroom work for many years, even when they didn't know they were doing so (I think they will get their money's worth). Kenneth Thimann, UC Santa Cruz, was an early inspiration to study secondary metabolites in plants (back in the days when fungi were still considered plants). The DOE Plant Research Lab and Michigan State University have provided an outstanding professional home for more than 30 years and supported my lab in many ways material and cultural. I would particularly like to express my gratitude to Hans Kende, a former director of the PRL, for his professional and personal support.

Through my good friend Elaine Taylor Jungleib (Portola Valley, CA) I made some important contacts with mushroom enthusiasts (and loathers) in the San Francisco Bay Area, including Diana Gerba (www.mushrooms911.blogspot.com). Giovanni Consiglio, Anne Pringle, Rod Tulloss, Hong Luo, Lucy Albertella, Oluna and Adolf Ceska, Dave Hood, Christopher Jeffree, Britney Wharton, Tamara Clark, George Barron, Dan Molter, T. Nakamori, Kathie Hodge, and Debbie Veiss generously allowed me to reprint their beautiful photographs. Many other authors graciously gave permission to reprint their journal figures and data.

And to the amateur mycologists of the world, selflessly and enthusiastically contributing to our knowledge of the Kingdom Mycota: All hail!

Contents

About the Author

Jonathan Walton was raised in upstate New York and the Bay Area of California. He attended high school in Ojai, California, and graduated from the University of California, Santa Cruz, in 1975. He received an MS degree in plant pathology from Cornell University (1978) and a PhD in biological sciences from Stanford University (1982). He was a NATO postdoctoral fellow at the University of Rome "La Sapienza" and worked at the ARCO Plant Cell Research Institute in Dublin, California. Since 1987, he has been a professor of plant biology at Michigan State University in the Department of Energy Plant Research Laboratory (DOE-PRL). In 1993, he spent a sabbatical year at the University of California, Berkeley, in the laboratory of John Casida. He has published more than 130 papers on the role of fungal toxins in plant disease, the mechanisms of plant disease resistance, biosynthesis of cell walls, enzymes for biomass conversion, and fungal cyclic peptides.

Chapter 1
Introduction

> *Birds, flowers, insects, stones delight the observant. Why not toadstools? A tramp after them is absorbing, study of them interesting, and eating of them health-giving and supremely satisfying.*
>
> *Charles McIlvaine (1840–1909)*

The concept of a "mushroom toxin" embraces a large chemical and taxonomic territory, since many mushrooms are at least mildly deleterious to some organism or another. This monograph focuses on the cyclic peptide toxins characteristic of, but not exclusive to, species of *Amanita* in section *Phalloideae*, whose fruiting bodies (mushrooms) are known by the common names "death cap" and "destroying angel." One group of the cyclic peptide toxins, the amatoxins, exemplified by α-amanitin, account for almost all fatal mushroom poisonings. Although primarily associated with the genus *Amanita* from which they were first isolated, the amatoxins are also found in the unrelated mushroom genera *Galerina*, *Lepiota*, and perhaps *Conocybe*.

Throughout this book, the bicyclic octapeptides related to α-amanitin are referred to as *amatoxins* and the bicyclic heptapeptides related to phalloidin as the *phallotoxins*. Whenever "cyclic peptide toxin" or just "toxin" is used, the reader can assume that the amatoxins and phallotoxins are being referred to and not the nonpeptidic toxins of *Amanita* such as ibotenic acid, bufotenine, and others yet to be characterized.

On the basis of early chemical work by Theodor Wieland and colleagues, and more recently from genome sequencing of amatoxin-producing species of *Amanita* and *Lepiota*, we know that these fungi can biosynthesize many, perhaps hundreds, of additional cyclic peptides. A few of these were structurally characterized in the 1960s and 1970s and named the cycloamanides (Wieland 1986). Recent studies indicate that the amatoxins, phallotoxins, and cycloamanides are all biosynthesized by the same ribosomal pathway. For nomenclatural simplicity and consistency, in this book the term *cycloamanide* is applied to all of the cyclic peptides with a common biogenesis that are made by species of *Amanita*, *Galerina*, and *Lepiota*.

J. Walton, *The Cyclic Peptide Toxins of Amanita and Other Poisonous Mushrooms*, https://doi.org/10.1007/978-3-319-76822-9_1

1.1 The Kingdom Fungi: Overview of Fungi and the Fungal Lifestyle

Like plants and animals, fungi are eukaryotic, i.e., they have nuclei and organelles such as mitochondria (Fig. 1.1). However, the fungi are now recognized as an evolutionarily distinct group with their own kingdom (sometimes also called the Mycota) distinct from the plants (kingdom Planta) and the animals (kingdom Animalia) (Bauldauf et al. 2000; Woese et al. 1990) (Fig. 1.2). The true fungi are also distinct from the oomycetes and "water molds" (superphylum Heterokonta or Stramenopiles), which are fungus-like organisms traditionally studied by mycologists but actually more closely related to the brown algae (Phaeophyceae) (Fig. 1.2). Like plants, fungi have a rigid polysaccharide cell wall and, like animals, are heterotrophic, depending on other organisms for their food. Most fungi obtain their nutrients from the decomposition of dead organisms, but some are pathogens (of plants, animals, and each other) and others form symbiotic (i.e., mutually advantageous) associations with plants. Common terms for different types of fungi include *yeast*, *mold*, and *mushroom*.

For further information on fungi and mycologists, see the suggested reading list at the end of the chapter.

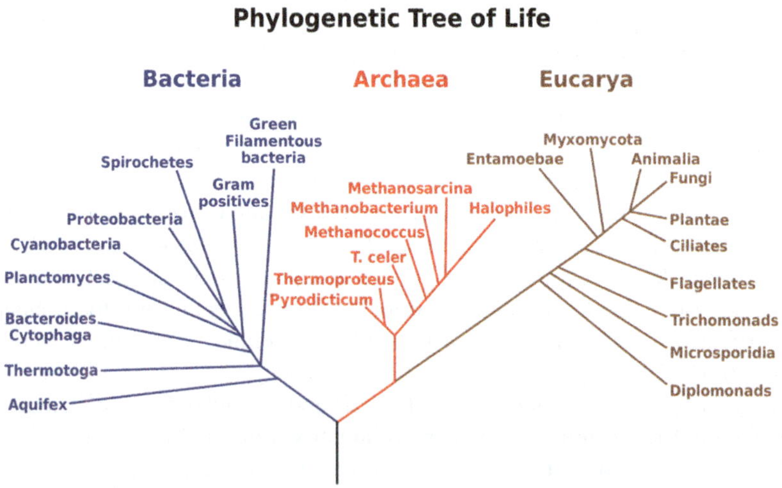

Fig. 1.1 The classic "tree of life" based on the small subunit ribosomal RNA, showing the evolutionary relationship of fungi to other living things. Redrawn from Woese et al. 1990 https://en.wikipedia.org/wiki/Tree_of_life_(biology). In light of our current appreciation of the importance of horizontal gene transfer among prokaryotes, newer trees for the Bacteria and Archaea are more complex and less dichotomous

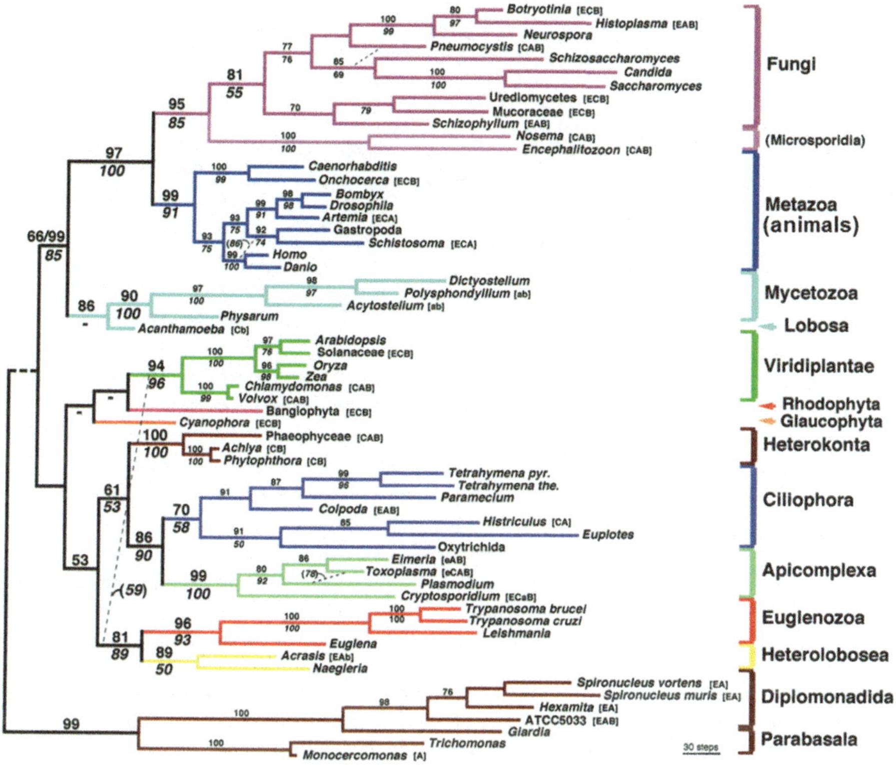

Fig. 1.2 A kingdom-level phylogeny of eukaryotes based on combined protein sequences. (From Bauldauf et al. (2000). Reprinted with permission from AAAS)

1.1.1 Classification of the Fungi

Like other members of the tree of life, the kingdom Mycota is taxonomically subclassified into phyla, classes, families, genera, and species. The two major phyla of higher fungi are the Ascomycota (ascomycetes) and the Basidiomycota (basidiomycetes), which together form the subkingdom Dikarya (Fig. 1.3). The ascomycetes include the common yeasts and molds and some fungi that can produce macroscopic (i.e., large) reproductive structures superficially resembling true mushrooms, such as morels and cup fungi. Ascomycetes produce many natural products (also known as secondary metabolites or specialized metabolites) with a wide range of chemistries, biological activities, and medicinal and agricultural uses. Some have been "domesticated" to grow efficiently to high densities for industrial-scale fermentations.

The phrase *filamentous fungus* is often used to refer to the ascomycetes in the subphylum Pezizomycotina (i.e., ascomycetes excluding unicellular yeasts) (Fig. 1.3), although the dominant growth habits of most basidiomycetes and many

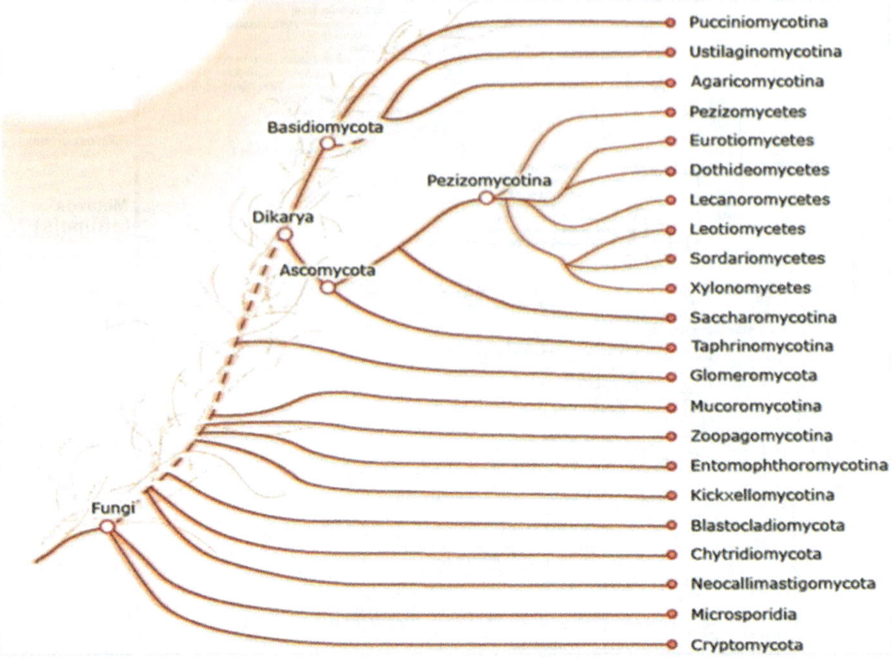

Fig. 1.3 The taxonomic tree of the fungi (Kingdom Fungi). (From: http://genome.jgi.doe.gov/programs/fungi/index.jsf. From Grigoriev et al. (2014), published by Oxford University Press. This work, written by US Government employees, is in the public domain)

other fungi are also filamentous, i.e., as threadlike *hyphae* (collectively known as *mycelium*) that elongate by tip growth and penetrate into solid substrates. Instead of filaments, some fungi grow as single-cell yeasts. The classic yeast, *Saccharomyces cerevisiae*, or baker's yeast, is an ascomycete. Some basidiomycetes also grow as single-cell yeasts, such as the human pathogen *Cryptococcus neoformans*. Furthermore, some fungi, e.g., *Candida albicans*, an ascomycete, and *Ustilago maydis*, a basidiomycete, can grow as either single-cell yeasts or as filamentous hyphae depending on environmental conditions.

The basidiomycetes include the rusts (Pucciniomycotina, smuts (Ustilaginomycotina), and Agaricomycetes (Fig. 1.3). Within the class Agaricomycetes, the order Agaricales contains the majority of organisms commonly known as mushrooms. *Agaric* is a common term for gilled mushrooms, which largely but not completely overlaps with the order Agaricales. The genera *Amanita*, *Galerina*, *Lepiota*, and *Conocybe*, which are the fungi of particular relevance to this book, all produce mushrooms with gills and belong to the order Agaricales within the class Agaricomycetes. However, at the next finer level of the phylogenetic tree, these four genera of fungi belong to different families. Details of the taxonomy of the toxin-producing fungi and the relationship between toxin production and taxonomy are discussed in Chaps. 3 and 6. For further information on

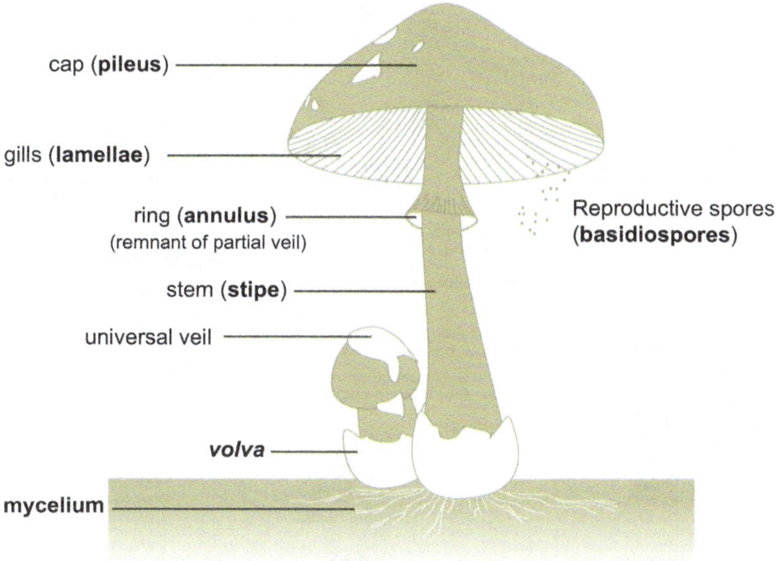

Fig. 1.4 Parts of a gilled mushroom such as *Amanita*, *Galerina*, or *Lepiota*. Scientific terms are in bold. (Figure courtesy of John Harris (john@mushroomdiary.co.uk) and used with permission)

fungal taxonomy, see http://genome.jgi-psf.org/programs/fungi/index.jsf, the NCBI taxonomy browser (http://www.ncbi.nlm.nih.gov/guide/taxonomy/), and the primary literature (Hibbett et al. 2007; Matheny et al. 2006).

Mushroom is the common term for the spore-producing structures of macrofungi in the order Agaricales, although it is also applied to some fungi in the ascomycetes, such as the morels and truffles (Fig. 1.4). Mushrooms are generally found above ground and are readily visible to the naked eye (~1 to 30 cm in height), but they emerge from the mass of mycelium that remains below ground or embedded in a plant structure such as a living tree or a rotten log. The mycelium of a mushroom-forming fungus is the permanent, perennial embodiment of the organism, whereas the mushroom itself is a transient, reproductive structure analogous to a fruit. Alternate names for mushroom in this sense are *basidiocarp*, *basidiome*, *fruiting body*, *carpophore*, and *sporocarp*. Common and scientific names for the parts of a typical mushroom are the cap (*pileus*), stem (*stipe*), gills (*lamellae*), and partial veil. Some mushrooms have a universal veil which originally covers the whole mushroom and remains as fragments on the cap and as a *volva* at the base (Fig. 1.4). The partial veil is derived from a transient membrane connecting the edge of the pileus to the stipe and often remains as a ring (*annulus*) on the stem. The presence/absence and shapes of the volva and annulus are distinguishing features of many of the mushrooms of relevance to this book.

Some mushrooms produce their spores on the surfaces of pores within the mushroom (e.g., the boletes) and some have gills. Both pores and gills serve to increase the magnitude of the spore-producing surface, which is technically known as the

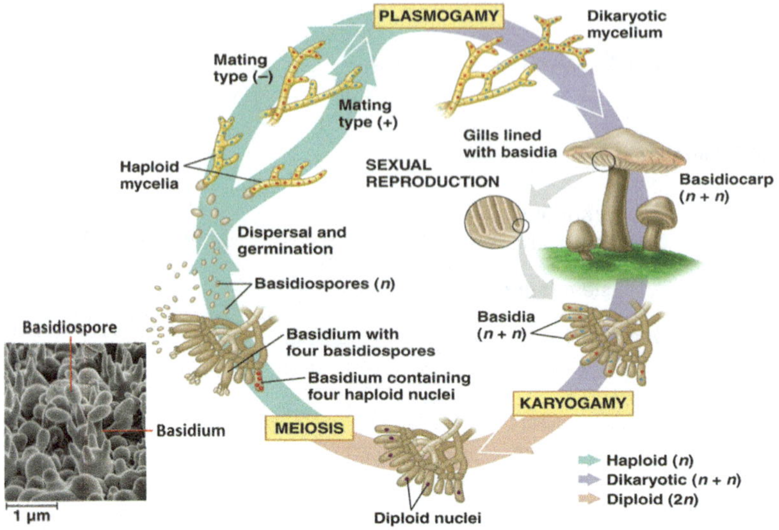

Fig. 1.5 Life cycle of a typical agaric (mushroom). *Plasmogamy* is cell fusion and *karyogamy* is nuclear fusion; see Fig. 1.6. *Inset*: scanning electron micrograph of basidia on the gills of a mushroom (*Polyporus squamosus*). Note that each basidium produces four stalks (sterigmata) and the basidiospores develop on the ends of these stalks. (Drawing taken from NA Campbell and JB Reece, Biology, 6th Ed., ©2002. Reprinted with permission of Pearson Education, Inc., New York, NY. Photo credit: Dr. Christopher Jeffree, University of Edinburgh, used by permission)

hymenium. The mushrooms with gills are commonly referred to as agarics, i.e., belonging to the order Agaricales of the class Agaricomycetes of the phylum Basidiomycota. All fungi that produce the amatoxin and phallotoxin cyclic peptides – the focus of this book – are gilled mushrooms.

1.1.2 Life Cycle of the Agarics (Mushrooms)

Agarics are typically *dikaroytic* for most of their lives, meaning that each cell contains two haploid nuclei, each with one copy of the genetic information (Fig. 1.5). (Mammals and many higher plants, in contrast, are *diploid*, containing two copies of the genetic information in a single nucleus). Ascomycetes, on the other hand, are usually *monokaryotic* for most of their lives, i.e., each cell contains one nucleus with one complement of the genetic information. Ascomycetes are thus functionally *haploid*. Although agarics are normally dikaryotic, it is possible to establish and maintain healthy monokaryotic strains in the laboratory, at least for those agarics that can be cultured.

In agarics, fusion of the two nuclei to form a transient true diploid state (a process called *karyogamy*) precedes meiosis during basidiospore formation within the fruiting body (Figs. 1.5 and 1.6). The net result of meiosis is typically the formation

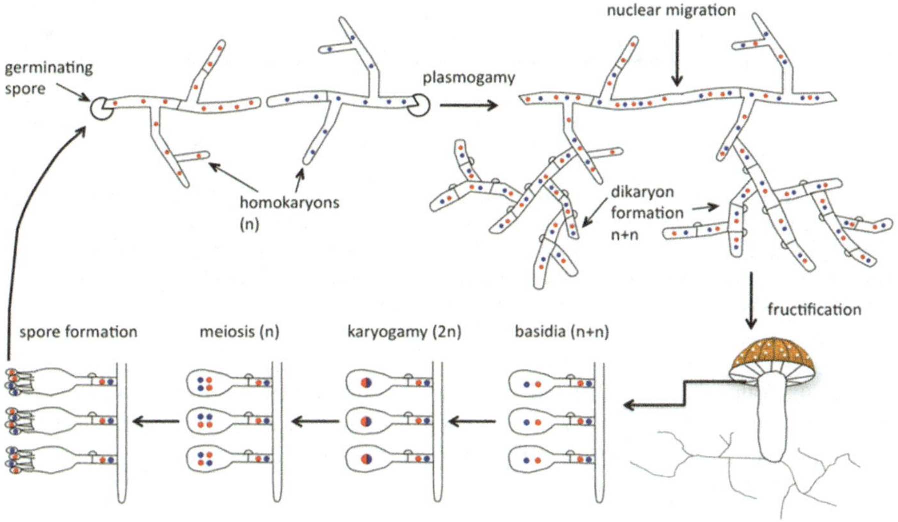

Fig. 1.6 Nuclear behavior during the life cycle of an agaric (mushroom-forming fungus). Haploid basidiospores, the products of meiosis, are produced on the gills of the mushroom. Each spore grows into a haploid mycelium (known as a *monokaryon* or *homokaryons*). Two haploid mycelia of compatible mating types fuse (i.e., undergo plasmogamy) to form a dikaryotic mycelium (*dikaryon*), the dominant growth form of mushroom-forming fungi. In the dikaryon, the nuclei remain distinct (n + n) and do not undergo karyogamy until basidiospore formation occurs in the gills of the next generation's mushroom. The true diploid phase (2n) is restricted to a single cell type, the basidium. (Figure courtesy of Tim James, University of Michigan. http://www.umich.edu/~mycology/research.html)

of four haploid basidiospores, although some fungi such as *A. bisporigera* produce only two basidiospores. Diagrams and photographs showing the tremendous morphological and chemical diversity of basidiocarp mycelia can be seen in Clémençon's classic monograph (Clémencon 2012). Confocal micrographs showing the internal structures of mushrooms including their reproductive hymenia can be seen in Luo et al. (2010).

Taxonomic classification of fungi and resolution of their evolutionary relationships has traditionally relied on morphological and chemical staining characteristics. The fungal fossil record is sparse. Modern fungal taxonomy relies heavily on molecular sequence data. The most widely used tool for establishing phylogenetic (evolutionary) relationships is the DNA sequence of the ITS (internal transcribed spacer) between the genes for the large and small subunits of ribosomal RNA (https://en.wikipedia.org/wiki/Internal_transcribed_spacer) (Fig. 1.7). This sequence, which is usually about 500–800 bp, can be obtained readily by PCR amplification using standardized primers (Gardes and Bruns 1993). The sequence of the ITS region can frequently, but not always, yield taxonomic resolution at the level of species. Taxonomic ambiguity can be resolved by sequencing additional genes including entire genomes (ascomycete and basidiomycete genomes range from 30 to 80 MB). When there is unresolved uncertainty about species boundaries

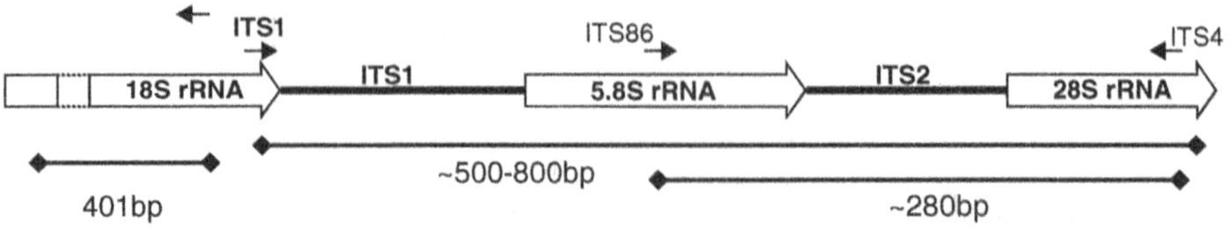

Fig. 1.7 Representation of the fungal ribosomal 18S rRNA gene and ITS (internal transcribed spacer) regions with representative primer (ITS1, ITS86, and ITS4) binding locations (From Embong et al. 2008 (Creative Commons License). See http://sites.biology.duke.edu/fungi/mycolab/primers.htm and https://nature.berkeley.edu/brunslab/tour/primers.html for primer sequences and additional primers useful for identification of fungi)

among a group of closely related fungi, the group can be referred to as a *species complex*. In light of the power of molecular tools for resolving phylogenetic relationships, fungal taxonomy is currently in a very active phase of expansion and revision (Hibbett and Taylor 2013).

1.1.3 The Genus Amanita

The name *Amanita* dates from antiquity. Its first scientifically accepted use within the Linnaean system of binomial classification is attributed to the South African mycologist Christian Persoon in 1797. *Amanita muscaria*, the fly agaric, is the type species. Mushrooms in this genus are some of the most attractive and easily recognized of all mushrooms (Figs. 1.8 and 1.9). Currently ~600 species of *Amanita* are recognized.

A comprehensive and continuously updated description of the species in the family *Amanitaceae* is available on the web site curated by Rod Tulloss (www.amanitaceae.org). Most but not all species in the *Amanitaceae* are *agaricoid* (having gills and a central stem). Two defining features of the family are gill tissue with *bilateral, divergent lamella trama* and stem tissue with vertically aligned inflated cells shaped like clubs, i.e., *longitudinally acrophysalidic* (www.amanitaceae.org). Two genera, *Amarrendia* and *Torrendia*, are sequestrate (i.e., the spores remain enclosed in the fruiting body at maturity). *Amarrendia* is now placed in the genus *Amanita*.

By the latest taxonomic classification, the family *Amanitaceae* contains three genera, *Amanita*, *Limacella*, and *Torrendia*. Genus *Amanita* is subdivided into two subgenera, *Amanita* and *Lepidella*. *Lepidella* spores are *amyloid*, i.e., they turn dark with iodine indicating the presence of starch, whereas the spores of subgenus *Amanita* are inamyloid.

The genus *Amanita* is further divided into seven sections (Bas 1969). The sections belonging to subgenus *Amanita* are *Amanita*, *Caesareae*, and *Vaginatae*, whereas the sections belonging to subgenus *Lepidella* are *Lepidella*, *Amidella*, *Phalloideae*, and *Validae*. It was recently proposed to split the genus *Amanita* into

Fig. 1.8 Examples of mushrooms (basidiocarps) in the genus *Amanita*. *Top row*: *Amanita muscaria* subsp. *flavivolvata*, *Amanita frostiana*, and *Amanita jacksonii*. *Bottom row*: an undescribed *Amanita* species, *Amanita manicata*, and *Amanita phalloides*. (From Wolfe et al. 2012. Reprinted with permission from John Wiley and Sons)

Fig. 1.9 Photogenic mushrooms, poisonous and edible, are popular subjects for postage stamps throughout the world

two (Redhead et al. 2016; Vizzini et al. 2012), but the issue is unresolved as of this writing (Tulloss et al. 2016).

Section *Phalloideae* of *Amanita* contains all of the peptide toxin-producing species and is thus the taxonomic group within the family *Amanitaceae* that is of most relevance to this book. Section *Phalloideae* contains ~60 species. Our understanding of the evolutionary relationships within *Amanita*, and between *Amanita* and other agarics, is still in a state of flux. The relationship between phylogenetics and toxin production is considered in more detail in Chaps. 3 and 6.

Section *Phalloideae* is characterized by a stipe (stem) with a bulbous volva, a persistent annulus (partial veil), and a pileus (cap) margin that is not appendiculate (having hanging remnants of the partial veil). According to Tulloss (www.amanitaceae.org/?Sections+of+Amanita), the oldest, basal taxa in sect. *Phalloideae*, which grow in Asia, do not make the cyclic peptide toxins and are, in fact, edible. This is an important point in regard to the origin of the cycloamanide gene family but needs solid confirmation before *any* mushroom in sect. *Phalloideae* should be considered safe to eat. For additional information on taxonomic relationships within *Amanita*, see Tulloss and Yang (2016).

Geographically, species in *Amanita* sect. *Phalloideae* occur on every continent except Antarctica. Section *Phalloideae* has been estimated to have diverged from the other sections of *Lepidella* ~64 million years ago (Cai et al. 2014, 2016). *A. phalloides* has spread on the roots of trees from its native range in Europe to North America, Africa, and Australia within the last 100 years, and its range is continuing to spread (Pringle et al. 2009; Vellinga et al. 2009; Wolfe et al. 2010) (Chap. 6). *A. marmorata* has also travelled intercontinentally, from Australia to Hawaii and Africa on the roots of *Eucalyptus* and *Casuarina* (Hallen et al. 2002; Miller et al. 1988).

A. phalloides is the type species of sect. *Phalloideae*. It is responsible for more deaths than any other mushroom, reflected in its common name "death cap" (frontispiece; Figs. 1.8–1.13 and Chap. 3). Its association with a high rate of fatality is due to a number of factors discussed in detail in Chap. 7.

Species of *Amanita*, especially *A. muscaria*, have been used as model organisms to study aspects of general mushroom biology, e.g., ethnobotanical uses, medicinal properties, mycorrhizal associations, horizontal gene transfer, organismal development, invasive species biogeography, ecology of symbiosis, and heavy metal remediation (Chap. 6). For a representative sample of papers on these topics, see Chaib De Mares et al. (2014), Dunk et al. (2012), Hrynkiewicz and Baum (2013), Michelot and Melendez-Howell (2003), Nagendran et al. (2009), Nehls et al. (1998), Schrey et al. (2007), Wasson (1971, 1979), Willmann et al. (2007) Wolfe et al. (2012), and Yoshida et al. (1996).

1.1.4 Other Toxin-Producing Fungi: Galerina *and* Lepiota

Some species of *Galerina* and *Lepiota* are well-documented to make amatoxins, the deadly poisonous peptides found in *Amanita* sect. *Phalloideae*. Similar to *Amanita*, only certain species in *Galerina* and *Lepiota* have the genetic capacity to biosynthesize amatoxins (Chap. 4). The taxonomic and evolutionary relationships between *Amanita*, *Lepiota*, and *Galerina* are discussed in Chap. 6. The evidence for amatoxins in other fungi is discussed in Chap. 3.

Whereas the genus *Amanita* occurs on six of the seven continents, *Galerina* has the distinction of also growing in Antarctica, where it was reported to be common in the vicinity of Palmer Station (Anvers Island) (http://antarcticsun.usap.gov/science/contenthandler.cfm?id=2723). The evidence that Antarctic *Galerina* mushrooms make α-amanitin is discussed in Chap. 3.

1.2 Ecology of Toxin-Producing Fungi

Fungi play key roles in terrestrial ecosystems as recyclers of organic material, releasing carbon, nitrogen, sulfur, phosphorus, and other essential elements back into the environment. Some of the toxin-producing fungi, such as *Galerina* and *Lepiota*, are saprobic, i.e., they obtain their energy from decomposition of dead plant material. In contrast, all toxin-producing species of *Amanita* are ectomycorrhizal, that is, they form obligatory symbiotic relationships on the roots of host plants. The ecology of *Amanita* and other cyclic peptide-producing fungi in relation to their toxicogenic potential is considered in more detail in Chap. 6, as are possible adaptive rationales for the evolution of toxicogenesis.

1.3 Poisonous Mushrooms in History and Culture

Among the historical personages that are associated with poisonous mushrooms are Greek playwright Euripides, Roman Emperor Claudius, Pope Clement VII, Holy Roman Emperor Charles VI, Daniel Fahrenheit (of thermometer fame), and Tsaritsa Natalia Naryshkina (second spouse of Tsar Alexei I and mother of Peter the Great). In the case of Claudius, thanks to Tacitus' "Annals of Imperial Rome" we even know the name of the witch (Locusta) who supplied the poisonous mushrooms and the name of the servant (eunuch Halotus) who served them on the orders of Claudius's wife (Agrippina). Poisonous mushrooms appear in several children's books (e.g., Tintin, Babar) as well as numerous murder mysteries. For authoritative and entertaining discussions of mushrooms and mycophagy in culture and history, the books by Benjamin (1995) and Money (2011) are highly recommended.

Despite extensive public service notices and other sources of information (Figs. 1.10, 1.11, 1.12, and 1.13), fatal poisonings from mushrooms still occur on a regular basis throughout the world. A geographical survey of poisonings from amatoxin-containing mushrooms in the past 5 years includes Canada, California, and New York in North America; China, India, and Japan in Asia; Turkey, Bulgaria, Spain, France, Austria, Italy, and Poland in Europe; Argentina in South America; Australia; and South Africa and Morocco in Africa. Many additional episodes are never publicized in either the popular or scientific literature, especially if the patients recover. Once ingested, the high concentrations of toxins in the mushrooms, their rapid absorption and recirculation in the enterohepatic system, and their high affinity for an essential enzyme come into play (Chaps. 5 and 7).

Fig. 1.10 Poisonous mushroom warning poster published by North American Mycological Association (NAMA). (Reprinted with permission of Martha Gottlieb, NAMA)

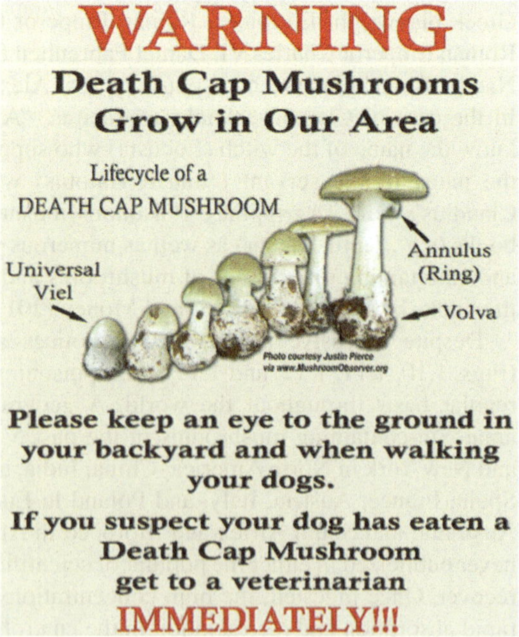

Fig. 1.11 Poster distributed by Diana Gerba in the San Francisco Bay Area to veterinarians and dog owners. After losing a dog to "death cap" poisoning, she maintains a blog to warn dog owners of the dangers of poisonous mushrooms (http://www.mushrooms911.blogspot.com/). (Used by permission. Photo credit: Archenzo, Italy (Creative Commons Attribution License))

Fig. 1.12 Information card distributed by Diana Gerba (http://www.mushrooms911.blogspot.com)

Not all mushroom poisonings in people and animals are attributable to amatoxins. Ibotenic acid, muscimol, and other, still unknown compounds are present in toxic levels in some species of *Amanita* and other fungi (Benjamin 1995; Bresinsky and Besl 1990; Spoerke and Rumack 1994) (Chaps. 2, 3, and 7).

1.4 History of Research on *Amanita* Cyclic Peptide Toxins

Major advances in methods for the extraction, purification, and structural elucidation of chemicals in the nineteenth and twentieth centuries gave rise to rapid progress on the identification of the "active ingredients" from living organisms that account for their manifold effects on other living organisms. So-called natural products, also known as secondary metabolites or specialized metabolites, are widespread in certain groups of organisms, especially particular taxa of bacteria, fungi, and plants. Ecologically, these chemical compounds act as attractants, repellents, behavioral modulators, and self-regulators. Many have been adapted to human uses as pharmaceutical agents, such as morphine and penicillin.

Attempts to identify the chemical agents responsible for the toxicity of poisonous mushrooms date to the early 19th century. In the early 20th century, William Ford at Johns Hopkins University made active ethanolic extracts. Almost all of our knowledge of the chemistry of the *Amanita* toxins comes from sustained efforts started by Heinrich Wieland and continued by colleagues at the Universities of Munich, Frankfurt, and Heidelberg (Wieland, 1983; Wieland and Faulstich, 1991).

Initially, Wieland and colleagues identified two types of toxin, one of which killed injected mice quickly and the other slowly. In 1937, Lynen purified the major rapid toxin, phalloidin (Lynen and Wieland 1938) (Chap. 2). The major slow-acting toxin, named amanitin, was purified and crystallized a few years later (Wieland and Hallermayer 1941).

Fig. 1.13 Multi-lingual warning sign for poisonous *Amanita* and *Galerina* mushrooms, San Mateo County park system, California. (Courtesy Midpeninsula Regional Open Space District. Photo credit: Dave Hood, used by permission)

Since the original chemical structural work, many major advances have been made, including isolation and characterization of the virotoxins from *A. virosa* and the identification of several nontoxic cyclic peptides such as the cycloamanides and antamanide (Chap. 2). On the biological side, we now know in great detail the modes of action of the major *Amanita* toxins, the amatoxins and the phallotoxins (Chap. 5). The "molecular revolution" that transformed biology in the late twentieth century made possible the isolation of the genes encoding the toxins and, to date, partial elucidation of their biosynthetic pathway (Chap. 4). The use of DNA as a taxonomic tool (i.e., ITS sequencing, phylogenetics, and genomics) is rapidly clarifying the evolutionary relationships between toxin-producing species, and between toxin producers and nonproducers (Chaps. 3 and 6).

1.5 Suggested Reading

Recommended textbooks on fungi include Carlile et al. (2001) and Webster and Weber (2007), and entertaining popular accounts include Bone (2011), Marley (2010), and Money (2011). Benjamin (1995) gives an excellent account of mushrooms in culture and the history of mushroom eating and poisoning.

Other interesting books on mycology include Moore (2013) and Petersen (2013). Online resources, including extensive photographs of mushrooms and descriptions of their life cycles, uses, natural history, and edibility, include Tom Volk's mushroom-of-the-month (http://botit.botany.wisc.edu/toms_fungi/), Nathan Wilson's Mushroom Observer (http://mushroomobserver.org/), and Michael Kuo's MushroomExpert (http://www.mushroomexpert.com).

References

Abel JJ, Ford WW (1907) On the poisons of *Amanita phalloides*. J Biol Chem 2:273–288

Bas C (1969) Morphology and subdivision of *Amanita* and a monograph on its section *Lepidella*. Persoonia 5:285–579

Bauldauf SL, Roger AJ, Wenk-Siefert I, Doolittle WF (2000) A kingdom-level phylogeny of eukaryotes based on combined protein data. Science 290:972–977. https://doi.org/10.1126/science.290.5493.972

Benjamin DR (1995) Mushrooms: poisons and panaceas. WH Freeman, New York

Bone E (2011) Mycophilia: revelations from the weird world of mushrooms. Rodale, New York

Bresinsky A, Besl H (1990) A colour atlas of poisonous fungi: a handbook for pharmacists, doctors, and biologists. Wolfe Publishing, London

Cai Q, Tulloss RE, Tang LP, Tolgor B, Zhang P, Chen ZH, Yang ZL (2014) Multi-locus phylogeny of lethal amanitas: implications for species diversity and historical biogeography. BMC Evol Biol 14:143. https://doi.org/10.1186/1471-2148-14-143

Cai Q, Cui YY, Yang ZL (2016) Lethal *Amanita* species in China. Mycologia 108:993–1009. https://doi.org/10.3852/16-008

Carlile MJ, Watkinson SC, Gooday GW (2001) The fungi, 2nd edn. Elsevier, Amsterdam

Chaib De Mares M, Hess J, Floudas D, Lipzen A, Choi C, Kennedy M, Grigoriev IV, Pringle A (2014) Horizontal transfer of carbohydrate metabolism genes into ectomycorrhizal *Amanita*. New Phytol 205:1552–1564. https://doi.org/10.1111/nph.13140

Clémencon H (2012) Cytology and plectology of the Hymenomycetes, 2nd edn. J. Cramer, Stuttgart

Dunk CW, Lebel T, Keane PJ (2012) Characterisation of ectomycorrhizal formation by the exotic fungus *Amanita muscaria* with *Nothofagus cunninghamii* in Victoria, Australia. Mycorrhiza 22:135–147. https://doi.org/10.1007/s00572-011-0388-9

Embong Z, Wan Hitam WH, Yean CY, Rashid NH, Kamarudin B, Abidin SK, Osman S, Zainuddin ZF, Ravichandran M (2008) Specific detection of fungal pathogens by 18S rRNA gene PCR in microbial keratitis. BMC Ophthalmol 8:7. https://doi.org/10.1186/1471-2415-8-7

Gardes M, Bruns TD (1993) ITS primers with enhanced specificity for basidiomycetes – application to the identification of mycorrhizae and rusts. Mol Ecol 2:113–118

Grigoriev IV, Nikitin R, Haridas S, Kuo A, Ohm R, Otillar R, Riley R, Salamov A, Zhao X, Korzeniewski F, Smirnova T, Nordberg H, Dubchak I, Shabalov I (2014) MycoCosm portal: gearing up for 1000 fungal genomes. Nucleic Acids Res 42:D699–D704. https://doi.org/10.1093/nar/gkt1183

Hallen HE, Adams GC, Eicker A (2002) Amatoxins and phallotoxins in indigenous and introduced South African *Amanita* species. S African J Bot 68:322–326

Hibbett DS, Taylor JW (2013) Fungal systematics: is a new age of enlightenment at hand? Nat Rev Microbiol 11:129–133. https://doi.org/10.1038/nrmicro2963

Hibbett DS, Binder M, Bischoff JF, Blackwell M, Cannon PF, Eriksson OE, Huhndorf S, James T, Kirk PM, Lücking R, Thorsten Lumbsch H, Lutzoni F, Matheny PB, McLaughlin DJ, Powell MJ, Redhead S, Schoch CL, Spatafora JW, Stalpers JA, Vilgalys R, Aime MC, Aptroot A, Bauer R, Begerow D, Benny GL, Castlebury LA, Crous PW, Dai YC, Gams W, Geiser DM, Griffith GW, Gueidan C, Hawksworth DL, Hestmark G, Hosaka K, Humber RA, Hyde KD, Ironside JE, Kõljalg U, Kurtzman CP, Larsson KH, Lichtwardt R, Longcore J, Miadlikowska J, Miller A, Moncalvo JM, Mozley-Standridge S, Oberwinkler F, Parmasto E, Reeb V, Rogers JD, Roux C, Ryvarden L, Sampaio JP, Schüssler A, Sugiyama J, Thorn RG, Tibell L, Untereiner WA, Walker C, Wang Z, Weir A, Weiss M, White MM, Winka K, Yao YJ, Zhang N (2007) A higher level phylogenetic classification of the fungi. Mycol Res 111:509–547. https://doi.org/10.1016/j.mycres.2007.03.004

Hrynkiewicz K, Baum C (2013) Selection of ectomycorrhizal willow genotype in phytoextraction of heavy metals. Environ Technol 34:225–230. https://doi.org/10.1080/09593330.2012.689369

Luo H, Hallen-Adams HE, Scott-Craig JS, Walton JD (2010) Co-localization of amanitin and a candidate toxin-processing prolyl oligopeptidase in *Amanita* basidiocarps. Eukaryot Cell 9:1891–1900. https://doi.org/10.1128/EC.00161-10

Lynen F, Wieland U (1938) Über die Giftstoffe des Knollenblätterpilzes. IV. Kristallisation von Phalloidin. Liebigs Ann Chem 533:93–117

Marley GA (2010) Chanterelle dreams, Amanita nightmares: the love, lore, and mystique of mushrooms. Chelsea Green, Vermont

Matheny PB, Curtis JM, Hofstetter V, Aime MC, Moncalvo JM, Ge ZW, Slot JC, Ammirati JF, Baroni TJ, Bougher NL, Hughes KW, Lodge DJ, Kerrigan RW, Seidl MT, Aanen DK, DeNitis M, Daniele GM, Desjardin DE, Kropp BR, Norvell LL, Parker A, Vellinga EC, Vilgalys R, Hibbett DS (2006) Major clades of Agaricales: a multilocus phylogenetic overview. Mycologia 98:982–995. https://doi.org/10.1080/15572536.2006.11832627

Michelot D, Melendez-Howell LM (2003) *Amanita muscaria*: chemistry, biology, toxicology, and ethnomycology. Mycol Res 107:131–146

Miller OK Jr, Hemmes DE, Wong G (1988) *Amanita marmorata* subsp. *myrtacearum*, a new subspecies in *Amanita* section *Phalloideae* from Hawaii. Mycologia 88:140–145

Money NP (2011) Mushroom. Oxford University Press, New York

Moore D (2013) Fungal biology in the origin and emergence of life. Cambridge University, Cambridge, UK. ISBN: 978- 11-07652-77-4

Nagendran S, Hallen-Adams HE, Paper JM, Aslam N, Walton JD (2009) Reduced genomic potential for secreted plant cell-wall-degrading enzymes in the ectomycorrhizal fungus *Amanita bisporigera*, based on the secretome of *Trichoderma reesei*. Fung Genet Biol 46:427–435

Nehls U, Wiese J, Guttenberger M, Hampp R (1998) Carbon allocation in ectomycorrhizas: identification and expression analysis of an *Amanita muscaria* monosaccharide transporter. Mol Plant-Microbe Interact 11:167–176

Petersen JH (2013) The kingdom of fungi. Princeton University Press, Princeton

Pringle A, Adams RI, Cross HB, Bruns TD (2009) The ectomycorrhizal fungus *Amanita phalloides* was introduced and is expanding its range on the west coast of North America. Mol Ecol 18:817–833. https://doi.org/10.1111/j.1365-294X.2008.04030.x

Redhead SA, Vizzini A, Drehmel DC, Contu M (2016) *Saproamanita*, a new name for both *Lepidella* E.-J. Gilbert and *Aspidella* E.-J. Gilbert (*Amaniteae*, *Amanitaceae*). IMA Fungus 7:119–129. https://doi.org/10.5598/imafungus.2016.07.01.07

Schrey SD, Salo V, Raudaskoski M, Hampp R, Nehls U, Tarkka MT (2007) Interaction with mycorrhiza helper bacterium *Streptomyces* sp. AcH 505 modifies organisation of actin cytoskeleton in the ectomycorrhizal fungus *Amanita muscaria* (fly agaric). Curr Genet 52:77–85

Spoerke DG, Rumack BH (1994) Handbook of mushroom poisoning. CRC Press, Boca Raton

Tulloss RE, Yang ZL (2016) Family Amanitaceae. Tulloss RE, Yang ZL Amanitaceae studies. [http://www.amanitaceae.org?Family+Amanitaceae]. Accessed 6 Jul 2016

Tulloss RE, Kuyper TW, Vellinga EC, Yang ZL, Halling RE, Geml J, Sánchez-Ramírez S, Gonçalves SC, Hess J, Pringle A (2016) The genus *Amanita* should not be split. Amanitaceae 1(3):1–16. Available at www.amanitaceathejournal.org

Vellinga EC, Wolfe BE, Pringle A (2009) Global patterns of ectomycorrhizal introductions. New Phytol 181:960–973. https://doi.org/10.1111/j.1469-8137.2008.02728.x

Vizzini A, Contu M, Ercole E, Voyron S (2012) Rivalutazione e delimitazione del genere *Aspidella* (Agaricales, Amanitaceae), nuovamente separato da *Amanita*. Micol Veget Medit 27:75–90

Wasson RG (1971) Soma: divine mushroom of immortality. Harcourt Brace Jovanovich, New York

Wasson RG (1979) Traditional use in North America of *Amanita muscaria* for divinatory purposes. J Psychedelic Drugs 11:25–28

Webster J, Weber RWS (2007) Introduction to fungi, Third edn. Cambridge University Press, Cambridge

Wieland T (1986) Peptides of poisonous *Amanita* mushrooms. Springer, Berlin

Wieland H, Hallermayer R (1941) Über die Giftstoffe des grünen Knollenblätterpilzes, VI. Amanitin, das Hauptgift des Knollenblätterpilzes. Liebigs Ann Chem 548:1–18

Wieland T (1983) The toxic peptides from *Amanita* mushrooms. Int J Pept Protein Res 22:257–276

Wieland T, Faulstich H (1991) Fifty years of amanitin. Experientia 47:1186–1193

Willmann A, Weiss M, Nehls U (2007) Ectomycorrhiza-mediated repression of the high-affinity ammonium importer gene *AmAMT2* in *Amanita muscaria*. Curr Genet 51:71–78. doi:10.1007/s00294-006-0106-x

Woese CR, Kandler O, Wheelis ML (1990) Towards a natural system of organisms: proposal for the domains Archaea, Bacteria, and Eucarya. Proc Natl Acad Sci U S A 87:4576–4579

Wolfe BE, Richard F, Cross HB, Pringle A (2010) Distribution and abundance of the introduced ectomycorrhizal fungus *Amanita phalloides* in North America. New Phytol 185:803–816. https://doi.org/10.1111/j.1469-8137.2009.03097.x

Wolfe BE, Tulloss RE, Pringle A (2012) The irreversible loss of a decomposition pathway marks the single origin of an ectomycorrhizal symbiosis. PLoS One 7:e39597. https://doi.org/10.1371/journal.pone.0039597

Yoshida I, Kiho T, Usui S, Sakushima M, Ukai S (1996) Polysaccharides in fungi. XXXVII. Immunomodulating activities of carboxymethylated derivatives of linear (1→3)-α-D-glucans extracted from the fruiting bodies of *Agrocybe cylindracea* and *Amanita muscaria*. Biol Pharm Bull 19:114–121

Chapter 2
Chemistry of the *Amanita* Peptide Toxins

> *The thermostable body is the most important constituent of Amanita and the one causing the most profound disturbance of the organs and tissues (Ford 1906). [We] have shown that the amanita-toxin can be isolated and purified by certain well-defined methods, and in its pure state is one of the most powerful poisons of organic origin known... The exact character of this poison is still under investigation.*
>
> *(Ford 1909)*

Extraordinary advances in scientific methodologies during the nineteenth and twentieth centuries enabled chemists to isolate and characterize biologically active molecules from all branches of the tree of life. Biochemicals (such as DNA, RNA, proteins, and lipids) that are essential for the basic cellular processes of life are known as *primary metabolites*, whereas other chemicals found in living organisms that have subsidiary roles, often in mediating interactions *between* organisms, are known as *secondary metabolites*, *specialized metabolites*, or *natural products* (Demain and Fang 2000). There are tens of thousands of known natural products, which are especially prevalent in some taxa of bacteria, fungi, and plants. Ecologically, natural products commonly behave as attractants, repellents, and behavioral modulators. Many have been adopted for human use, for example, caffeine and penicillin. The possible ecological functions of the *Amanita* cyclic peptide toxins are discussed in Chap. 6.

The earliest attempts to identify the chemical agents responsible for the toxicity of poisonous mushrooms date to the early nineteenth century. In 1899, Kobert described a toxic ethanol extract of *A. phalloides* (cited in Abel and Ford 1907 and Wieland 1986). The first systematic scientific research on the toxic components of *Amanita* mushrooms were initiated by William Ford, professor of bacteriology at John Hopkins University, Maryland, USA, in the early twentieth century (Ford 1906, 1909). He showed that ethanol extracts of *A. phalloides* could kill guinea pigs and rabbits at a dose of 0.4 mg when delivered by subcutaneous injection. He called this material "*Amanita* toxin" to distinguish it from a hemolytic protein

J. Walton, *The Cyclic Peptide Toxins of Amanita and Other Poisonous Mushrooms*, https://doi.org/10.1007/978-3-319-76822-9_2

("*Amanita* hemolysin") that had earlier been studied by Kolbert (see "phallolysin" below). Ford collected his specimens of "*A. phalloides*", sometimes referred to as the "white variety" of *A. phalloides*, in North Carolina, Massachusetts, and New Hampshire. However, *A. phalloides* did not exist at that time in North America, being introduced only later in the 20th century (Wolfe et al. 2010). Ford's specimens were probably native American species closely related but not identical to *A. verna* and *A. virosa*. To this day the numbers of species in *Amanita* sect. *Phalloideae* and the relationships between the European and North American species have not been completely clarified (Chap. 3). Ford's compound was resistant to boiling and gastric juice, characteristics now known to pertain to the amatoxins.

The longest period of sustained research on the toxic constituents of *Amanita* mushrooms started in the 1930s at the University of Munich in the laboratory of Heinrich Wieland and continuing through the 1980s in the labs of several of his professional descendants, notably Theodor Wieland and Heinz Faulstich in Frankfurt and later Heidelberg (Wieland 1983; Wieland and Faulstich 1991). All of the early work was done with *A. phalloides*, which is native to Germany and presumably collected there. The early choice of bioassay, injection into mice, had a strong influence on the development of the field, because the two major classes of toxins (amatoxins and phallotoxins) behave differently when injected versus delivered orally. In the early years, two conclusions stand out. One was the recognition that poisonous *Amanita* mushrooms contain multiple toxic chemicals, sub-divided into two classes, the "slow acting" toxins and the "rapid acting" toxins based on the speed with which they killed mice when injected. Second, each type of toxin itself comprised different chemical species with different toxicity profiles.

Initial work focused on the toxicologically more dramatic "rapid" toxins, which kill mice in a few hours when injected. In 1937, Lynen purified the major rapid toxin, phalloidin (Lynen and Wieland 1938; Fig. 2.1). The major slow-acting toxin, α-amanitin, was purified and crystallized a few years later (Wieland and Hallermayer 1941). It is now known that mammalian toxicity of poisonous mushrooms when consumed orally (i.e., eaten) is due exclusively to the amatoxins because only they, and not the phallotoxins, are absorbed by the digestive tract (Chap. 5). The amatoxins act relatively slowly compared to the phallotoxins because inhibition of RNA polymerase II (pol II) causes gradual depletion of essential proteins in cells (Chap. 5).

Major advances in the chemistry of the *Amanita* cyclic peptides since the 1960s include identification of a number of additional forms of the amatoxins and phallotoxins, isolation and characterization of the cyclic peptide virotoxins from *A. virosa*, the identification of several nontoxic cyclic peptides, including the cycloamanides and antamanide, and a large number of chemical synthetic studies.

Über die Giftstoffe des Knollenblätterpilzes.

Inaugural-Dissertation
zur Erlangung der Doktorwürde
der
Hohen Naturwissenschaftlichen Fakultät
der Ludwig-Maximilians-Universität
München

vorgelegt
am 1. Februar 1937
von
Feodor L y n e n
aus München

Berichterstatter: Geheimrat Prof. Dr. H. W i e l a n
Tag der mündlichen Prüfung: 12. Februar 1937.

Fig. 2.1 Title page of Feodor Lynen's Ph.D. dissertation from the University of Munich, 1937, written under the guidance of Nobel laureate Prof. Dr. Heinrich Wieland. Dr. Lynen himself won the Nobel Prize in 1964. The title "Uber die Giftstoffe des Knollenblatterpilzes" translates as "On the toxins in the death cap mushroom." In his dissertation, Dr. Lynen reported the first purification of phalloidin

2.1 Purification of the Toxins

2.1.1 Historical Methods

Early methods of cyclic peptide toxin purification and detection are covered by Wieland (1986). These included solvent extraction, differential precipitation, paper chromatography, thin-layer chromatography (TLC), and low-pressure column chromatography on various media including silica, alumina, size exclusion (gel filtration), and Sephadex LH-20. Amatoxins are soluble in water, methanol, ethanol, DMF, and DMSO and insoluble in ether, benzene, gasoline, chloroform, ethyl acetate, and other organic solvents. They have been crystallized from various solvents (Wieland 1986).

Historically, purification was guided by bioassay (typically toxicity when injected into mice), and detection was by cinnamaldehyde/HCl (paper and TLC chromatography) or ultraviolet absorption (column chromatography). Using silica gel TLC and cinnamaldehyde detection, the detection limit is about ~0.5 μg of α-amanitin (Fig. 2.2).

The current method of choice for analysis of the peptide toxins and for purification of low-mg quantities is high-pressure liquid chromatography (HPLC), whose advantages include high speed, resolution, reproducibility, and sensitivity. TLC has the advantages of simplicity, low capital cost, and simultaneous analysis of multiple samples (Stijve and Seeger 1979). The major toxins can be separated in a single

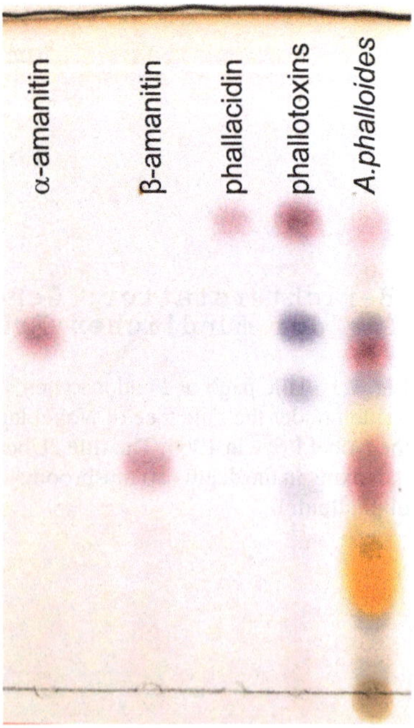

Fig. 2.2 Thin-layer chromatography (TLC) of representative peptide toxins and a methanolic extract of *A. phalloides*. The stationary phase was Merck silica gel 60. Mobile phase was chloroform/methanol/acetic acid/water (70:38:5:7.5) (Stijve and Seeger 1979). Visualization was with cinnamaldehyde/HCl. (Courtesy of Richard Ransom, Funite, Inc)

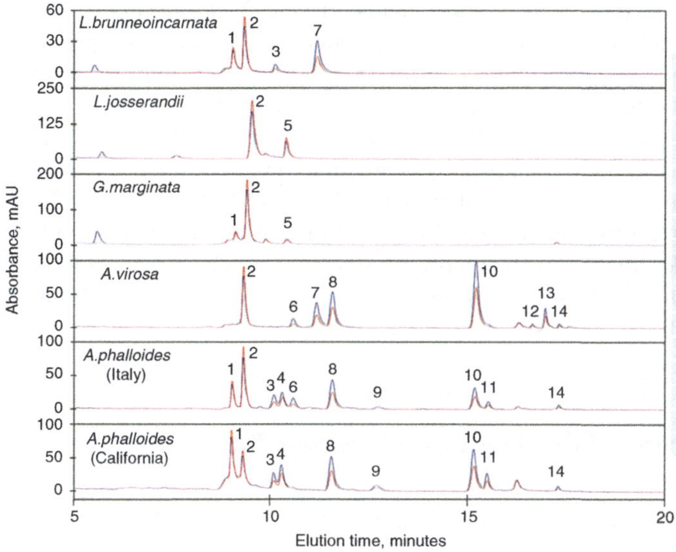

Peak number	Identification
1	β-amanitin
2	α-amanitin
3	amanin
4	phallisacin
5	γ-amanitin
6	phallisin II
7	amaninamide
8	phallacidin
9	phallisin I
10	phalloidin
11	unknown
12	unknown
13	unknown
14	unknown

Fig. 2.3 HPLC UV absorption profiles of mushroom extracts from six amatoxin- and phallotoxin-producing mushrooms. Blue, absorbance at 295 nm; red, absorbance at 305 nm. Peaks are labeled in order of elution time and shared numbers among extracts indicate the same compound. Note that compounds with hydroxy-Trp (#'s 1, 2, and 5) have stronger absorbance at 305 nm than 295 nm, whereas compounds lacking hydroxy-Trp (#'s 3, 4, 6, 7, 8, and 10) have the opposite pattern. The shift in retention time of α-amanitin (peak 2) in the *L. josserandii* extract is due to column performance and is within the deviations observed for standards. For structures see Figs. 2.6, 2.7, 2.8, and 2.9. (From Sgambelluri et al. 2014, reprinted under the Creative Commons License)

HPLC run using a standard reverse-phase column, and identifications can be made with reasonable certainty on the basis of their retention times and comparative UV absorption spectra (Fig. 2.3). HPLC methods for analyzing the *Amanita* toxins in mushrooms are described in numerous publications; recent examples include Clarke et al. (2012), Deng et al. (2011), Enjalbert et al. (1992), Garcia et al. (2015a), Muraoka et al. (1999), and Sgambelluri et al. (2014).

2.1.2 Current Protocols for Analysis of Peptide Toxins

Mushrooms can be extracted in their fresh state or after drying with low heat or lyophilization (freeze-drying). The toxins are hydrophilic (polar and/or anionic), and therefore the first extraction step typically utilizes 50–90% methanol. In our experience, 100% methanol or 90% ethanol do not efficiently extract the phallotoxins. α-Amanitin has been detected in dried mushrooms more than 20 years old (Brandon Landry, unpublished results from the author's lab).

The toxins can be extracted from mushrooms by simply soaking in 50% methanol for 30 min to overnight, but grinding of freeze-dried mushrooms to a fine powder in liquid nitrogen gives the highest yields. Friable dried mushroom tissues (5–10 mg) can be effectively pulverized with a wooden stick in a 1.5 ml Eppendorf tube. Acidifying the extraction solution with 0.001 N HCl is thought to improve recovery of the acidic toxins such as β-amanitin and phallacidin (Enjalbert et al. 1992). The methanol-insoluble materials, including cell walls, proteins, nucleic acids, and polysaccharides, are removed by centrifugation (10,000 × g for 10 min) or filtration using a methanol-resistant filter material such as Whatman #1 or PVDF.

For a cleaner preparation, the methanol can be removed by evaporation and the aqueous residue back-extracted with chloroform (the major toxins are not soluble in chloroform). This eliminates some interfering UV peaks. The detection limit by HPLC (using either UV_{295} or mass detection) is ~0.1 μg.

2.1.3 HPLC Protocol for Analysis of Peptide Toxins

- Mushrooms can be stored at −80 °C indefinitely, before or after drying.
- Grind between 20 mg and 1 g dried mushroom in a mortar and pestle with liquid N_2. Prechill the mortar and pestle with liquid N_2. Add pieces of mushroom and just enough liquid N_2 to cover the sample. Allow the N_2 to evaporate completely before grinding the sample to a fine powder. Do not add liquid N_2 to material once it is reduced to powder to avoid spattering the toxic dust, and always wear eye protection and a dust mask over your nose and mouth. The powder can be stored at −20 °C or −80 °C indefinitely.
- Transfer the powder to a 50-ml disposable polypropylene centrifuge tube, and add 50 ml extraction solution per gm dried tissue (alternatively, 1 ml per 20 mg in a 1.5 ml Eppendorf centrifuge tube). One preferred extraction solution is methanol + water + 0.01 N HCl (70:20:10). (Methanol concentrations from 30% to 80% work satisfactorily for the major toxins, although some of the more hydrophobic cycloamanides require less polar solvents such as acetone or ethyl acetate). Mix thoroughly by vortexing and incubate for 1 h with gentle rocking at room temperature.
- Remove the insoluble material by centrifugation (~10,000 × g for 5 min).
- Transfer the supernatant to a fresh tube. Remove the methanol and water by rotary evaporation under vacuum. Taking to complete dryness is satisfactory. Redissolve the pellet in 5.0 ml water per gm original material (100 μl for 20 mg starting material). At this point, the material can be extracted with an equal volume of chloroform. If a large emulsion forms, the two phases can be separated by centrifugation. Discard the chloroform phase.
- HPLC analysis: If an autosampler is being used with vial inserts, add 15 μl of sample and 50 μl water to each insert and inject the entire 65 μl. This is equivalent to 3 mg dry weight of the original mushroom. If the mushroom contains 4 mg α-amanitin per gm dry weight, the α-amanitin loading is 12 μg, approximately 100 times the detection limit.
- Any good reverse phase (C18 or ODS) column should give satisfactory separation, e.g., a 250 × 4.5 mm column with a 5 μ packing (Sgambelluri et al. 2014).

For gradient elution, solution A is 0.02 M ammonium acetate, adjusted to pH 5 with glacial acetic acid, plus acetonitrile in the ratio of 90:10 (v/v), and solution B is acetonitrile. Ammonium formate can be used instead of ammonium acetate. A suitable elution protocol is a linear gradient from 100% solution A to 60% solution B in 30 min at a flow rate of 1 ml/min. For the major toxins, the eluant is monitored at 280, 295, and/or 305 nm. With an in-line quadrupole mass analyzer, the eluant is typically monitored in the range of *m/z* ~600 to ~1200 in positive ion mode. Results are reanalyzed after the run by selected ion monitoring (SIM) at *m/z* 919 ($M+H^+$ for α-amanitin), *m/z* 920 (β-amanitin), *m/z* 847 (phallacidin), and *m/z* 789 (phalloidin). In our experience, it is common to see significant *m/z* peaks corresponding to the Na^+ and K^+ adducts of the major toxins (Sgambelluri et al. 2014).

In a single HPLC run, no one criterion is sufficient to provide unambiguous identification and quantitation of the peptide toxins, especially the minor ones. In the author's lab, the criteria for positive identification are a combination of retention time relative to standards, relative UV absorbance at 295 nm and 305 nm, and nominal mass (*m/z*). Retention time compared to standards can vary day to day due to column aging and minor differences in elution solution composition. Furthermore, as of this writing, commercial standards are available only for α-amanitin, β-amanitin, γ-amanitin, phalloidin, and phallacidin. With the above gradient protocol, β-amanitin elutes before α-amanitin (~10 and ~12 min, respectively), phallacidin elutes later (~21 min), and phalloidin elutes last at ~26 min (Enjalbert et al. 1992). A faster separation with a modified gradient program is described in Sgambelluri et al. (2014); the four major toxins elute in the same order between ~8 and ~18 min (Fig. 2.3).

All of the major *Amanita* peptide toxins absorb in the range 280–305 nm due to the presence of Trp or hydroxy-Trp (Fig. 2.4 and Table 2.1), but UV absorption alone is not sufficiently specific for unambiguous identification. The absorption ratios at 295 and 305 nm can be used to classify a UV peak as a probable phallotoxin (295 > 305 nm) or amatoxin (305 > 295 nm). Exceptions to this rule are amanin and amaninamide, which, although amatoxins, lack hydroxy-Trp and therefore have stronger absorption at 295 than 305 nm (Figs. 2.3 and 2.4). UV absorption is the most reliable method to quantitate the toxins, assuming that standards of known concentration are available for construction of standard curves and that the toxin preparations are free of co-eluting UV-absorbing compounds.

In-line single quadropole mass analyzers provide a suitable method to identify with high certainty compounds eluting from an HPLC (LC/MS). The mass spectra of the eluting compounds must be examined to ensure that the observed masses correspond to the major compound present and not to a minor compound or isotope peak. For example, if the eluant is monitored at *m/z* 919 ($M+H^+$ for α-amanitin) and 920 ($M+H^+$ for β-amanitin), the +1 isotope peak from α-amanitin will overlap with the parent β-amanitin peak. Ultimate identification of the peptide toxins requires high-resolution mass spectrometry to demonstrate that the compounds have the correct elemental formulae.

These extraction and analysis methods have also proved suitable for analyzing the toxins in *Galerina*, *Conocybe*, and *Lepiota* (Enjalbert et al. 2004; Hallen et al. 2003; Luo et al. 2012; Sgambelluri et al. 2014).

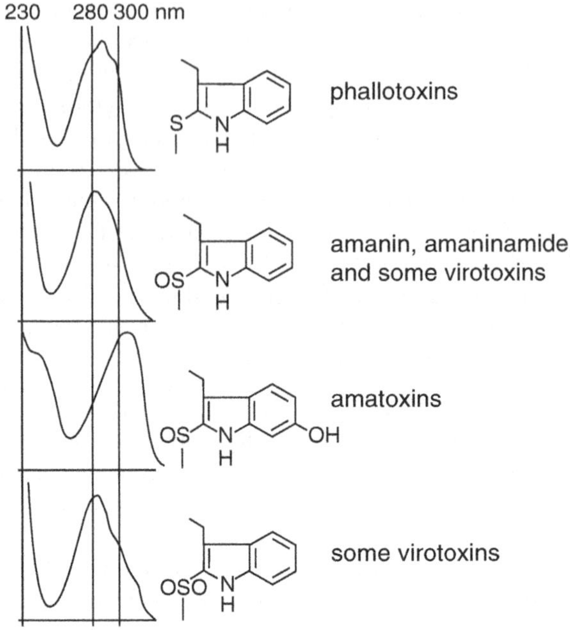

Fig. 2.4 Characteristic UV absorption spectra of the modified indole moieties found in amatoxins, phallotoxins, and virotoxins. (Redrawn from Wieland 1986 with permission of Springer Nature)

Table 2.1 UV absorption maxima and molar extinction coefficients of *Amanita* cyclic peptide toxins

Toxin family	Chromophore	λ_{max} (nm)	Molar extinction coefficient ($mol^{-1}\ l^{-1}$)
Phallotoxins	2′-(SR)Trp	292	ε_{292} = 12,600
			ε_{300} = 10,200
Amanin, amaninamide, and some virotoxins	2′-(SOR)Trp	287	ε_{287} = 12,500
Amatoxins	2′-(SOR), 6′-(OH)Trp	305	ε_{305} = 14,600
			ε_{305} = 14,000 (methanol)
			ε_{310} = 13,900
			ε_{310} = 14,600 (methanol)
Virotoxins	2′-(SO_2R)Trp	276	ε_{276} = 13,400

Data taken from Wieland (1986)
Extinction coefficients are in water at pH 7 unless otherwise indicated

2.1.4 Alternative Methods for Detection and Quantitation

Alternate methods of separation and detection include capillary electrophoresis and electrochemical detection. Methods have also been developed for quantitating amatoxins in clinically relevant matrices such as urine, serum, bile, gastric juice, and mammalian tissues (e.g., Brüggemann et al. 1996; Feng et al. 2015; Garcia et al. 2015b; Gicquel et al. 2014; Leite et al. 2013; Tomková et al. 2015). A method for extracting and analyzing amatoxins from a bonafide "mushroom stew" is described by Jansson et al. (2012).

Radioimmunoassay (RIA) and enzyme-linked immunosorbent assay (ELISA) for detection and quantitation of amatoxins have been developed using antibodies against α-amanitin (Andres et al. 1986; Andres and Frei 1987; Chen et al. 1993; Faulstich et al. 1975, 1982; Fiume et al. 1975). An ELISA test kit has been developed by Bühlmann Laboratories AG (Switzerland) and as of this writing is commercially available (www.apollo.com). The ELISA antibodies can detect α-amanitin and γ-amanitin (both of which have the core amino acid sequence IWGIGCNP), but not β-amanitin (which has the sequence IWGIGCDP) (differences underlined). Although β-amanitin can be a major constituent in poisonous *Amanita* mushrooms, the author's laboratory has never seen it in the absence of α-amanitin in *Amanita*, *Lepiota*, or *Galerina*. Therefore, the ELISA test should be reliable for detecting the presence of toxic peptides but perhaps not for quantitation of the total load of clinically relevant compounds. The anti-α-amanitin antibodies also do not detect the phallotoxins, which, however, are clinically irrelevant. Antibodies have also been raised in sheep specifically against β-amanitin; these showed some cross-reactivity with α-amanitin (Abuknesha and Maragkou 2004). A monoclonal antibody has been developed that detects α, β, and γ-amanitin but not amanin, amaninamide, or the phallotoxins (He et al. 2017).

^{15}N-labeled α-amanitin has been produced using *Galerina* fermentation (Luo et al. 2015). A quantity of this material was produced in the author's laboratory for the United States Centers for Disease Control (CDC). One potential use of this material is as an internal standard to quantitate amanitin recovery from complex matrices such as blood, urine, or bile.

For future work, methods orthogonal to reverse-phase chromatography, such as ion exchange, hold promise at least for the acidic toxin variants (e.g., Ritorto et al. 2013). An alternative eluant gradient for HPLC is 0.1% formic acid to 60% acetonitrile. Under these conditions, some of the peptide retention times shift relative to each other, making it a useful adjunct to the ammonium acetate-containing eluants described above.

2.2 The Wieland-Meixner Test

The juice of an amatoxin-containing mushroom yields a characteristic blue color when rubbed on crude paper and treated with acid. This color reaction was first described by F. Wieland, W. Dilger, and co-workers (Wieland et al. 1949) and later was the subject of a paper by Meixner (1979). It is now often referred to as simply the Meixner test, but to acknowledge Wieland's precedence, we here refer to it as the Wieland-Meixner test. It is a reasonably reliable first-pass test for the presence of amatoxins but should never be relied on by itself to prove or disprove the toxicity of a mushroom. The test is also described in Olson et al. (1982) and Duffy (2008), and see also Vergeer (1983).

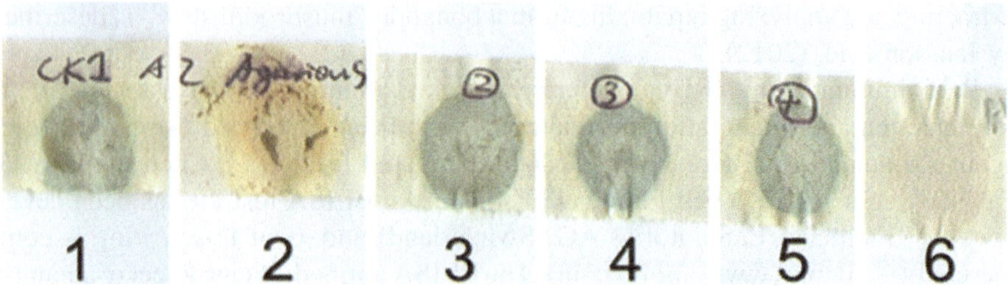

Fig. 2.5 The Wieland-Meixner test. The matrix was ordinary newsprint from the United States. (1) Positive control: *Amanita bisporigera*, a known amatoxin producer. (2) Negative control: *Agaricus bisporus*. (3, 4) Two unknowns, both showing positive reaction for amatoxins. (5) *Galerina marginata*, which produces α-amanitin. (6) Concentrated HCl alone. (Photo credit: Hong Luo, Kunming Institute of Botany, Chinese Academy of Sciences, used with permission)

2.2.1 Protocol for the Wieland-Meixner Test

Rub a small piece of fresh (or rehydrated) mushroom on a ~1 cm^2 piece of lignin-containing paper, such as a newsprint or the white pages of a United States telephone book (pure cellulose does not work). Moisten with a minimum amount of concentrated hydrochloric acid (HCl). Keep the sample away from high heat or direct sunlight. The presence of amatoxins is indicated by the appearance within a few minutes of a blueish-green color (Fig. 2.5). The color fades in a day or two. As a positive control, a known amatoxin-containing mushroom or a crude or pure amatoxin preparation can be used. A negative control can be any amatoxin-non-containing mushroom such as *Agaricus bisporus*, which will give at best a brownish-yellow color (Fig. 2.5). A spot of hydrochloric acid alone is another important negative control (Fig. 2.5).

The Wieland-Meixner test detects hydroxylated indoles, and therefore it cannot distinguish between amatoxins and compounds such as psilocin (N,N-dimethyl-4-hydroxytryptamine, the active form of psilocybin), serotonin (5-hydroxytryptamine), 5-hydroxytryptophan, or bufotenine (N,N-dimethyl-5-hydroxytryptamine). *Amanita citrina* and *A. porphyria* give positive results with the Wieland-Meixner test despite lacking amatoxins. *A. citrina* contains bufotenine, and *A. porphyria* contains 5-hydroxytryptophan (Beutler and Vergeer 1980). Phallotoxins, amaninamide, and amanin, lacking a hydroxy-substituted indole, are not detected. This is potentially significant because amanin and amaninamide are toxic to mammals (LD_{50} ~0.5 mg/kg), so in theory a mushroom could test negative for the Wieland-Meixner test and still be orally poisonous. Amaninamide, which is α-amanitin without the 6-hydroxyl on the Trp residue, was originally found in *A. virosa* (Buku et al. 1980b) and is also present in *Lepiota brunneoincarnata* (Fig. 2.3).

In a blind test using emergency room physicians, 2 μg of amanitin could be detected by the Wieland-Meixner test with 100% agreement among the evaluators (Beuhler et al. 2004). False positives with psilocin were also detected 100% of the time, consistent with psilocin being a hydroxylated indole (Beutler and Vergeer 1980). In regard to the rate of false positives among mushrooms, Seeger (1984; cited in Duffy 2008) reported that 62 of 335 nontoxic mushrooms gave a false positive.

2.3 Structures of the Toxins

The structures of the toxic *Amanita* cyclic peptides have been studied by all modern methods, including mass spectrometry (MS), nuclear magnetic resonance (NMR), circular dichroism (CD), and X-ray crystallography (e.g., Henderson 2010; Isernia et al. 1996; Luo et al. 2015; Pehk et al. 1989; Zanotti et al. 2001). The major toxins comprise two families, the amatoxins and the phallotoxins (Figs. 2.6, 2.7, 2.8, and 2.9). The ten known amatoxins are bicyclic octapeptides and the seven known

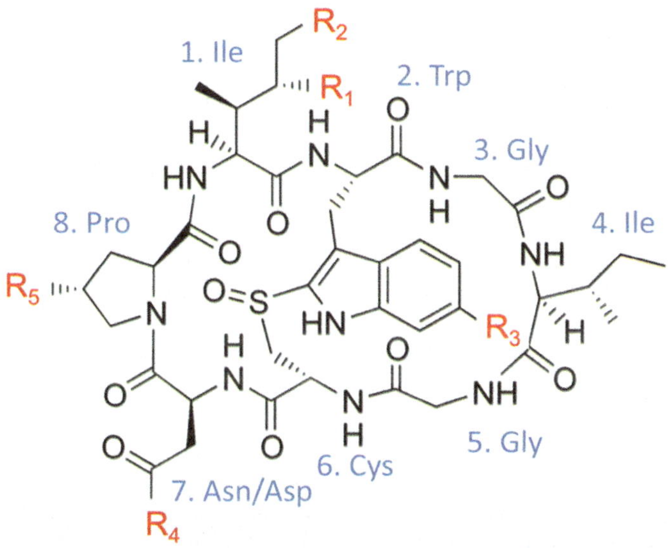

Compound	R1	R2	R3	R4	R5	Mono-isotopic mass	Elemental formula	LD_{50} (mg/kg)	K_i, pol II inhibition (M)
α-amanitin	OH	OH	OH	NH_2	OH	918.4	$C_{39}H_{54}N_{10}O_{14}S$	0.3-0.6	3×10^{-8}
β-amanitin	OH	OH	OH	OH	OH	919.4	$C_{39}H_{53}N_9O_{15}S$	0.5	3×10^{-8}
λ-amanitin	OH	H	OH	NH_2	OH	902.4	$C_{39}H_{54}N_{10}O_{13}S$	0.2-0.5	3×10^{-8}
ε-amanitin	OH	H	OH	OH	OH	903.3	$C_{39}H_{53}N_9O_{14}S$	0.3-0.6	
amanin	OH	OH	H	OH	OH	903.3	$C_{39}H_{53}N_9O_{14}S$	0.5	3×10^{-8}
amaninamide	OH	OH	H	NH_2	OH	902.4	$C_{39}H_{54}N_{10}O_{13}S$	0.5	
amanullin	H	H	OH	NH_2	OH	886.4	$C_{39}H_{54}N_{10}O_{12}S$	>20	$10^{-6} - 10^{-7}$
amanullinic acid	H	H	OH	OH	OH	887.3	$C_{39}H_{53}N_9O_{13}S$	>20	
proammanullin	H	H	OH	NH_2	H	870.4	$C_{39}H_{54}N_{10}O_{11}S$	>20	6×10^{-4}
θ-amanitin	OH	H	OH	NH_2	H	887.3	$C_{39}H_{54}N_{10}O_{12}S$	unknown	

Fig. 2.6 The amatoxin family of bicyclic octapeptides. LD_{50} values are based on white mouse injection assay. (Taken from Wieland 1986). K_i values (i.e., concentrations for half-maximal inhibition in vitro) are for *Drosophila* pol II, which is ~10-fold less sensitive than mammalian pol II. (Taken from Wieland and Faulstich 1991; Zanotti et al. 1987). Also see Shoham et al. (1989). θ-Amanitin has been found only in an engineered mutant of *Galerina marginata* (unpublished results from the author's laboratory)

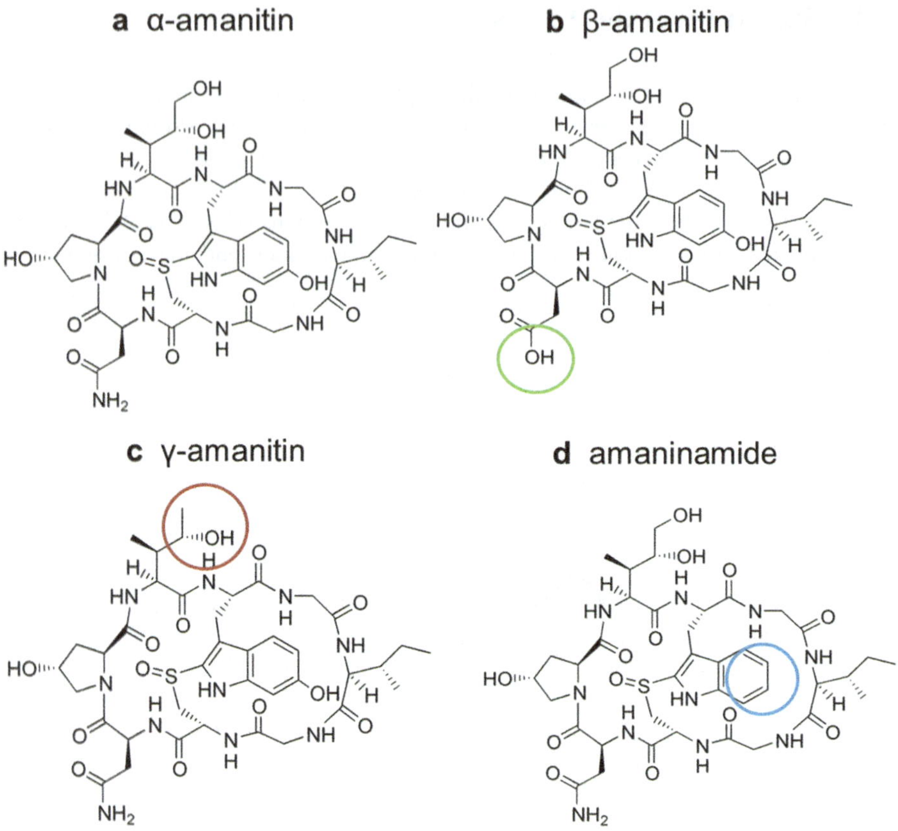

Fig. 2.7 The major amatoxins. (**a**) α-Amanitin, (**b**) β-amanitin, (**c**) γ-amanitin, (**d**) amaninamide. *Colored circles* highlight points of difference from α-amanitin. (Source: Edgar181, https://commons.wikimedia.org/w/index.php?curid= 5909292, 5909480, 5909650, and 5914713)

phallotoxins are cyclic heptapeptides. *Amanita* mushrooms also make other cyclic peptides, the virotoxins (related to the phallotoxins) and the nontoxic cycloamanides including antamanide (see below). Within the two main families, members differ from each other in amino acid sequence and/or pattern of hydroxylation (Figs. 2.6, 2.7, 2.8, and 2.9). The molecular genetic evidence indicates that the differences in amino acid sequences are DNA-encoded, whereas the different hydroxylation patterns are due to posttranslational modifications (Chap. 4). Therefore, four genes can encode all of the known amatoxins and phallotoxins.

Historically, there has been little consistency in the numbering of the amino acids in the *Amanita* cyclic peptides. The numbering scheme used in this book follows the order of the amino acids in the linear toxin precursor peptides. Numbering schemes used in other publications have been converted to this system in this book. For

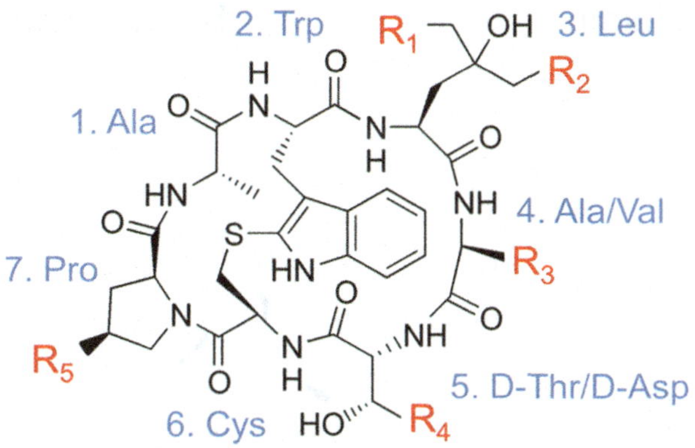

Compound	R1	R2	R3	R4	R5	Mono-isotopic mass	Elemental formula	LD_{50} (mg/kg)
phalloidin	OH	H	CH_3	CH_3	OH	788.3	$C_{35}H_{48}N_8O_{11}S$	2
phallacidin	OH	H	$CH(CH_3)_2$	COOH	OH	846.3	$C_{37}H_{50}N_8O_{13}S$	1.5
phallisacin	OH	OH	$CH(CH_3)_2$	COOH	OH	862.3	$C_{37}H_{50}N_8O_{14}S$	4.5
phallisin	OH	OH	CH_3	CH_3	OH	804.3	$C_{35}H_{48}N_8O_{12}S$	2
phallacin	H	H	$CH(CH_3)_2$	COOH	OH	830.3	$C_{37}H_{50}N_8O_{12}S$	1.5
phalloin	H	H	CH_3	CH_3	OH	772.3	$C_{35}H_{48}N_8O_{10}S$	1.5
prophalloin	H	H	CH_3	CH_3	H	756.3	$C_{35}H_{48}N_8O_9S$	>20

Fig. 2.8 The phallotoxin family of bicyclic heptapeptides. LD_{50} values are based on white mouse injection assay. (Data taken from Wieland 1986)

α-amanitin, the order is Ile^1-Trp^2-Gly^3-Ile^4-Gly^5-Cys^6-Asn^7-Pro^8 in standard three-letter code or IWGIGCNP in single-letter code. Phalloidin has the sequence Ala^1-Trp^2-Leu^3-Ala^4-Thr^5-Cys^6-Pro^7 or AWLATCP in single-letter code.

2.3.1 *Structural Features Common to the Amatoxins and Phallotoxins*

Both families of compounds, with the exception of amanullin, amanullinic acid, and proamanullin, contain a hydroxyl group at the 4-carbon (also known as the γ position) of Ile #1 in the amatoxins or Leu #3 in the phallotoxins. The 4-hydroxylation of Ile or Leu has an influence on the chemical reactivity of amatoxins and phallotoxins by making this amino acid prone to cleavage under acidic conditions via a γ-lactone intermediate (Wieland 1986). Ile #1 and Leu #3 often contain an additional one or two hydroxylations in the amatoxins and phallotoxins, respectively (Figs. 2.6, 2.7, 2.8, and 2.9).

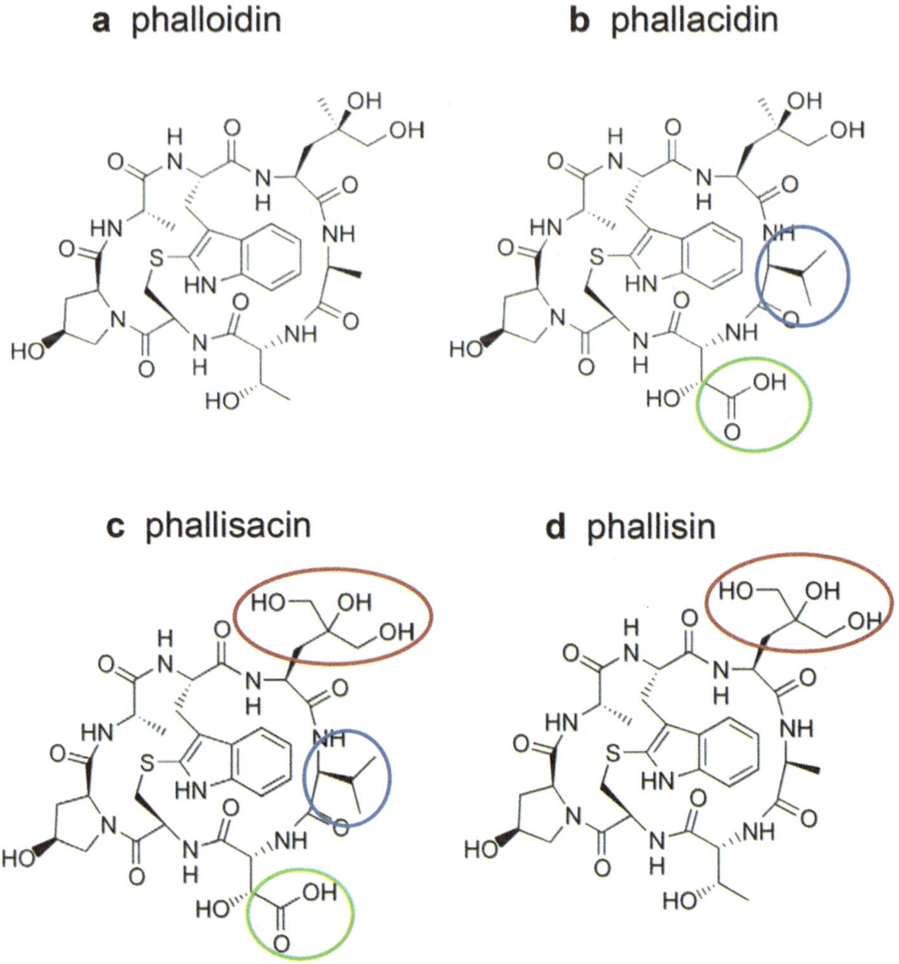

Fig. 2.9 The major phallotoxins. (**a**) Phalloidin, (**b**) phallacidin, (**c**) phallisacin, (**d**) phallisin. *Colored circles* highlight points of difference from phalloidin. (Source: Edgar181 – https://commons.wikimedia.org/w/index.php?curid=3395734, 5330289, 5330296, and 5330271)

A salient feature of both the amatoxins and phallotoxins is the presence of tryptathionine, the trivial name for the dipeptide formed by the thiol of Cys cross-linked to C2 of Trp #2 (May and Perrin 2007). In the amatoxins the sulfur is further oxidized to the (*R*)-sulfoxide, whereas in the phallotoxins, the sulfur remains as the thioether (sulfide). Both families also contain 4-hydroxy-L-Pro (Hyp). In the amatoxins, Hyp has the *trans* configuration (2*S*,4*R*), whereas in the phallotoxins, it has the *cis* configuration (2*S*,4*S*). Hyp in collagen and plant cell wall Hyp-rich glycoproteins has the *trans* configuration. Several sources state that the *cis* isomer has not been found in nature apart from the *Amanita* cyclic peptide toxins; however, there are bacterial α-ketoglutarate-dependent dioxygenases that catalyze *cis*-4-hydroxylation of Pro (Hara and Kino 2009).

2.3.2 *Amatoxins*

The bioassay used as the basis of the original purification of the toxic principles of *A. phalloides* was mouse mortality by injection. The use of this assay allowed for discrimation between "slow-acting" and "fast-acting" toxins, which were subsequently named the amatoxins and phallotoxins, respectively. Although the amatoxins are slower acting than phallotoxins when injected (several days vs. a few hours), we now know that the toxicity of *A. phalloides* to humans, dogs, and other mammals when consumed orally is due solely to the amatoxins because the phallotoxins are not absorbed (Chap. 5).

There are ten characterized amatoxins (Figs. 2.6 and 2.7). All but θ-amanitin are natural compounds extracted from *A. phalloides* and sometimes other species of *Amanita*. θ-Amanitin is produced by a mutant of *Galerina marginata* (unpublished work from the author's laboratory; Chap. 4). The backbone amino acid sequences of the amatoxins can be encoded by just two genes, typified by α- and β-amanitin. α-Amanitin has the amino acid sequence IWGIGCNP and β-amanitin has the sequence IWGIGCDP. The sole difference between these two (Asn or Asp at position #7) is probably not due to posttranslational modification (e.g., deamidation of Asn to Asp or amidation of Asp to Asn) because genes encoding both are present in the genome of *A. phalloides* (Chap. 4). The other amatoxins (γ-amanitin, ε-amanitin, amanin, etc.) are variants in posttranslational hydroxylations and therefore can be encoded by the same two genes.

2.3.2.1 Structure/Activity Relationships in the Amatoxins

The amatoxins have been the subject of many structure/activity studies, based on either chemical modification of natural toxins or de novo synthesis. Some studies have looked at the structural elements involved in toxicity to organisms (typically by injection into mice but also by oral delivery), whereas others have studied structural features that influence the in vitro inhibition of pol II or binding to monoclonal antibodies (Baumann et al. 1993, 1994). Because of the complexity of pharmacokinetics, i.e., the path through an organism from ingestion to the site of action, a perfect correlation between the two types of assay is not to be expected.

The overall bicyclic ring structure, which contributes overall shape and rigidity to the molecule, has been shown to be critical, because breaking the cycle at either the bond between Ile #1 and Trp #2 or between Trp #2 and Cys #5 (the cross-bridge) destroys activity. Based on 24 α-amanitin derivatives, a β-turn was concluded to be an important structural feature (Baumann et al. 1994).

2.3.2.2 The Individual Amino Acids of the Amatoxins

#1: Isoleucine α-Amanitin and β-amanitin both contain 4,5-dihydroxy-L-Ile, also known as γ,δ-dihydroxy-L-Ile. γ-Amanitin is α-amanitin with only the 4-hydroxyl (γ-hydroxyl) group and lacking the 5-hydroxyl (δ-hydroxyl) group on Ile. Analogously, ε-amanitin is β-amanitin without the 5-hydroxylation (Figs. 2.6 and 2.7). Amatoxins with only 5-hydroxylation have not been reported, but 5-hydroxy-Ile does occur in the orbitide yunnanin B; Chap. 4). In γ-amanitin, the 4-hydroxyl group has the *S* configuration, so the overall stereochemistry of the hydroxy-Ile moiety is 2*S*,3*R*,4*S*. In α-amanitin, due to the rules of nomenclature, the 4-hydroxy group has the *R* configuration, making the overall stereochemistry (2*S*,3*R*,4*R*) (Wieland 1986).

The 4-hydroxy group of Ile #1 is required for full activity, as evidenced by the much lower toxicity of amanullin and related compounds lacking this hydroxyl group (Fig. 2.6). The 4-hydroxyl group appears to be less important for in vitro inhibition of pol II, indicating that its major role is in amanitin pharmacokinetics (Shoham et al. 1989; Fig. 2.6). The 5-hydroxyl group is not essential, because γ-amanitin is fully toxic in vivo when injected into mice and in vitro against purified pol II (Wieland 1986; Wieland and Faulstich 1991). The 5-hydroxyl group forms a hydrogen-bond to Gln763 in subunit 2 of RNA polymerase II (Rpb2) (the only contact between α-amanitin and Rbp2); this interaction is apparently not critical for pol II inhibition because γ-amanitin is fully potent.

Periodiate cleavage of the bond between carbons 4 and 5 (i.e., the diol) in α-amanitin destroys its activity, but activity in vivo and in vitro is partially restored by reintroduction of a primary alcohol (equivalent to replacing hydroxy-Ile with hydroxy-Val). An impact on pol II inhibition of shortening the chain length is consistent with the co-crystal structure, in which the Ile side chain fits into a hydrophobic pocket in pol II (Bushnell et al. 2002).

Wieland (1986) presents additional activity data for a number of other derivatives with modifications at this amino acid. For example, the 3-methyl group of Ile #1 (i.e., the β-branch) is critical for toxicity and pol II inhibition (Shoham et al. 1989). This methyl group forms hydrophobic contacts with Gly819 and Gly820 of subunit 1 of RNA polymerase II (Rpb1) (Brueckner and Cramer 2008; Kaplan et al. 2008).

Regarding stereochemistry, various diastereomers of Ile #1 were tested as amaninamide derivatives for inhibition of pol II, from which it was concluded that a methyl group in the (*R*) configuration at the 3-carbon is essential (Zanotti et al. 1992). Other studies have expanded on these earlier studies to make a number of amatoxin derivatives (Zanotti et al. 1987, 1992). In summary, the 4-hydroxyl group is essential for mammalian toxicity but not absolutely for inhibition of pol II, the presence of the correct stereoisomer of the 3-carbon methyl group is critical for both toxicity and enzyme inhibition, and chain length is not essential but contributes to both toxicity and enzyme inhibition.

The backbone carbonyl of 4,5-dihydroxy-Ile forms a hydrogen bond with residue Gln768 and the 4-hydroxyl group with Gln760 in Rpb1 (Kaplan et al. 2008; Brueckner and Cramer 2008). The interactions between pol II and amanitin are discussed further in Chap. 5.

#2: Tryptophan In the major amatoxins, but not the phallotoxins, the Trp is hydroxylated at C6. The amatoxin exceptions are amaninamide and amanin, which are α-amanitin and β-amanitin, respectively, without the 6-hydroxyl group (Figs. 2.6 and 2.7). The hydroxyl substitution on Trp gives the amatoxins their characteristic UV absorption profiles compared to the phallotoxins (Figs. 2.3 and 2.4 and Table 2.1). It also accounts for the characteristic CD spectra of the amatoxins and their specific color reactions with acidified cinnamaldehyde and in the Wieland-Meixner test (Wieland 1986).

The 6-hydroxyl group on Trp is not essential for either toxicity or inhibition of pol II, evidenced by the high potency of amanin and amaninamide (Fig. 2.6). Methylation of the 6-hydroxyl group to form O-methyl-α-amanitin also does not reduce activity (Wieland 1986). This position has therefore attracted attention for introducing modifications to amanitin such as tritiation and coupling to fluorescent dyes, affinity column matrices, and proteins including antibodies (Anderl et al. 2013; Liu et al. 2015; Lutter and Faulstich 1984). α-Amanitin labeled with [^{125}I] at C7 of the indole ring is as inhibitory to pol II as the parent compound (Morris et al. 1978).

A number of alkyl ether substitutions at the 6-hydroxyl group of Trp have been made and tested for toxicity and pol II inhibition. Divergence from a simple positive correlation between toxicity and enzyme inhibition in the series of alkyl derivatives could be explained by lipophilicity promoting cell penetration (i.e., some substitutions affect pol II accessibility rather than pol II binding) (Wieland 1986).

The backbone carbonyl of Trp #2 forms hydrogen bonds with Arg726 and Gln767 of Rpb1. The benzyl ring of the Trp indole forms hydrophobic contacts with Gln760, Arg726, Ile756, and/or Ala759. Consistent with the chemical structure/activity studies, there is no evidence for contact between the 6-hydroxyl of Trp #2 and Rpb1 or Rpb2 (Bushnell et al. 2002; Brueckner and Cramer 2008; Chap. 5).

#3, 4, 5: Glycine, Isoleucine, and Glycine The amatoxins contain two Gly residues at positions #3 and #5. Precursor peptide precursors that are substituted with bulky amino acids at these positions are generally not cyclized efficiently by POPB, the amanitin biosynthetic macrocyclase (Chap. 4). The Ile at position #4 is not modified in any of the amatoxins.

Substitution with Ala at positions #3, 4, or 5 in amanullin (which is α-amanitin with no hydroxylations on Ile #1 – Fig. 2.6) is less inhibitory to pol II than native amanullin (Wieland 1986; Shoham et al. 1989). Substitution of Gly #3 with Ala reduces activity more than substitution of Gly #5 with Ala. Therefore these amino acids, especially Gly #5, were predicted to form critical contacts with pol II (Wieland et al. 1981).

In a study of the binding of 24 α-amanitin derivatives to three monoclonal antibodies and pol II, Ile #4 was the single most consistently critical residue. Ile #4 is part of a β-turn (Baumann et al. 1994; Kostansek et al. 1978). Amanullin is not toxic when injected into mice, so the contribution of these amino acids to toxicity as opposed to pol II inhibition could not be tested.

From the co-crystal structure of α-amanitin and yeast pol II, the backbone nitrogen of Ile #4 forms a hydrogen bond with Gln767 of Rpb1 (Bushnell et al. 2002), and its backbone carbonyl forms a hydrogen bond with Asn723 (Brueckner and Cramer 2008). The co-crystal structure also indicates that the side chain of Ile #4 inserts into a hydrophobic pocket on pol II which includes Gly772 and Val719 (Bushnell et al. 2002; Brueckner and Cramer 2008). The backbone nitrogen of Gly #5 forms a hydrogen bond with His1085 of yeast Rpb1 (Kaplan et al. 2008). Overall, the structure/activity studies and the co-crystal structure appear to be consistent with each other in regard to the importance of these three amino acids.

#6: Cysteine The Cys residue forms a cross-bridge to C2 of Trp, thereby making the amatoxins bicyclic. The dipeptide Cys-Trp has the trivial name tryptathionine; to date it has not been found in any natural product other than the *Amanita* cyclic peptides. In the amatoxins (but not phallotoxins), the sulfur atom is oxidized to the sulfoxide with the (*R*) configuration.

Reduction of the (*R*)-sulfoxide to the thioether (sulfide) or oxidation to the sulfone did not reduce the toxicity of O-methyl-α-amanitin to mice. However, replacing the (*R*)-sulfoxide with the (*S*)-sulfoxide reduced toxicity ~20-fold (Matinkhoo et al. 2018; Wieland 1986). In binding to pol II, the (*S*)-sulfoxide form of Ile[1]-amaninamide (i.e., amaninamide with unmodified Ile at position #1 instead of 4,5-dihydroxy-Ile; Fig. 2.6) bound five times less strongly than the (*R*)-sulfoxide (K_i = 1 μM vs. 5 μM) (Zanotti et al. 1981).

#7: Asparagine or Aspartic Acid Amino acid #7 is Asn in α-amanitin and Asp in β-amanitin. Thus, there are amatoxins that are chemically neutral and those that are acidic. Amatoxins containing Asp, such as β-amanitin and ε-amanitin, and those containing Asn, such as α-amanitin and γ-amanitin, are fully biologically active in vivo and in vitro. A number of derivatives of β-amanitin with modifications to the carboxyl group besides amidation are active (Wieland 1986). The backbone carbonyl of Asn #7 forms a hydrogen bond with Ser769 in Rpb1 (Bushnell et al. 2002).

Wieland and Boehringer (1960) published a protocol to amidate β-amanitin with ethoxycarbonyl chloride (ethyl chloroformate) and ammonia and thereby convert it to α-amanitin. This protocol could be useful since β-amanitin is a large percentage of the total amatoxins made by *A. phalloides* (Fig. 2.3), yet commercial demand for α-amanitin is stronger.

#8: Proline The Pro residue in both the phallotoxins and the amatoxins is hydroxylated at C4 (abbreviated as 4-L-Hyp or just Hyp). In the amatoxins, Hyp has the (2*S*,4*R*) configuration, also known as *trans*-Hyp. This assignment is based on the crystal structures of β-amanitin (Kostansek et al. 1978) and α-amanitin (Brueckner and Cramer 2008) and on chemical synthesis of active amanitin derivatives containing *trans*-4-Hyp (Matinkhoo et al. 2018; Zhao et al. 2015). *Trans*-Hyp is also found in collagen and plant Hyp-rich glycoproteins. The Hyp in phallotoxins has the *cis* configuration (see below).

From a comparison of the activities of amanullin and proamanullin, which differ only in the presence or absence of the 4-hydroxyl group, this functional group is important for amatoxin activity (Fig. 2.6). Neither compound is toxic in mouse assays, but proammanullin is ~1000-fold less inhibitory than amanullin to pol II in vitro (Fig. 2.6). This is consistent with the co-crystal structure of amanitin and yeast pol II, which shows hydrogen bonds between the hydroxyl group of Hyp and Glu822 and His1085 in Rpb1 (Bushnell et al. 2002; Brueckner and Cramer 2008). It is apparently not known if amatoxins with *cis*-Hyp are active either in vivo or in vitro.

2.3.3 Phallotoxins

Toxicity of phallotoxin derivatives in general corresponds well with their ability to bind F-actin (Chap. 5), with the usual caveat that pharmacokinetics (e.g., stability and cell membrane permeability) can strongly influence toxicity by influencing uptake and target access. Like the amatoxins, overall molecular shape, which has been studied by circular dichroism (CD) and NMR, is critical for the phallotoxins. Cleaving the amino acid backbone or removing the sulfur bridge (tryptathionine) abolishes toxicity and F-actin binding. Bicyclic hexapeptides and octapeptides based on phalloidin are nontoxic.

There are seven described phallotoxins (Figs. 2.8 and 2.9). *Aci* in the name of a phallotoxin, e.g., phallacidin and phallacin, indicates it is acidic due to substitution of Asp for Thr at position #5. The suffix *din* indicates the presence of dihydroxy-Leu at position #3 (e.g., phalloidin), whereas *sin* indicates the presence of trihydroxy-Leu (e.g., phallisin). The core amino acid sequences of the phallotoxins vary at two amino acids, positions #4 (Ala or Val) and #5 (Thr or Asp). Genes encoding both of these, Ala-Trp-Leu-<u>Ala-Thr</u>-Cys-Pro and Ala-Trp-Leu-<u>Val-Asp</u>-Cys-Pro (differences underlined), are present in the genome of *A. phalloides* (Pulman et al. 2016). Since all of the other differences between the members of the phallotoxin family are posttranslational modifications, all of the phallotoxins can be encoded by just two genes (Chap. 4).

Regarding important overall characteristics of the phallotoxins, based on many structure/activity studies of derivatives containing modified side chains, it appears that amino acids #1, 2, 6, and 7 (Ala, Trp, Cys, and Pro), which form one face of the molecule, are more important than amino acids #3, 4, and 5 (Leu, Val/Ala, and Asp/Thr) on the other side (Wieland 1986). The phallotoxins have not been crystallized in a form suitable for high-resolution X-ray crystallographic analysis. Their three-dimensional structures have been deduced from NMR studies of phallotoxin derivatives (Patel et al. 1973; Zanotti et al. 2001).

2.3.3.1 The Individual Amino Acids of the Phallotoxins

Like the amatoxins, the amino acid numbering in this volume follows the order in the linear precursor peptides. Therefore, phalloidin has the sequence Ala^1-Trp^2-Leu^3-Ala^4-Thr^5-Cys^6-Pro^7 or AWLATCP in single-letter code. Most of the literature citations relevant to the structure/activity studies can be found in Wieland (1986).

#1: Alanine This amino acid is unmodified in all of the natural phallotoxins. Substitution with Gly reduces potency, so the methyl group is predicted to form a critical interaction with F-actin.

#2: Tryptophan Like the amatoxins, the natural phallotoxins contain Trp and Cys linked as tryptathionine. The bicyclic nature of the phallotoxins is critical for activity. Spectrophotometric analysis indicates that the indole chromophore interacts with F-actin. Unlike the amatoxins, none of the phallotoxins are hydroxylated on Trp #2, which affects their UV absorbance and reaction with cinnamaldehyde/HCl. The lack of hydroxy-Trp also renders phallotoxins negative with the Wieland-Meixner test.

#3: Leucine This amino acid is hydroxylated once, twice, or three times in both the neutral and acidic phallotoxins. 4-Hydroxy-L-Leu is found in phalloin and phallacin, 4,5-dihydroxy-L-Leu is found in phalloidin and phallacidin, and 4,5,5′-trihydroxy-L-Leu is found in phallisacin and phallisin (Figs. 2.8 and 2.9). All hydroxylation variants are about equally toxic when injected into mice (Fig. 2.8). Wieland (1986) calls this amino acid side "the most versatile," in that it can tolerate many modifications without affecting toxicity or binding to F-actin. For example, the 4-hydroxyl group in phalloidin can be acetylated without loss of activity. Phalloidin has been modified at this position with fluorescent dyes such as coumarin, tetramethyl rhodamine B isothiocyanate, and fluorescein isothiocyanate for visualization of the actin cytoskeleton (Wulf et al. 1979; Chap. 5).

A series of phalloidin derivatives with modified chain lengths of amino acid #3 were prepared and tested by Falcigno et al. (2001). All were active in actin binding; a branched side chain appears to be the only requirement.

Due to the vicinal hydroxyl groups, the phallotoxins containing 4,5-dihydroxy-Leu are cleaved by periodate. Norphalloin, which contains norvaline at position #3 of phalloin instead of Leu, is toxic and binds F-actin, confirming that, unlike Ile #1 of the amatoxins, not even a single hydroxylation is required on this amino acid for activity of the phallotoxins. However, replacement of dihydroxy-L-Leu with the D isomer in phalloidin or phallasin eliminates toxicity, presumably due to steric hindrance within the F-actin binding site.

#4: Valine or Alanine The acidic phallotoxins, i.e., those containing Asp instead of Thr at position #5, contain Val at this position, whereas the neutral phallotoxins contain Ala. Phalloin with D-Ala at position #4 instead of L-Ala is nontoxic. A number of other phallotoxin derivatives with altered side chain lengths at position #4 have also been synthesized and tested for toxicity and F-actin binding (Wieland

1986). The overall conclusion is that the side chain at position #4 appears to be only slightly important for toxicity and binding to F-actin.

#5: Aspartic Acid or Threonine Phallacidin (an acidic phallotoxin) contains Asp at this position and phalloidin (a neutral phallotoxin) contains Thr. For activity the side chain must have at least two carbon atoms (Falcigno et al. 2001). Whereas Thr is a naturally hydroxylated amino acid, an additional posttranslational hydroxylation must occur in phallotoxins containing hydroxy-Asp (i.e., phallacidin, phallisacin, and phallacin). The hydroxyl group is not essential because position #5 can be replaced with aminoisobutyric acid (Abu) with only a two- to threefold reduction in actin binding (Falcigno et al. 2001). Acetylation of the Asp side chain carboxyl group also does not affect activity, consistent with the high potency of the neutral phallotoxins.

Alone among all of the amino acids in all of the *Amanita* peptide toxins, amino acid #5 has the D configuration at its α-carbon. The chirality of this amino acid was determined by optical rotation of the isolated individual amino acids and by digestion with specific L-Thr deaminase (Wieland and Schnabel 1962a, b). The D configuration is important for actin binding (Falcigno et al. 2001).

#6: Cysteine The phallotoxins contain the same tryptathionine cross-bridge as the amatoxins, but unlike the amatoxins, the Cys sulfur exists as the thioether (sulfide) and is not oxidized. Evidence from difference spectra suggests the thioether interacts with the amino acid side chains of actin (Wieland 1986). Molecular modeling indicates an interaction between the S atom and Asp180 and Lys270 of the B chain of actin (Skillman et al. 2011). Phalloidin with the sulfur oxidized to the (*R*) configuration was as toxic as native phalloidin, as was the sulfone. Phalloidin oxidized to the (*S*) configuration was 100-fold less active (Wieland 1986).

#7: Proline Pro in the phallotoxins is 4-(*S*)-hydroxylated (i.e., the *cis* isomer), whereas Hyp has the *trans* configuration in the amatoxins (see above). The *cis* isomer of Hyp is also known as *allo*-Hyp. The hydroxyl group makes a major contribution to toxicity of the phallotoxins, evidenced by comparing phalloin with prophalloin (Fig. 2.8). Phallotoxins with 4-(*R*)-Hyp or 4-(*S*)-acetyl-Pro are also inactive (Wieland 1986).

2.3.4 Virotoxins

Amanita virosa is a white species of *Amanita* in sect. *Phalloideae* originally described from Europe. From mushrooms identified as "*A. virosa*" growing in West Virginia (but now designated as its own species; Chap. 3), Faulstich et al. (1980) extracted novel compounds that they named virotoxins (Fig. 2.10). Chemically, the virotoxins are closely related to the phallotoxins, and they also bind and stabilize F-actin. LD_{50}s for mice for the virotoxins are in the range 1–5 mg/kg, comparable to

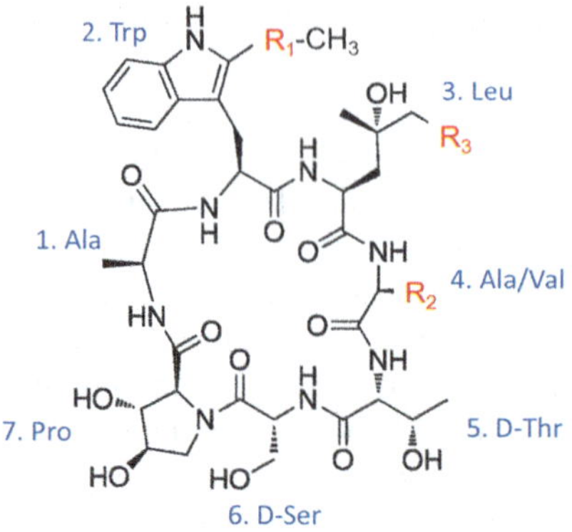

	R_1	R_2	R_3	LD_{50} (mg/kg)
viroidin	SO_2	$CH(CH_3)_2$	CH_3	1.0
desoxoviroidin	SO	$CH(CH_3)_2$	CH_3	5.1
Ala⁴-viroidin	SO_2	CH_3	CH_3	3.7
Ala⁴-desoxoviroidin	SO	CH_3	CH_2OH	
viroisin	SO_2	$CH(CH_3)_2$	CH_2OH	1.7
desoxovirosin	SO	$CH(CH_3)_2$	CH_2OH	3.4

Fig. 2.10 Structures of the known virotoxins. LD_{50} values are for intraperitoneal injection into mice. (From Loranger et al. 1985). In the same study, the LD_{50} for phalloidin was 1.4 mg/kg

phallotoxins (Loranger et al. 1985; Fig. 2.10). *A. virosa* also produces amatoxins and phallotoxins (Faulstich et al. 1974; Fig. 2.3).

Desoxoviroidin was reported from *A. exitialis* collected in China (Deng et al. 2011), but this result was questioned by Wei et al. (2017). Ala-viroidin, viroisin, and viroidin were detected in *A. subpallidorosea*, a species that grows in China but that is closely related to the European "true" *A. virosa* (Wei et al. 2017; Chap. 3).

2.3.4.1 Structure/Activity Studies in the Virotoxins

There are six known virotoxins (Fig. 2.10). Assuming they are derived from phallotoxin-like molecules, all of the virotoxins are variants of the amino acid sequences AWLATCP or AWLVTCP, i.e., phalloidin in which position #4 can be Ala or Val. In *A. virosa*, viroisin is the most abundant, followed by viroidin. The virotoxins are all monocyclic due to lack of the tryptathionine cross-bridge found in the phallotoxins. Amino acid #1 is always Ala (in our numbering scheme, which follows the precursor peptide sequences of the phallotoxins). Amino acid #2 is L-Trp modified at position #2 with methylsulfinyl ($-SO-CH_3$) or methylsulfonyl ($-SO_2-CH_3$) groups, i.e., a sulfoxide or sulfone, respectively. Amino acid #3 is L-Leu that is hydroxylated either two or three times, always on the 4-carbon and the 5-carbon and then optionally again on the 5′-carbon, similar to the phallotoxins. Amino acid #4 is either unmodified L-Ala or L-Val, like in phalloidin or phallacidin, respectively. Amino acid #5 is D-Thr, like in phalloidin, and amino acid #6 is D-Ser, instead of L-Cys as in the phallotoxins. The chiralities of these and the other amino acids were determined by optical rotation of the isolated amino acids and by selective digestion with L- and D-amino acid oxidases (Faulstich et al. 1980). Amino acid #7 is 3,4-dihydroxy-L-Pro. Stereochemically, this amino acid is the 2,3-*trans*-3,4-*trans* isomer, also known as (2*S*,3*R*,4*R*)-diHyp (Buku et al. 1980a). In regard to the nomenclature of the virotoxins, those that end in *din* have dihydroxylated Leu #3, and those that end in *sin* are trihydroxylated at this amino acid.

By CD and NMR, the virotoxins have a stable three-dimensional conformation more similar to the bicyclic phallotoxins than to the monocyclized phallotoxins, which exist as a population of conformers. Furthermore, the hydrophobic residues of the virotoxins tend to be oriented in one direction, similar to the phallotoxins and consistent with structure/activity studies on the individual amino acids (Bhaskaran and Yu 1994).

Because possession of a defined overall shape is critical for binding to F-actin, one or more of the features that distinguishes the virotoxins from the phallotoxins must play a critical role in shape and binding. Unique features include the methylsulfinyl-Trp (or methylsulfonyl-Trp), the dihydroxyHyp, and the D-Ser moieties. Based on synthetic derivatives of viroisin, Zanotti et al. (1999) concluded that the hydroxyl group on D-Ser is not important for conformational stability and F-actin binding, but that the D configuration at the α-carbon is.

The second hydroxyl group on the Pro residue contributes to binding to F-actin, because derivatives with monohydroxy-Pro (Hyp) are fivefold less active (Kahl et al. 1984). Like the phallotoxins, hydroxylation of Leu is not essential. The sulfur-containing side chain of Trp has likewise been shown to be unimportant for F-actin binding, whereas, based on UV difference spectroscopy, the indole ring does interact with F-actin (Wieland 1986).

A stereoisomer of viroidin, called alloviroidin, has been purified from *A. suballiacea* (Little et al. 1986). Alloviroidin is viroidin containing 2*S*,4*S*-dihydroxy-Leu in place of the 2*S*,4*R* isomer. It was as active as viroidin in binding to actin.

2.3.5 Antamanide and Cycloamanides

In addition to the amatoxins, phallotoxins, and virotoxins, other cyclic peptides have been isolated from *A. phalloides* and *A. exitialis*. They are lipophilic compared to the amatoxins and phallotoxins, being sparingly soluble in water but soluble in acetone, alcohols, and pyridine. They can be extracted with ethyl acetate. At the time of their isolation, these compounds were considered to be biologically inert other than the protective effect of antamanide against phalloidin (see below). However, since then several of them have been shown to have diverse biological activities (Chap. 5).

2.3.5.1 Cycloamanides

The original cycloamanides were all monocyclic peptides of six to eight amino acids isolated from *A. phalloides* collected in Germany (Gauhe and Wieland 1977) (Table 2.2). At the time of their initial discovery, they had no known biological activities. Some of them have subsequently been shown to be immunosuppressive and have other activities, but they are not toxic in any context (Chap. 5). Wieland and co-workers originally identified four cycloamanides, CyA-A, CyA-B, CyA-C, and CyA-D, which we here simplify to CylA through CylD to eliminate the

Table 2.2 Structures of known cycloamanides

Name	Structure	Number of amino acids	Reference
Cycloamanide A (CylA)	Cyclo(VFFAGP)	6	Wieland (1986)
Cycloamanide B (CylB)	Cyclo(SFFFPIP)	7	Wieland (1986)
Cycloamanide C (CylC)	Cyclo(MLGFLVLP)	8	Wieland (1986)
Cycloamanide D (CylD)	Cyclo(MLGFLPLP)	8	Wieland (1986)
Cycloamanide E (CylE)	Cyclo(SFFFPVP)	7	Pulman et al. (2016)
Cycloamanide F (CylF)	Cyclo(IVGILGLP)	8	Pulman et al. (2016)
Antamanide	Cyclo(FFVPPAFFPP)	10	Wieland (1986)
Amanexitide	Cyclo(VFFPVFSLP)	9	Xue et al. (2011)

All are from *A. phalloides* except amanexitide, which is from *A. exitialis*. All are homodetic (head-to-tail) monocyclic peptides synthesized as precursor peptides on ribosomes (Chap. 4). All amino acids have the L configuration. As isolated, cycloamanides C and D contain oxidized Met (Wieland 1986)

syntactically awkward hyphen (Table 2.2). The cycloamanides are unmodified by hydroxylations or other posttranslational modifications (other than cyclization), and all of their amino acids have the L configuration. CylC and CylD have oxidized Met, although it is not clear if this is their natural state or an artifact that occurs during isolation (Wieland 1986). Lacking Cys and/or Trp, the cycloamanides cannot be bicyclized through tryptathionine. The crystal structure of CylA has been solved (Chiang et al. 1982).

In light of the discovery of the gene family encoding the *Amanita* cyclic peptide toxins, it can now be assumed that the cycloamanides are also synthesized on ribosomes as small (<40 amino acid) precursor peptides and cyclized by prolyl oligopeptidase B (POPB) (Chap. 4). To reflect their common biogenesis, one can apply the more general term *cycloamanide* to all of the *Amanita* cyclic peptides including amatoxins, phallotoxins, virotoxins, classic cycloamanides, and antamanide.

The sequenced genome of *A. phalloides* has a gene encoding CylB, cyclo(SFFFPIP), but none encoding CylA, CylC, or CylD (Table 2.2) (Pulman et al. 2016). *A. bisporigera* has no genes for any of the chemically characterized cycloamanides (Pulman et al. 2016). However, *A. phalloides* contains at least two novel cycloamanide genes that are expressed at the chemical level, which have been named CylE and CylF (Pulman et al. 2016). Fungi in *Amanita* sect. *Phalloideae* probably produce additional cycloamanides, which can be predicted from the encoding gene sequences (Li et al. 2014; Pulman et al. 2016; Chap. 4). A novel cycloamanide, amanexitide, and its encoding gene have been described from *A. exitialis* (Li et al. 2013; Xue et al. 2011) (Table 2.2).

Differences in the complement of cycloamanide genes and cyclic peptides found in *A. phalloides* in different studies are probably due to natural variation in the complement of genes in the cycloamanide superfamily between individual mushrooms, populations, or subspecies (Wieland 1986; Pulman et al. 2016) (Chap. 6).

An operationally important chemical attribute shared by the cycloamanides that distinguishes them from the amatoxins, phallotoxins, and virotoxins is their relatively greater lipophilicity. The cycloamanides generally contain a higher proportion of hydrophobic amino acids and lack polar posttranslational hydroxylations. Traditionally, the cycloamanides were extracted from mushrooms with solvents such as acetone and ethyl acetate instead of aqueous methanol used for the amatoxins, phallotoxins, and virotoxins. Such differences in solubility can be critical for successfully finding novel cycloamanides predicted from genome sequences (Pulman et al. 2016).

G. marginata makes α-amanitin, and its genome has no additional related genes capable of encoding cycloamanides (Riley et al. 2014). Some species of *Lepiota* make α-amanitin (Chap. 3). Five genes predicted to encode novel cycloamanides were found in the genome of a specimen of *L. subincarnata* from the U.S. Pacific Northwest. Three of these predicted cycloamanides, LsCyl2, LsCyl3, and LsCyl4, with structures cyclo(LFFPLPIPP), cyclo(IALVLSLFP), and cyclo(FVLLLIVPP), respectively, were chemically identified by LC/MS/MS in mushroom extracts (unpublished results from the author's lab; Chap. 4). Note that these *Lepiota* cycloamanides are dominated by hydrophobic amino acids.

2.3.5.2 Antamanide

Antamanide is a monocyclic decapeptide (Table 2.2) that has been the subject of extensive structural and biological research (Wieland 1986). Antamanide is nontoxic, but it was purified on the basis of its ability to protect mice against simultaneous injection with phalloidin. A molecular explanation for the protective effect of antamanide is that it is a competitive inhibitor of the uptake of phalloidin by liver cells through the OATP1B1 transporter, similar to the competitive inhibition of antamanide for amanitin through the transporter OATP1B3 (Letschert et al. 2006; Chap. 5). Derivatives of antamanide have been synthesized and the conformations of their sodium complexes studied by NMR (Amodeo et al. 1998). The biosynthesis and biological activities of antamanide are considered in Chaps. 4 and 5, respectively.

2.3.6 Other Toxic Peptidic and Nonpeptidic Natural Products from Amanita *and Other Agarics*

Agarics and related fungi make many other compounds with interesting biological activities (Abraham 2001; Agger et al. 2009; Alves et al. 2012; Schneider et al. 2008; Sliva 2004; Thomas et al. 2001). Although it is beyond the scope of this book to discuss the full gamut of natural products from agarics, those of particular interest or relevance are discussed here. Bresinsky and Besl (1990), Benjamin (1995), and Spoerke and Rumack (1994) provide excellent illustrated guides to the full range of known mushroom toxins.

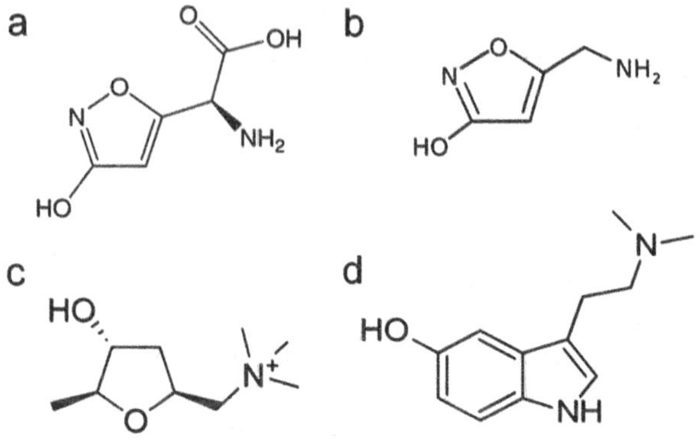

Fig. 2.11 Nonpeptidic toxins from species of *Amanita*. (**a**) Ibotenic acid. Credit: Elbreapoly, https://commons.wikimedia.org/w/index.php?curid=4767783; (**b**) Muscimol. Credit: Edgar181, https://commons.wikimedia.org/w/index.php?curid=11418162; (**c**) Muscarine. (**d**) Bufotenine. Credit: Edgar181, https://commons.wikimedia.org/w/index.php?curid=11418162

Muscimol and Ibotenic Acid These two compounds are related isoxazoles (Fig. 2.11a, b). They are found in *Amanita muscaria*, the fly agaric (Fig. 1.8), and *Amanita pantherina*, the panther cap. Neither of these two species make amatoxins or phallotoxins. Ibotenic acid, (*S*)-2-amino-2-(3-hydroxyisoxazol-5-yl) acetic acid, is converted to muscimol (5-(aminomethyl)-isoxazol-3-ol) by decarboxylation during cooking and/or in the gastrointestinal tract. Muscimol is a mimic of γ-aminobutyric acid (GABA) and therefore a potent agonist of the $GABA_A$ neurotransmitter receptors. It is almost certainly the compound responsible for the psychotropic effects of consuming *A. muscaria* (Krogsgaard-Larsen and Hansen 1992; Michelot and Melendez-Howell 2003; Stebelska 2013). Mycophagous species of *Drosophila* are less inhibited by ibotenic acid and muscimol than frugiverous species (Tuno et al. 2007). The ecological ramifications of fly resistance to α-amanitin and other mushroom toxins are discussed in Chap. 6.

Muscarine Some mushrooms, in particular species of *Inocybe* and *Clitocybe*, contain significant levels (~1.6%) of L-(+)-muscarine (Kosentka et al. 2013). Despite its name, *A. muscaria* itself only contains low levels of muscarine. Structurally, muscarine is 2,5-anhydro-1,4,6-trideoxy-6-(trimethylammonio)-D-ribo-hexitol (Fig. 2.11c). It is a mimic of acetylcholine and defines the muscarinic acetylcholine subclass of neurotransmitter receptors. It stimulates the peripheral parasympathetic nervous system and as such is an important tool in neurobiology (Jin 2013).

Bufotenine This tryptamine derivative (*N*,*N*-dimethylserotonin; Fig. 2.11d) differs from psilocin (the active form of psilocybin) in having a hydroxyl group on C5 of the indole ring instead of C4. It was originally isolated from the *Bufo* genus of toads. The psychoactivity of bufotenine is disputed, but it is almost certainly not toxic when consumed orally (Bresinsky and Besl 1990). It is found in *Amanita citrina*, which is not in sect. *Phalloideae* and therefore does not contain amatoxins. Beutler and Der Marderosian (1981) estimated the concentrations to be in the range of 1.6–7.5 mg/g dry weight. It reacts positively with the Wieland-Meixner assay (Beutler and Vergeer 1980).

Orellanine More than 34 species of *Cortinarius*, including *C. rubellus* (= *C. speciosissimus*), *C. armillatus*, and *C. orellanus*, produce orellanines, which are glycosylated bipyridyl N-dioxides. In Europe, many deaths have been attributed to these compounds, which are highly nephrotoxic (Benjamin 1995; Herrmann et al. 2012).

Cortinarins Three cyclic peptides, cortinarins A, B, and C, were reported from *Cortinarius speciosissimus* collected in Scotland (Tebbett and Caddy 1984). Most other species of surveyed *Cortinarius* species also contained compounds with a similar fluorescence profile. The compounds were nephrotoxic when injected intraperitoneally into mice. The compounds were characterized as bicyclic decapeptides of sequence cyclo(Phe-Trp-Val-Orn-Leu-Ile-Cys-Thr-Gly-Lys) in which the Trp and Cys are cross-bridged by a thioether, like the phallotoxins. The Trp is substituted at C4 with a methoxyl group in cortinarin A and with a hydroxyl group in cortinarin B. Cortinarin C is cortinarin A with Ala in place of Cys (and therefore lacking tryptathionine). All amino acids have the L configuration except Thr.

Existence of the cyclic peptide cortinarins was questioned by Matthies and Laatsch (1991), who attemped unsuccessfully to replicate key aspects of the work including detection of any cyclic peptides with the postulated chemical properties. The fluorescent compounds detected by Tebbett and Caddy (1984) were attributed by Matthies and Laatsch (1991) to degradation products of steroids and the previously described nephrotoxic orellanines, which are structurally bipyridines.

Because agarics outside the genus *Amanita* are known to produce cyclic peptides that are structurally related to the amatoxins and phallotoxins, it is plausible that *Cortinarius* might also produce related compounds. However, an unusual feature of the cortinarins in light of our current understanding of the biosynthesis of the *Amanita* cyclic peptides in *Amanita*, *Galerina*, and *Lepiota* is their lack of proline (Pro). Pro is found in all of the *Amanita* cyclic peptides, including not only the amatoxins and phallotoxins but also the cycloamanides, virotoxins, and antamanide, as well as all members of the predicted cycloamanide superfamily in *A. phalloides* and *A. bisporigera* (Chap. 4). Pro must be present in order for the amatoxin and phallotoxin precursor peptides to be hydrolyzed and cyclized by prolyl oligopeptidase B (POPB). Furthermore, Tebbett and Caddy (1984) found the noncanonical amino acid ornithine in the cortinarins. Ornithine is common in nonribosomal peptides but not in ribosomally encoded peptides such as the cycloamanide family. Despite having significant other chemical features in common with the *Amanita* cyclic peptides, the absence of Pro and the presence of ornithine in the presumptive cortinarins would require that they have a strikingly different biogenesis. Taking all of the evidence into consideration, the existence of cortinarins with the claimed structures should be considered doubtful in the absence of confirmatory data (Chilton 1994; Laatsch and Matthies 1992; Tebbett 1992).

Other Bioactive Molecules from Agarics: Preliminary Reports Bioactive compounds have been isolated from species of *Amanita* outside sect. *Phalloideae*, some of which are also poisonous. For many of these compounds, their contributions to poisoning symptoms are possible but still tentative (Michelot and Melendez-Howell 2003).

Some species of *Amanita*, mainly in sect. *Lepidella*, growing in Asia, North America, and Europe, have been implicated in poisonings in which renal failure is the major symptom (Apperley et al. 2013; West et al. 2009; Yang et al. 2006). In North America, the main culprit is *A. smithiana*. Kirchmair et al. (2012) describe the "*Amanita* nephrotoxic syndrome" as "characterized by an early onset of gastrointestinal symptoms, mild hepatic damage and severe but reversible acute renal failure due to acute interstitial nephritis." The responsible agents have been detected by TLC but not yet purified or fully characterized (Apperley et al. 2013). Kirchmair et al. (2012) discuss the evidence against the active agent being allenic norleucine ((2*S*)-2-amino-4,5-hexadienoic acid), which is found in several species of *Amanita* and is nephrotoxic in mice (Chilton et al. 1973; Kirchmair et al. 2012; Pelizzari et al. 1994). European relatives of *A. smithiana*, such as *A. proxima*, *A. boudieri*, *A. gracilior*, and *A. echinocephala*, can also cause nephrotoxicity (Kirchmair et al. 2012). Because none of these species are in sect. *Phalloideae*, the responsible nephrotoxins are probably not chemically or biogenically related to the *Amanita* cyclic peptides. Yamaura et al. (1986) reported three toxic compounds, L-2-amino-4-pentynoic acid, D,L-propargylglycine, and L-2-amino-4,5-hexadienoic acid, from *Amanita abrupta* in Japan. The chemical basis of many mushroom poisonings remains unknown (Saviuc and Danel 2006).

The mushroom toxins responsible for sudden unexplained death syndrome in China caused by eating *Trogia venenata* are aminohexynoic acids (Zhou et al. 2012). Cycloprop-2-ene carboxylic acid from *Russula subnigricans* causes fatal rhabdomyolysis (Matsuura et al. 2009).

A number of agarics make compounds toxic to animals that are still unidentified. *Conocybe albipes* makes a nematicidal compound (Hutchinson et al. 1996). *Russula bella* makes an unknown compound toxic to *Collembola* (springtails) (Nakamori and Suzuki 2007). *Inocybe* makes compounds toxic to dogs (Lee et al. 2009). In all of these cases, the compounds accumulate in specialized structures such as cystidia. The relationship between mushroom cellular structures and toxin biosynthesis is discussed in more detail in Chap. 3.

For the sake of completeness, mention must be made of the myriamanins, which are discussed by Litten (1975). These are high molecular weight compounds or conjugates from which the amatoxins and phallotoxins were proposed to be derived by fragmentation during purification. There is apparently but a single scientific article on these conjugates, which was unobtainable by the author, and apparently no follow-up in the primary scientific literature (Courtillot and Staron 1970). Wieland (1986) states that he and co-workers were unable to replicate the myriamanin results, and they are therefore probably spurious.

In the early days of *Amanita* chemistry, a toxic hemolytic protein, phallolysin, was discovered in *A. phalloides* and later purified (Faulstich et al. 1983). Phallolysin, a mixture of two or three 34-kDa proteins, is toxic to rabbits at ~40 μg/kg and disrupts mammalian erythrocyte cell membranes in vitro. The encoding gene(s) have not yet been identified but are presumably present in the sequenced genome of *A. phalloides* (Pulman et al. 2016). Wieland (1986) thoroughly covers the work on phallolysin up to 1986; little has been published since then. A cytolytic protein isolated from *Flammulina velutipes* mushrooms might be related to phallolysin (Bernheimer and Oppenheim 1987). Toxophallin, a distinct, cytotoxic protein, has been isolated from *A. phalloides* and shown to have L-amino acid oxidase activity (Stasyk et al. 2010). Based on partial amino acid sequences from the purified protein, genes for toxophallin are present in the genomes of *A. phalloides* (NCBI MEHY01000000.1) and *A. bisporigera* (MIPV01000000.1) (Pulman et al. 2016). Because of the intrinsic instability of enzymatic proteins in the mammalian gastrointestinal tract, neither phallolysin nor toxophallin are likely to be involved in mushroom poisonings.

2.4 Chemical Synthesis of the Toxins

2.4.1 Synthesis of Amatoxins

Work on the complete de novo synthesis of the amatoxins began shortly after the initial structural studies. Synthetic work has been motivated by three considerations: first, to prove the structures; second, to make derivatives with desirable modifications; and third, to provide a toxin source that is more reliable than extraction from wild mushrooms.

Although over the years a number of biologically active amatoxins have been synthesized, complete synthesis of α-amanitin was accomplished only recently (Matinkhoo et al. 2018). Methods for solid-phase synthesis of peptides, including cyclic peptides, are well-established, but the amatoxins present unique difficulties due to their special posttranslational modifications. A major obstacle to synthesizing amatoxins has been the hydroxylated Ile (position #1); both mono-hydroxylated and di-hydroxylated Ile have three chiral centers (Zhao et al. 2015). 4-Hydroxy-Ile with the correct stereochemistry (2*S*,3*R*,4*S*) has been synthesized by itself (De Lamo Marin et al. 2010; Wang et al. 2002). Another route to this compound was developed by modifying 4-hydroxy-Ile dehydrogenase, which is normally non-stereoselective, to efficiently produce (2*S*,3*R*,4*S*)-4-hydroxy-Ile at >99% stereoselectivity (Shi et al. 2017).

6-Hydroxy-Trp can be avoided by synthesizing versions of the native amatoxin amanin, which lacks the Trp hydroxyl group but is highly active in vivo and in vitro (Fig. 2.6). However, the hydroxyl group of Trp is a good attachment site for modifying reagents including antibodies (Anderl et al. 2013). Since sulfoxidation of the Cys is also not necessary for activity, this step can also be omitted in the synthesis of active amatoxins.

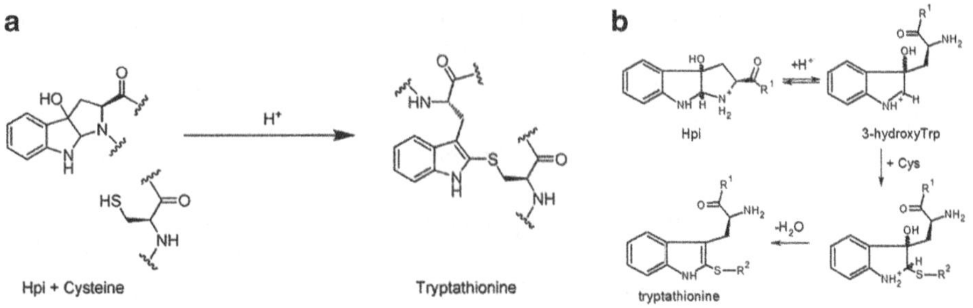

Fig. 2.12 (**a**) The acid-catalyzed Savige-Fontana reaction for tryptathionine biosynthesis. (Reprinted with permission from May et al. 2005. Copyright 2005 American Chemical Society). (**b**) Proposed reaction pathway of the Savige-Fontana synthesis of tryptathionine. Hpi is 3a-hydroxy-pyrrolo[2,3-b]-indole

Routes to the synthesis of tryptathionine (the cross-bridge between Trp and Cys) have attracted attention because it is a prominent and unique feature of the amatoxins and phallotoxins. An account of the various synthetic approaches that have been used to generate a thioether bridge between Trp and Cys can be found in May and Perrin (2007). The early approaches depended on reaction of the indole of Trp with the S-chloride derivative of Cys. An effective alternate procedure is the Savige-Fontana reaction which involves pre-oxidation of Trp and subsequent reaction with Cys (Savige and Fontana 1980). In this process, the Trp is activated by hydroxylation at C3, to make L-3*a*-hydroxy-pyrrolo[2,3-*b*]-indoline-2-carboxylic acid (Hpi). Under acidic conditions, Hpi reacts with Cys to form the 2-thioether (Fig. 2.12). An analogous reaction may occur in the biosynthetic pathway to tryptathionine (Chap. 4).

The Savige-Fontana reaction has been used to make derivatives of amanitin, some of which show activity despite lacking hydroxylated Ile or Trp (summarized in Wieland 1986). May et al. (2008) and May and Perrin (2008) showed that the Savige-Fontana reaction can also be used to synthesize bicyclic peptides such as Pro^8-Ile^1-*S*-deoxo-amaninamide and Pro^8-D-*allo*-Ile^1-*S*-deoxo-amaninamide (revised amino acid numbering), which are epimers of each other that form concomitantly during the final macrocyclization step. Structurally, these two compounds are similar to α-amanitin but lack hydroxyl groups on Pro, Ile, and Trp (Fig. 2.6). Both are only weakly toxic to hepatocytes. These studies exploited a robust route to Hpi synthesis and its incorporation into dipeptides ("synthons") suitable for making larger peptides (May et al. 2005).

Another alternate approach to the synthesis of tryptathionine-containing bicyclic peptides starts with trityl-protected Cys incorporated into a monocyclic peptide. In the presence of I_2, the thioether forms. This has been used to make derivatives of phalloidin (Schuresko and Lokey 2007).

An amanitin derivative with the amide of Asn #7 modified with an alkyne (propargyl group) was synthesized by Zhao et al. (2015). All of their amanitin derivatives lacked both the hydroxyl group on Trp and the sulfoxide but did contain the native *trans*-Hyp (2*S*,4*R*-Hyp). The derivative was subsequently coupled with rhodamine and RGD peptide through the alkyne. Rhodamine is fluorescent, and RGD peptide facilitates cellular uptake through attachment to integrins in mammalian cells (Dal Pozzo et al. 2010). The derivatized compounds were almost as active as native α-amanitin, killing HeLa cells at ~2 μM. To synthesize amanitin derivatives containing 4,5-dihydroxy-Ile, Zhao et al. (2015) also synthesized a mixture of the four diastereomers of dihydroxy-Ile using the Claisen rearrangement. The diastereomers could not be chromatographically resolved from each other and were therefore incorporated together synthetically into the amanitin derivatives. After incorporation, the resulting amanitin-like diastereomers could be partially resolved by HPLC, and only one showed biological activity, presumably the one containing the native stereoisomer, 2*S*,3*R*,4*R*-dihydroxy-L-Ile. Zhao et al. (2015) found that the 4-hydroxy-Ile-containing propargyl derivative was inactive against mammalian (Chinese hamster ovary) cells, whereas the 4,5-dihydroxy-Ile derivative was active. This is puzzling insofar as native γ-amanitin (which only has the 4-hydroxylation) is as active as α-amanitin (which is identical to γ-amanitin except with an additional hydroxyl group at C5). A patent application for synthesis of γ, δ-dihydroxy-Ile was filed by Lutz et al. (2014), but further details are not available.

The eventual successful synthesis of α-amanitin had to overcome several challenges (Matinkhoo et al. 2018). First, the tryptathionine crosslink had to be introduced using 6-hydroxy-Trp, which is unstable. Second, the correct stereoisomer of dihydroxy-Ile, which has three chiral centers, had to be obtained and incorporated into the peptide. Third, the sulfoxidation had to be stereoselectively synthesized. Successfully overcoming these hurdles led to the first synthesis of bonafide α-amanitin with an IC_{50} against Chinese hamster ovary (CHO) cells of 0.34 μM, identical to natural α-amanitin.

2.4.2 Synthesis of Phallotoxins

Active phallotoxins are easier to synthesize than amatoxins because they do not require any hydroxyl groups on Leu #3, and they naturally lack the hydroxyl group on Trp #2 and the sulfoxide on Cys #6. The first active phallotoxin to be synthesized was norphalloin, which is phalloin with norvaline at position #3 (Fahrenholz et al. 1971). Phalloin, one of the naturally occuring phallotoxins (Fig. 2.8), has been synthesized de novo (Munekata et al. 1977; Munekata 1981). Derivatives of phalloidin with Ala or Glu in position #3 (revised numbering scheme; Fig. 2.8) have been synthesized and shown to bind actin (Schuresko and Lokey 2007). Some synthetic routes to bicyclic phallotoxins lead to two atropisomers (Anderson et al. 2005), whereas others give what is presumed to be the natural isomer based on circular dichroism (CD) (Schuresko and Lokey 2007). Phalloin with L-Lys at position #3 and D-aminobutyric acid (D-Abu) at position #5 is active at binding actin, whereas substitution at positions #3 and #5 with L-Leu or D-Ala, respectively, results in an inactive product (Wieland et al. 1983). Substitutions at position #5 (which is D-Thr in native phalloin; Fig. 2.9) indicate that the side chain at this position must be at least two carbons in length in order to interact with actin.

Phallacidin, phalloidin, and some virotoxins contain dihydroxy-Leu with the 2*S*,4*R* stereochemistry, whereas alloviroidin contains the 2*S*,4*S* isomer. The synthesis of a dipeptide containing the latter compound in optically pure was accomplished by Edagwa and Taylor (2009).

2.4.3 Synthesis of Virotoxins

Kahl et al. (1984) synthesized four virotoxin-like molecules containing 4-hydroxy-Pro (4-Hyp) in place of diHyp and tested them for actin binding. Viroisin with L-Ala replacing L-Val, Hyp replacing 3,4-dihydroxy-Pro (diHyp), and L-Leu replacing 4,5-dihydroxy-L-Leu was synthesized de novo by Zanotti et al. (1999). The resulting compound bound F-actin with 20% of the efficiency of the native compound.

2.4.4 Synthesis of Cycloamanides

Being unmodified cyclic peptides, the cycloamanides including antamanide pose no particular barriers to chemical synthesis. Early synthetic work on the cycloamanides is summarized in Wieland (1986). More recently, a number of additional cycloamanides have been synthesized (e.g., Wieczorek et al. 1993). Antamanide and derivatives including the octapeptide precursor to α-amanitin, cyclo(IWGIGCNP), have been synthesized both chemically (Azzolin et al. 2011; Luo et al. 2014; Ruzza et al. 1999) and with prolyl oligopeptidase B (POPB) (Sgambelluri et al. 2018; Chaps. 4 and 7).

References

Abel JJ, Ford WW (1907) On the poisons of *Amanita phalloides*. J Biol Chem 2:273–288

Abraham WR (2001) Bioactive sesquiterpenes produced by fungi: are they useful for humans as well? Curr Med Chem 8:583–606

Abuknesha RA, Maragkou A (2004) A highly sensitive and specific enzyme immunoassay for detection of beta-amanitin in biological fluids. Anal Bioanal Chem 379:853–860. https://doi.org/10.1007/s00216-004-2663-5

Agger S, Lopez-Gallego F, Schmidt-Dannert C (2009) Diversity of sesquiterpene synthases in the basidiomycete *Coprinus cinereus*. Mol Microbiol 72:1181–1195. https://doi.org/10.1111/j.1365-2958.2009.06717.x

Alves MJ, Ferreira IC, Dias J, Teixeira V, Martins A, Pintado M (2012) A review on antimicrobial activity of mushroom (Basidiomycetes) extracts and isolated compounds. Planta Med 78:1707–1718. https://doi.org/10.1055/s-0032-1315370

Amodeo P, Saviano G, Borin G, Calderan A, Ruzza P, Tancredi T (1998) Solution conformational analysis of sodium complexed [Gly^6]- and [Gly^9]-antamanide analogs. J Pept Res 51:180–187

Anderl J, Faulstich H, Hechler T, Kulke M (2013) Antibody–drug conjugate payloads. Meth Mol Biol 1045:51–70. https://doi.org/10.1007/978-1-62703-541-5_4

Anderson MO, Shelat AA, Guy RK (2005) A solid-phase approach to the phallotoxins: total synthesis of [Ala^7]-phalloidin. J Org Chem 70:4578–4584. https://doi.org/10.1021/jo0503153
Andres RY, Frei W (1987) [^{125}I]amatoxin and anti-amatoxin for radioimmunoassay prepared by a novel approach: chemical and structural considerations. Toxicon 25:915–922
Andres RY, Frei W, Gautschi K, Vonderschmitt DJ (1986) Radioimmunoassay for amatoxins by use of a rapid, ^{125}I-tracer-based system. Clin Chem 32:1751–1755
Apperley S, Kroeger P, Kirchmair M, Kiaii M, Holmes DT, Garber I (2013) Laboratory confirmation of *Amanita smithiana* mushroom poisoning. Clin Toxicol 51:249–251. https://doi.org/10.3109/15563650.2013.778995
Azzolin L, Antolini N, Calderan A, Ruzza P, Sciacovelli M, Marin O, Mammi S, Bernardi P, Rasola A (2011) Antamanide, a derivative of *Amanita phalloides*, is a novel inhibitor of the mitochondrial permeability transition pore. PLoS One 28:e16280. https://doi.org/10.1371/journal.pone.0016280
Baumann K, Münter K, Faulstich H (1993) Identification of structural features involved in binding of a-amanitin to a monoclonal antibody. Biochemistry 32:4043–4050
Baumann K, Zanotti G, Faulstich H (1994) A beta-turn in alpha-amanitin is the most important structural feature for binding to RNA polymerase II and three monoclonal antibodies. Protein Sci 3:750–756. https://doi.org/10.1002/pro.5560030504
Benjamin DR (1995) Mushrooms: poisons and panaceas. A handbook for naturalists, mycologists, and physicians. WH Freeman, New York
Bernheimer AW, Oppenheim JD (1987) Some properties of flammutoxin from the edible mushroom *Flammulina velutipes*. Toxicon 25:1145–1152
Beuhler M, Lee DC, Gerkin R (2004) The Meixner test in the detection of alpha-amanitin and false-positive reactions caused by psilocin and 5-substituted tryptamines. Ann Emerg Med 44:114–120. https://doi.org/10.1016/S0196064404002860
Beutler JA, Der Marderosian AH (1981) Chemical variation in *Amanita*. J Nat Prod 44:422–444
Beutler JA, Vergeer PP (1980) Amatoxins in American mushrooms: evaluation of the Meixner test. Mycologia 72:1142–1149
Bhaskaran R, Yu C (1994) NMR spectra and restrained molecular dynamics of the mushroom toxin viroisin. Int J Pept Protein Res 43:393–401
Bresinsky A, Besl H (1990) A colour atlas of poisonous fungi. Wolfe Publishing, Regensburg
Brueckner F, Cramer P (2008) Structural basis of transcription inhibition by alpha-amanitin and implications for RNA polymerase II translocation. Nat Struct Mol Biol 15:811–818. https://doi.org/10.1038/nsmb.1458
Brüggemann O, Meder M, Freitag R (1996) Analysis of amatoxins alpha-amanitin and beta-amanitin in toadstool extracts and body fluids by capillary zone electrophoresis with photodiode array detection. J Chromatogr A 744:167–176
Buku A, Faulstich H, Wieland T, Dabrowski J (1980a) 2,3-*trans*-3,4-*trans*-3,4-Dihydroxy-L-proline: an amino acid in toxic peptides of *Amanita virosa* mushrooms. Proc Natl Acad Sci U S A 77:2370–2371
Buku A, Wieland T, Bodenmüller H, Faulstich H (1980b) Amaninamide, a new toxin of *Amanita virosa* mushrooms. Experientia 36:33–34
Bushnell DA, Cramer P, Kornberg RD (2002) Structural basis of transcription: alpha-amanitin-RNA polymerase II cocrystal at 2.8 Å resolution. Proc Natl Acad Sci U S A 99:1218–1222. https://doi.org/10.1073/pnas.251664698
Chen N, Bowles MR, Pond SM (1993) Polyclonal amanitin-specific antibodies: production and cytoprotective properties in vitro. Biochem Pharmacol 46:327–329
Chiang CC, Karle IL, Wieland T (1982) Unusual intramolecular hydrogen bonding in cycloamanide A, cyclic (LPro-LVal-LPhe-LPhe-LAla-Gly). A crystal structure analysis. Int J Pept Protein Res 20:414–420
Chilton W (1994) The chemistry and mode of action of mushroom toxins. In: Spoerke DG, Rumack BH (eds) Handbook of mushroom poisoning: diagnosis and treatment. CRC Press, Boca Raton, pp 165–231

Chilton W, Decato L, Malone M (1973) The unsaturated norleucines of *Amanita solitaria*, chemical and pharmacologic studies. Lloydia 36:169–173
Clarke DB, Lloyd AS, Robb P (2012) Application of liquid chromatography coupled to time-of-flight mass spectrometry separation for rapid assessment of toxins in *Amanita* mushrooms. Anal Meth 4:1298–1309
Courtillot M, Staron T (1970) *Amanita virosa* Fr. Précision sur l'espèce mise en evidence de sa toxine principale (virosine). Ann Phytopathol 2:561–584
Dal Pozzo A, Esposito E, Ni M, Muzi L, Pisano C, Bucci F, Vesci L, Castorina M, Penco S (2010) Conjugates of a novel 7-substituted camptothecin with RGD-peptides as $\alpha_v\beta_3$ integrin ligands: an approach to tumor-targeted therapy. Bioconjug Chem 21:1956–1967. https://doi.org/10.1021/bc100097r
De Lamo Marin S, Catala C, Kumar SR, Valleix A, Wagner A, Mioskowski C (2010) A practical and efficient total synthesis of potent insulinotropic (2*S*,3*R*,4*S*)-4-hydroxyisoleucine through a chiral N-protected γ-keto-α-aminoester. Eur J Org Chem 2010:3985–3989
Demain AL, Fang A (2000) The natural functions of secondary metabolites. Adv Biochem Eng Biotechnol 69:1–39
Deng WQ, Li TH, Xi PG, Gan LX, Xiao ZD, Jiang ZD (2011) Peptide toxin components of *Amanita exitialis* basidiocarps. Mycologia 103:946–949. https://doi.org/10.3852/10-319
Duffy TJ (2008) Toxic fungi of western North America. Available online at http://www.mykoweb.com/TFWNA/index.html. Retrieved Dec 2016
Edagwa BJ, Taylor CM (2009) Peptides containing γ,δ-dihydroxy-L-leucine. J Org Chem 74:4132–4136. https://doi.org/10.1021/jo900459f
Enjalbert F, Gallion C, Jehl F, Monteil H, Faulstich H (1992) Simultaneous assay for amatoxins and phallotoxins in *Amanita phalloides* Fr. by high-performance liquid chromatography. J Chromatogr 598:227–236
Enjalbert F, Cassanas G, Rapior S, Renault C, Chaumont JP (2004) Amatoxins in wood-rotting *Galerina marginata*. Mycologia 96:720–729
Fahrenholz F, Faulstich H, Wieland T (1971) Über die Giftstoffe des grünen Knollenblätterpilzes, XLII: Über Peptidsynthesen, XLVIII. Synthese des Norphalloins und eines Monocyclus mit 18 gliedrigem Ring. Liebigs Ann Chem 743:83–94
Falcigno L, Costantini S, D'Auria G, Bruno BM, Zobeley S, Zanotti G, Paolillo L (2001) Phalloidin synthetic analogues: structural requirements in the interaction with F-actin. Chemistry 7:4665–4673
Faulstich H, Georgopoulos D, Bloching M, Wieland T (1974) Analysis of the toxins of amanitin-containing mushrooms. Z Naturforsch 29c:86–88
Faulstich H, Trischmann H, Zobeley S (1975) A radioimmunoassay for amanitin. FEBS Lett 56:312–315
Faulstich H, Buku A, Bodenmüller H, Wieland T (1980) Virotoxins: actin-binding cyclic peptides of *Amanita virosa* mushrooms. Biochemistry 19:3334–3343. https://doi.org/10.1021/bi00555a036
Faulstich H, Zobeley S, Trischmann H (1982) A rapid radioimmunoassay, using a nylon support, for amatoxins from *Amanita* mushrooms. Toxicon 20:913–924
Faulstich H, Bühring HJ, Seitz J (1983) Physical properties and function of phallolysin. Biochemistry 22:4574–4580
Feng L, Tan L, Li H, Xu Z, Shen G, Tang Y (2015) Selective fluorescent sensing of α-amanitin in serum using carbon quantum dots-embedded specificity determinant imprinted polymers. Biosens Bioelectron 69:265–271. https://doi.org/10.1016/j.bios.2015.03.005
Fiume L, Busi C, Campadelli-Fiume G, Franceschi C (1975) Production of antibodies to amanitins as the basis for their radioimmunoassay. Experientia 31:1233–1234
Ford WW (1906) The toxicological constitution of *Amanita phalloides*. J Exp Med 8:437–450
Ford WW (1909) The distribution of poisons in mushrooms. Science 30:97–108. https://doi.org/10.1126/science.30.760.97
Garcia J, Oliveira A, de Pinho PG, Freitas V, Carvalho A, Baptista P, Pereira E, de Lourdes Bastos M, Carvalho F (2015a) Determination of amatoxins and phallotoxins in *Amanita phalloides*

mushrooms from northeastern Portugal by HPLC-DAD-MS. Mycologia 107:679–687. https://doi.org/10.3852/14-253

Garcia J, Costa VM, Baptista P, de Lourdes Bastos M, Carvalho F (2015b) Quantification of alpha-amanitin in biological samples by HPLC using simultaneous UV- diode array and electrochemical detection. J Chromatogr B Anal Technol Biomed Life Sci 997:85–95. https://doi.org/10.1016/j.jchromb.2015.06.001

Gauhe A, Wieland T (1977) Die cycloamanide, monocyclische peptide; isolierung und strukturaufklärung eines cyclischen heptapeptids (CyA B) und zweir cyclischer oktapeptide (CyA C und CyA D). Liebigs Ann Chem 1977:859–868

Gicquel T, Lepage S, Fradin M, Tribut O, Duretz B, Morel I (2014) Amatoxins (α- and β-amanitin) and phallotoxin (phalloidin) analyses in urines using high-resolution accurate mass LC-MS technology. J Anal Toxicol 38:335–340. https://doi.org/10.1093/jat/bku035

Hallen HE, Watling R, Adams GC (2003) Taxonomy and toxicity of *Conocybe lactea* and related species. Mycol Res 107:969–979

Hara R, Kino K (2009) Characterization of novel 2-oxoglutarate dependent dioxygenases converting L-proline to *cis*-4-hydroxy-L-proline. Biochem Biophys Res Commun 379:882–886. https://doi.org/10.1016/j.bbrc.2008.12.158

He K, Mao Q, Zang X, Zhang Y, Li H, Zhang D (2017) Production of a broad-specificity monoclonal antibody and application as a receptor to detection amatoxins in mushroom. Biologicals 49:57–61. PMID: 28688778. https://doi.org/10.1016/j.biologicals.2017.06.008

Henderson TJ (2010) Feasibility study for the rapid screening of target molecules using translational diffusion coefficients: diffusion-ordered NMR spectroscopy of biological toxins. Anal Bioanal Chem 396:1465–1471. https://doi.org/10.1007/s00216-009-3349-9

Herrmann A, Hedman H, Rosén J, Jansson D, Haraldsson B, Hellenäs KE (2012) Analysis of the mushroom nephrotoxin orellanine and its glucosides. J Nat Prod 75:1690–1696. https://doi.org/10.1021/np300135k

Hutchinson LJ, Madzia SE, Barron GL (1996) The presence and antifeedant function of toxin-producing secretory cells on hyphae of the lawn-inhabiting agaric *Conocybe lactea*. Can J Bot 74:431–434

Isernia C, Falcigno L, Macura S, Paolillo L, Pastore AL, Zanotti G (1996) Elucidation of the structure of constrained bicyclopeptides in solution by two-dimensional cross-relaxation spectroscopy: amatoxin analogues. J Pept Sci 2:3–13. https://doi.org/10.1002/psc.44

Jansson D, Fredriksson SÅ, Herrmann A, Nilsson C (2012) A concept study on identification and attribution profiling of chemical threat agents using liquid chromatography-mass spectrometry applied to *Amanita* toxins in food. Forensic Sci Int 221:44–49. https://doi.org/10.1016/j.forsciint.2012.03.023

Jin Z (2013) Muscarine, imidazole, oxazole and thiazole alkaloids. Nat Prod Rep 30:869–915. https://doi.org/10.1039/c3np70006b

Kahl JU, Vlasov GP, Seeliger A, Wieland T (1984) Analogs of viroidin. Synthesis of four virotoxin-like F-actin binding heptapeptides with one less hydroxyl group in the dihydroxy-proline ring. Int J Pept Protein Res 23:543–550

Kaplan CD, Larsson KM, Kornberg RD (2008) The RNA polymerase II trigger loop functions in substrate selection and is directly targeted by alpha-amanitin. Mol Cell 30:547–556. https://doi.org/10.1016/j.molcel.2008.04.023

Kirchmair M, Carrilho P, Pfab R, Haberl B, Felgueiras J, Carvalho F, Cardoso J, Melo I, Vinhas J, Neuhauser S (2012) Amanita poisonings resulting in acute, reversible renal failure: new cases, new toxic *Amanita* mushrooms. Nephrol Dial Transplant 27:1380–1386. https://doi.org/10.1093/ndt/gfr511

Kosentka P, Sprague SL, Ryberg M, Gartz J, May AL, Campagna SR, Matheny PB (2013) Evolution of the toxins muscarine and psilocybin in a family of mushroom-forming fungi. PLoS One 8:e64646. https://doi.org/10.1371/journal.pone.0064646

Kostansek EC, Lipscomb WN, Yocum RR, Thiessen WE (1978) Conformation of the mushroom toxin β-amanitin in the crystalline state. Biochemistry 17:3790–3795

Krogsgaard-Larsen P, Hansen JJ (1992) Naturally-occurring excitatory amino acids as neurotoxins and leads in drug design. Toxicol Lett 64-65:409–416

Laatsch H, Matthies L (1992) Letter to the editor. Experientia 48:533–534

Lee S, Nam S-J, Hyun C (2009) Mushroom poisoning by *Inocybe fastigiata* in a maltese dog. J Anim Vet Adv 8:708–710

Leite M, Freitas A, Azul AM, Barbosa J, Costa S, Ramos F (2013) Development, optimization and application of an analytical methodology by ultra performance liquid chromatography-tandem mass spectrometry for determination of amanitins in urine and liver samples. Anal Chim Acta 799:77–87. https://doi.org/10.1016/j.aca.2013.08.044

Letschert K, Faulstich H, Keller D, Keppler D (2006) Molecular characterization and inhibition of amanitin uptake into human hepatocytes. Toxicol Sci 91:140–149. https://doi.org/10.1093/toxsci/kfj141

Li P, Deng WQ, Li TH, Song B, Shen YH (2013) Illumina-based de novo transcriptome sequencing and analysis of *Amanita exitialis* basidiocarps. Gene 532:63–71. https://doi.org/10.1016/j.gene.2013.09.014

Li P, Deng W, Li T (2014) The molecular diversity of toxin gene families in lethal *Amanita* mushrooms. Toxicon 83:59–68. https://doi.org/10.1016/j.toxicon.2014.02.020

Litten W (1975) The most poisonous mushrooms. Sci Am 232:90–101. https://doi.org/10.1038/scientificamerican0375-90

Little MC, Preston JF 3rd, Jackson C, Bonetti S, King RW, Taylor LC (1986) Alloviroidin, the naturally occurring toxic isomer of the cyclopeptide viroidin. Biochemistry 25:2867–2872

Liu Y, Zhang X, Han C, Wan G, Huang X, Ivan C, Jiang D, Rodriguez-Aguayo C, Lopez-Berestein G, Rao PH, Maru DM, Pahl A, He X, Sood AK, Ellis LM, Anderl J, Lu X (2015) TP53 loss creates therapeutic vulnerability in colorectal cancer. Nature 520:697–701. https://doi.org/10.1038/nature14418

Loranger A, Tuchweber B, Gicquaud C, St-Pierre S, Côté MG (1985) Toxicity of peptides of *Amanita virosa* mushrooms in mice. Fundam Appl Toxicol 5:1144–1152. https://doi.org/10.1016/0272-0590(85)90151-4

Luo H, Hallen-Adams HE, Scott-Craig JS, Walton JD (2012) Ribosomal biosynthesis of α-amanitin in *Galerina marginata*. Fungal Genet Biol 49:123–129. https://doi.org/10.1016/j.fgb.2011.12.005

Luo H, Hong SY, Sgambelluri RM, Angelos E, Li X, Walton JD (2014) Peptide macrocyclization catalyzed by a prolyl oligopeptidase involved in α-amanitin biosynthesis. Chem Biol 21:1610–1617. https://doi.org/10.1016/j.chembiol.2014.10.015

Luo H, DuBois B, Sgambelluri RM, Angelos ER, Li X, Holmes D, Walton JD (2015) Production of ^{15}N-labeled α-amanitin in *Galerina marginata*. Toxicon 103:60–64. https://doi.org/10.1016/j.toxicon.2015.06.018

Lutter LC, Faulstich H (1984) Affinity isolation of RNA polymerase II on amanitin-sepharose. Biochem Biophys Res Commun 119:42–48

Lutz C, Müller C, Simon W (2014) Methods for synthesizing amatoxin building block and amatoxins. Eur Patent App EP 2 684 865 A1

Lynen F, Wieland H (1938) Über die Giftstoffe des grünen Knollenblätterpilzes, IV: Kristallisation von Phalloidin. Liebigs Ann Chem 533:93–117

Matinkhoo K, Pryyma A, Todorovic M, Patrick BO, Perrin DM (2018) Synthesis of the death-cap mushroom toxin α-amanitin. J Amer Chem Soc, in press. https://doi.org/10.1021/jacs.7b12698

Matsuura M, Saikawa Y, Inui K, Nakae K, Igarashi M, Hashimoto K, Nakata M (2009) Identification of the toxic trigger in mushroom poisoning. Nat Chem Biol 5:465–467. https://doi.org/10.1038/nchembio.179

Matthies L, Laatsch H (1991) Cortinarins in *Cortinarius speciosissimus*? A critical revision. Experientia 47:634–640

May JP, Perrin DM (2007) Tryptathionine bridges in peptide synthesis. Biopolymers 88:714–724. https://doi.org/10.1002/bip.20807

May JP, Perrin DM (2008) Intraannular Savige–Fontana reaction: one-step conversion of one class of monocyclic peptides into another class of bicyclic peptides. Chemistry 14:3404–3409. https://doi.org/10.1002/chem.200701088

May JP, Fournier P, Pellicelli J, Patrick BO, Perrin DM (2005) High yielding synthesis of 3*a*-hydroxypyrrolo[2,3-*b*]indoline dipeptide methyl esters: synthons for expedient introduction of the hydroxypyrroloindoline moiety into larger peptide-based natural products and for the creation of tryptathionine bridges. J Org Chem 70:8424–8430. https://doi.org/10.1021/jo051105t

May JP, Fournier P, Patrick BO, Perrin DM (2008) Synthesis, characterisation, and in vitro evaluation of Pro^2-Ile^3-*S*-deoxo-amaninamide and Pro^2-D-*allo*-Ile^3-*S*-deoxo-amaninamide: implications for structure-activity relationships in amanitin conformation and toxicity. Chemistry 14:3410–3417. https://doi.org/10.1002/chem.200701297

Meixner A (1979) Amatoxin-nachweis in pilzen. Z Mykol 45:137–139

Michelot D, Melendez-Howell LM (2003) *Amanita muscaria*: chemistry, biology, toxicology, and ethnomycology. Mycol Res 107:131–146

Morris PW, Venton DL, Kelley KM (1978) Biochemistry of the amatoxins: preparation and characterization of a stable iodinated α-amanitin. Biochemistry 17:690–698

Munekata E (1981) New advances in phallotoxin chemistry. In: Eberle A, Geiger R, Wieland T (eds) Perspectives in peptide chemistry. Karger, Basel, pp 129–140

Munekata E, Faulstich H, Wieland T (1977) Über die Inhaltsstoffe des grünen knollenblätterpilzes. LIII. Totalsynthese von Phalloin und Leu^7-phalloin. Liebigs Ann Chem 1977:1758–1765

Muraoka S, Fukamachi N, Mizumoto K, Shinozawa T (1999) Detection and identification of amanitins in the wood-rotting fungi *Galerina fasciculata* and *Galerina helvoliceps*. Appl Environ Microbiol 65:4207–4210

Nakamori T, Suzuki A (2007) Defensive role of cystidia against *Collembola* in the basidiomycetes *Russula bella* and *Strobilurus ohshimae*. Mycol Res 111:1345–1351. https://doi.org/10.1016/j.mycres.2007.08.013

Olson KR, Pond SM, Seward J, Healey K, Woo OF, Becker CE (1982) *Amanita phalloides*-type mushroom poisoning. West J Med 137:282–289

Patel DJ, Tonelli AE, Pfaender P, Faulstich H, Wieland T (1973) Experimental and calculated conformational characteristics of the bicyclic heptapeptide phalloidin. J Mol Biol 79:185–196

Pehk T, Haga M, Vija H, Lippman E (1989) High-field 2D NMR spectroscopy of amanitin isomers. Mag Res Chem 27:173–183. https://doi.org/10.1002/mrc.1260270216

Pelizzari V, Feifel E, Rohrmoser MM, Gstraunthaler G, Moser M (1994) Partial purification and characterization of a toxic component of *Amanita smithiana*. Mycologia 86:555–560

Pulman JA, Childs KL, Sgambelluri RM, Walton JD (2016) Expansion and diversification of the MSDIN family of cyclic peptide genes in the poisonous agarics *Amanita phalloides* and *A. bisporigera*. BMC Genomics 17:1038. https://doi.org/10.1186/s12864-016-3378-7

Riley R, Salamov AA, Brown DW, Nagy LG, Floudas D, Held BW, Levasseur A, Lombard V, Morin E, Otillar R, Lindquist EA, Sun H, LaButti KM, Schmutz J, Jabbour D, Luo H, Baker SE, Pisabarro AG, Walton JD, Blanchette RA, Henrissat B, Martin F, Cullen D, Hibbett DS, Grigoriev IV (2014) Extensive sampling of basidiomycete genomes demonstrates inadequacy of the white-rot/brown-rot paradigm for wood decay fungi. Proc Natl Acad Sci U S A 111:9923–9928. https://doi.org/10.1073/pnas.1400592111

Ritorto MS, Cook K, Tyagi K, Pedrioli PGA, Trost M (2013) Hydrophilic strong anion exchange (hSAX) chromatography for highly orthogonal peptide separation of complex proteomes. J Proteome Res 12:2449–2457. https://doi.org/10.1021/pr301011r

Ruzza P, Calderan A, Biondi B, Carrara M, Tancredi T, Borin G (1999) Ion binding and pharmacological properties of Tyr^6 and Tyr^9 antamanide analogs. J Pept Res 53:442–452

Savige WE, Fontana A (1980) A novel synthesis of 2-thioether derivatives of tryptophan. Covalent binding of tryptophan to cysteine sulfhydryl groups in peptides and proteins. Int J Pept Protein Res 15:102–112

Saviuc P, Danel V (2006) New syndromes in mushroom poisoning. Toxicol Rev 25:199–209

Schneider P, Bouhired S, Hoffmeister D (2008) Characterization of the atromentin biosynthesis genes and enzymes in the homobasidiomycete *Tapinella panuoides*. Fungal Genet Biol 45:1487–1496. https://doi.org/10.1016/j.fgb.2008.08.009

Schuresko LA, Lokey RS (2007) A practical solid-phase synthesis of Glu^7-phalloidin and entry into fluorescent F-actin binding reagents. Angew Chem Int Ed 46:3547–3549. https://doi.org/10.1002/anie.200700017

Seeger R (1984) Zeitungspapiertest für Amanitine—falsch-positive Ergebnisse. Z Mykol 50:353–359

Sgambelluri RM, Epis S, Sassera D, Luo H, Angelos ER, Walton JD (2014) Profiling of amatoxins and phallotoxins in the genus *Lepiota* by liquid chromatography combined with UV absorbance and mass spectrometry. Toxins 6:2336–2347. https://doi.org/10.3390/toxins6082336

Sgambelluri RM, Smith MO, Walton JD (2018) Versatility of prolyl oligopeptidase B in peptide macrocyclization. ACS Synth Biol 7:145–152. https://doi.org/10.1021/acssynbio.7b00264

Shi X, Miyakawa T, Nakamura A, Hou F, Hibi M, Ogawa J, Kwon Y, Tanokura M (2017) Engineering a short-chain dehydrogenase/reductase for the stereoselective production of (2*S*,3*R*,4*S*)-4-hydroxyisoleucine with three asymmetric centers. Sci Rep 7:13703. https://doi.org/10.1038/s41598-017-13978-w

Shoham G, Lipscomb WN, Wieland T (1989) Conformations of amatoxins in the crystalline state. J Am Chem Soc 111:4791–4809. https://doi.org/10.1021/ja00195a036

Skillman KM, Diraviyam K, Khan A, Tang K, Sept D, Sibley LD (2011) Evolutionarily divergent, unstable filamentous actin is essential for gliding motility in apicomplexan parasites. PLoS Pathog 7:e1002280. https://doi.org/10.1371/journal.ppat.1002280

Sliva D (2004) Cellular and physiological effects of *Ganoderma lucidum* (Reishi). Mini Rev Med Chem 4:873–879

Spoerke DG, Rumack BH (1994) Handbook of mushroom poisoning. CRC Press, Boca Raton

Stasyk T, Lutsik-Kordovsky M, Wernstedt C, Antonyuk V, Klyuchivska O, Souchelnytskyi S, Hellman U, Stoika R (2010) A new highly toxic protein isolated from the death cap *Amanita phalloides* is an L-amino acid oxidase. FEBS J 277:1260–1269. https://doi.org/10.1111/j.1742-4658.2010.07557.x

Stebelska K (2013) Fungal hallucinogens psilocin, ibotenic acid, and muscimol: analytical methods and biologic activities. Ther Drug Monit 35:420–442. https://doi.org/10.1097/FTD.0b013e31828741a5

Stijve T, Seeger R (1979) Determination of α-, β-, and γ-amanitin by high performance thin-layer chromatography in *Amanita phalloides* (Vaill. ex Fr.) Secr. from various origin. Z Naturforsch 34C:1133–1138

Tebbett IR (1992) Letter to the editor. Experientia 48:532–533

Tebbett IR, Caddy B (1984) Mushroom toxins of the genus *Cortinarius*. Experientia 40:441–446

Thomas JP, Arzoomanian R, Alberti D, Feierabend C, Binger K, Tutsch KD, Steele T, Marnocha R, Smith C, Smith S, MacDonald J, Wilding G, Bailey H (2001) Phase I clinical and pharmacokinetic trial of irofulven. Cancer Chemother Pharmacol 48:467–472

Tomková J, Ondra P, Válka I (2015) Simultaneous determination of mushroom toxins α-amanitin, β-amanitin and muscarine in human urine by solid-phase extraction and ultra-high-performance liquid chromatography coupled with ultra-high-resolution TOF mass spectrometry. Forensic Sci Int 251:209–213. https://doi.org/10.1016/j.forsciint.2015.04.007

Tuno N, Takahashi KH, Yamashita H, Osawa N, Tanaka C (2007) Tolerance of *Drosophila* flies to ibotenic acid poisons in mushrooms. J Chem Ecol 33:311–317. https://doi.org/10.1007/s10886-006-9228-3

Vergeer PP (1983) The Meixner test. West J Med 138:576

Wang Q, Ouazzani J, Sasaki NA, Potier P (2002) A practical synthesis of (2*S*,3*R*,4*S*)-4-hydroxyleucine, a potent insulinotropic α-amino acid from fenugreek. Eur J Org Chem 2002:834–839

Wei J, Wu J, Chen J, Wu B, He Z, Zhang P, Li H, Sun C, Liu C, Chen Z, Xie J (2017) Determination of cyclopeptide toxins in *Amanita subpallidorosea* and *Amanita virosa* by high-performance liquid chromatography coupled with high-resolution mass spectrometry. Toxicon 133:26–32. https://doi.org/10.1016/j.toxicon.2017.04.012

West PL, Lindgren J, Horowitz BZ (2009) Amanita smithiana mushroom ingestion: a case of delayed renal failure and literature review. J Med Toxicol 5:32–38

Wieczorek Z, Siemion IZ, Zimecki M, Bolewska-Pedyczak E, Wieland T (1993) Immunosuppressive activity in the series of cycloamanide peptides from mushrooms. Peptides 14:1–5

Wieland T (1983) The toxic peptides from *Amanita* mushrooms. Int J Pept Protein Res 22:257–276

Wieland T (1986) Peptides of poisonous *Amanita* mushrooms. Springer, Berlin

Wieland T, Boehringer W (1960) Über die Giftstoffe des grünen Knollenblätterpilzes, XIX: Umwandlung von β-amanitin in α-amanitin. Liebigs Ann Chem 635:178–181

Wieland T, Faulstich H (1991) Fifty years of amanitin. Experientia 47:1186–1193

Wieland T, Götzendörfer C, Zanotti G, Vaisius AC (1981) The effect of the chemical nature of the side chains of amatoxins in the inhibition of eukaryotic RNA polymerase B. Eur J Biochem 117:161–164

Wieland H, Hallermayer R (1941) Über die Giftstoffe des grünen Knollenblätterpilzes, VI: Amanitin, das Hauptgift des Knollenblätterpilzes. Liebigs Ann Chem 548:1–18

Wieland T, Schnabel HW (1962a) Über die Giftstoffe des grünen Knollenblätterpilzes, XXI: Die Konstitution des Phallacidins. Liebigs Ann Chem 657:218–225

Wieland T, Schnabel HW (1962b) Über die Giftstoffe des grünen Knollenblätterpilzes, XXII: Neue Sequenzanalyse von Phalloidin und Phalloin. Liebigs Ann Chem 657:225–228

Wieland T, Wirth L, Fischer E (1949) Über die Giftstoffe des Knollenblätterpilzes VII: β-Amanitin, eine dritte Komponente des Knollenblätterpilzgiftes. Liebigs Ann Chem 564:152–160

Wieland T, Miura T, Seeliger A (1983) Analogs of phalloidin. D-Abu^{2}-Lys^{7}-phalloin, an F-actin binding analog, its rhodamine conjugate (RLP) a novel fluorescent F-actin-probe, and D-Ala^{2}-Leu^{7}-phalloin, an inert peptide. Int J Pept Protein Res 21:3–10

Wulf E, Deboben A, Bautz FA, Faulstich H, Wieland T (1979) Fluorescent phallotoxin, a tool for the visualization of cellular actin. Proc Natl Acad Sci U S A 76:4498–4502

Xue JH, Wu P, Chi YL, Xu LX, Wei XY (2011) Cyclopeptides from *Amanita exitialis*. Nat Prod Bioprospect 1:52–66

Yamaura Y, Fukuhara M, Takabatake E, Ito N, Hashimoto T (1986) Hepatotoxic action of a poisonous mushroom, *Amanita abrupta* in mice and its toxic component. Toxicology 38:161–173

Yang WS, Lin CH, Huang JW, Fang CC (2006) Acute renal failure caused by mushroom poisoning. J Formos Med Assoc 105:263–267. https://doi.org/10.1016/S0929-6646(09)60317-X

Zanotti G, Birr C, Wieland T (1981) Analogs of amanin: synthesis of Ile^{3}-amaninamide and its diastereoisomeric (*S*)-sulfoxide. Int J Pept Protein Res 18:162–168

Zanotti G, Möhringer C, Wieland T (1987) Synthesis of analogues of amaninamide, an amatoxin from the white *Amanita virosa* mushroom. Int J Pept Protein Res 30:450–459

Zanotti G, Petersen G, Wieland T (1992) Structure-toxicity relationships in the amatoxin series. Structural variations of side chain 3 and inhibition of RNA polymerase II. Int J Pept Protein Res 40:551–558

Zanotti G, Kobayashi N, Munekata E, Zobeley S, Faulstich H (1999) D-configuration of serine is crucial in maintaining the phalloidin-like conformation of viroisin. Biochemistry 38:10723–10729. https://doi.org/10.1021/bi982692y

Zanotti G, Falcigno L, Saviano M, D'Auria G, Bruno BM, Campanile T, Paolillo L (2001) Solid state and solution conformation of [Ala^{7}]-phalloidin: a synthetic phallotoxin analogue. Chemistry 7:1479–1485

Zhao L, May JP, Blanc A, Dietrich DJ, Loonchanta A, Matinkhoo K, Pryyma A, Perrin DM (2015) Synthesis of a cytotoxic amanitin for bio-orthogonal conjugation. Chembiochem 16:1420–1425. https://doi.org/10.1002/cbic.201500226

Zhou ZY, Shi GQ, Fontaine R, Wei K, Feng T, Wang F, Wang GQ, Qu Y, Li ZH, Dong ZJ, Zhu HJ, Yang ZL, Zeng G, Liu JK (2012) Evidence for the natural toxins from the mushroom *Trogia venenata* as a cause of sudden unexpected death in Yunnan Province, China. Angew Chem Int Ed Eng 51:2368–2370. https://doi.org/10.1002/anie.201106502

Chapter 3
Distribution and Taxonomic Variation in the *Amanita* Cyclic Peptide Toxins

So notorious were the whole proceedings that authors of the period have recorded that the poison was sprinkled on an exceptionally fine mushroom. (On the death of Emperor Claudius, from the Annals of Tacitus, Loeb Classical Library, trans. J. Jackson). doi: 10.4159/DLCL.tacitus-annals.1931

3.1 The Importance of Understanding Distribution and Variation in the Cyclic Peptide Toxins

The literature on the taxonomic and tissue distribution of the *Amanita* cyclic peptides is extensive. Dating from the earliest day of toxin research, many studies have surveyed different mushroom species, specimens of the same species, growing environments, mushroom development stages, and mushroom tissues. It would be an encyclopedic and not very informative exercise to tabulate all of the results, not just because of their number but because many of the different studies are not easily comparable due to the use of different methods of toxin extraction and quantitation (or lack thereof) and the frequent omission of sufficient descriptions of the samples to allow species identifications. For an earlier summary of knowledge about the distribution of toxins among species and tissues, see Wieland (1986).

Taxonomic uncertainty in particular has confounded studies of the *Amanita* toxins since the earliest days. Ford (1906, 1909) collected his specimens of "*A. phalloides*" in North Carolina, Massachusetts, and New Hampshire. However, *A. phalloides* did not exist at that time in North America, being introduced only later in the twentieth century (Wolfe et al. 2010; Yocum and Simons 1977). Ford's specimens were almost certainly American relatives of the European species *A. verna* and *A. virosa*. To this day the numbers of species in *Amanita* sect. *Phalloideae* and the relationships between the European, Asian, and North American species have not been thoroughly resolved. Especially in the older literature, it is impossible to know what species were actually being analyzed.

J. Walton, *The Cyclic Peptide Toxins of Amanita and Other Poisonous Mushrooms*, https://doi.org/10.1007/978-3-319-76822-9_3

Despite these uncertainties, it is important to attempt to make some sense of the myriad studies, for reasons both biological and medical. First, it is important to attempt to understand the level of natural variation that exists across taxa in order to address whether cyclic peptide production arose once or multiple times in evolutionary history. Multiple genera (i.e., *Amanita*, *Galerina*, and *Lepiota*) in different families make the toxins (i.e., the trait is *polyphyletic*, the possible result of convergent evolution, but see Chap. 6), but the picture is less clear *within* genera. Did toxin production arise once or multiple times within a particular genus, i.e., is it monophyletic within *Amanita*, *Galerina*, or *Lepiota*? Another way of stating this question is to ask whether the trait of toxin production is derived from a common ancestor within a genus or whether it evolved independently multiple times even within a genus.

Another important aspect of cyclic peptide toxinology that is addressed by understanding variation in toxin levels among and within species is how evolutionarily and developmentally stable the trait is. The existence of natural qualitative and quantitative variation is supported by substantial anecdotal evidence, but how pervasive is variation between specimens of the same species? The molecular genetic studies of toxin production (Chap. 4) indicate that the cycloamanide family (formerly known as the "MSDIN" family and including the amatoxins, phallotoxins, virotoxins, and classic cycloamanides) is diverging and expanding rapidly in the genus *Amanita*, suggesting that this gene family is under strong evolutionary pressure. Yet another question that is addressable by studying the specimen to specimen differences in toxin levels across geographical space is how significant the effect of environment is on the production of the toxins. A strong correlation with soil composition, for example, might reflect a requirement of particular organic or inorganic nutrients for toxin biosynthesis or accumulation.

Second, the existence of differences in toxin levels in different tissues might shed light on the ecological role(s) of the toxins. Anti-herbivore compounds in plants often show strong tissue dependencies, reflecting an evolutionary imperative to protect certain plant parts over others. Does the pattern of toxin distribution in different mushroom tissues reflect herbivore pressure on the different tissues? From a basic research point of view, if tissue toxin analyses suggested a stringent developmental regulation of toxin biosynthesis, this situation could be exploited to help identify the genes in the pathway by comparative gene expression (e.g., Brown et al. 2012). In *Galerina*, toxin biosynthesis and expression of the gene encoding the α-amanitin precursor peptide, *GmAMA1*, are up-regulated in vitro by low glucose concentration (Luo et al. 2012). Therefore, perhaps co-regulation with *GmAMA1* could be exploited to identify other pathway genes. The broader relationship between specialized cell types and the biosynthesis and accumulation of natural products in mushrooms is discussed at the end of this chapter. In the author's opinion there are many under-explored facets of this relationship in fungi, especially in the agarics.

Third, knowledge about the *subcellular* distribution of the toxins might be informative in regard to the mechanism by which the mushrooms protect themselves against their own potent inhibitors of RNA biosynthesis and of the actin cytoskeleton.

Perhaps the mushrooms sequester the toxins in particular organelles to prevent the toxins from coming into contact with their endogenous sites of action.

Fourth, knowledge about taxonomic, geographical, and tissue variation in toxin concentrations has clinical significance. In the context of the emergency room, it could be important to know which species are always toxic, sometimes toxic, or never toxic. In medical cases involving a mushroom species that is usually or always toxic, and if there is some idea of how much the patient ingested, a clinician would like to be able to estimate if the patient received a medically significant dose or not.

3.2 Taxonomic Distribution of the Cyclic Peptide Toxins

3.2.1 *The Genus* Amanita

There are an estimated 900–1000 species in the family Amanitaceae (kingdom Fungi, phylum Basidiomycota, class Agaricomycetes, order Agaricales) (www.amanitaceae.org). By current taxonomic classification, the family comprises three genera: *Amanita*, *Limacella*, and *Torrendia*. The genus *Amanita* is divided into two subgenera, *Amanita* and *Lepidella*. Subgenus *Lepidella* is itself divided into four sections: *Lepidella*, *Validae*, *Phalloideae*, and *Amidella*. Only mushrooms in sect. *Phalloideae* of subgenus *Lepidella* are known to make the amatoxins and phallotoxins. Some of the species in sections other than *Phalloideae* (e.g., sect. *Lepidella*) are edible (Cai et al. 2014). There are ~60 species in sect. *Phalloideae*, making it one of the smaller sections within *Amanita*. Many species are not yet officially described in the scientific literature or delimited from each other. Of the 60 species listed by Tulloss (http://www.amanitaceae.org/?section+Phalloideae), about half have been assayed for the presence of the cyclic peptide toxins.

All of the cyclic peptide toxin-producing species of *Amanita* are ectomycorrhizal on a broad range of woody plants (oak, pine, beech, birch, hornbeam, eucalyptus, chestnut, etc.). Whether there might be any biological connection between this ecological niche and toxin production is considered in Chap. 6.

Species of *Amanita* that are well-documented to produce cyclic peptide toxins and to have poisoned humans and other mammals include *Amanita phalloides*, *A. virosa*, *A. verna*, *A. ocreata*, *A. pallidorosea*, *A. suballiacea*, *A. exitialis*, and *A. bisporigera*. The common name for *A. phalloides* is the "death cap," whereas *A. ocreata*, *A. verna*, *A. virosa*, *A. exitialis*, and *A. bisporigera* are all known as "destroying angels" or simply "deadly whites" for their distinctive pure white cap color (Figs. 3.1 and 3.2). *A. verna* and *A. virosa* are European species, but these names have historically been applied to morphologically similar mushrooms growing in in North America. Furthermore, there exist pure white forms of *A. phalloides* called var. *alba* (Kaya et al. 2013), and some authorities suspect that *A. bisporigera* (the most common destroying angel in much of the eastern and midwestern United States) itself is really a "species complex" that will eventually be resolved into multiple species (see below). *A. ocreata* is native to western North America and *A. exitialis* to Asia. Cai

Fig. 3.1 Representative cyclic peptide toxin-producing species of *Amanita*. (**a**) *A. phalloides* from Italy (Photo credit: Giovanni Consiglio, Associazione Micologica Bresadola, Italy. Used by permission.) (**b**) *A. bisporigera* from Michigan. (Reprinted from Luo et al. (2010) with permission. Photo credit: Hong Luo, Kunming Institute of Botany, Chinese Academy of Sciences.) (**c**) *A. marmorata* from Australia. (Photo credit: Lucy Albertella. New South Wales, Australia. http://mushroomobserver.org/163271 Used with permission.) (**d**) *A. ocreata* from California. (Photo credit: Debbie Viess, Bay Area Mycological Society. Used with permission.) (**e**) *A. verna* from Italy. (Photo credit: Giovanni Consiglio, Associazione Micologica Bresadola, Italy. Used with permission.) (**f**) *A. virosa* from Sweden. (Photo credit: Ben DeRoy – http://commons.wikimedia.org/wiki/File:Destroying_Angel_02.jpg. Used with permission)

et al. (2014) concluded that the genus *Amanita* contains a high level of cryptic diversity, leading to strong underestimation of the species diversity within sect. *Phalloideae* as currently delimited.

Every phylogenetic analysis to date places all of the cyclic peptide toxin-producing species of *Amanita* in sect. *Phalloideae* (Cai et al. 2014; www.amanitaceae.org). However, not all species in sect. *Phalloideae* make the toxins. For example, *A. areolata* (=*A. zangii*) is a basal lineage in sect. *Phalloideae* that does not produce either amatoxins or phallotoxins (Cai et al. 2014). In other words, a robust phylogenetic tree based on alignments of multiple housekeeping genes predicts that these species split off from the other species in sect. *Phalloideae* before the evolution of the capacity to make the cyclic peptide toxins. Somewhere, somehow, a species of *Amanita* acquired this extraordinary trait, and it was carried along through vertical descent, parents to progeny, through all subsequent speciation events.

Another molecular phylogenetic analysis, using a different set of species, grouped several toxin nonproducers (i.e., *A. heslerii*, *A. mutabilis*, and *A. crassiconus*) with *A. marmorata*, a known cycloamanide (phallotoxin) producer (Table 3.1), on a different branch from other toxin producers (*A. virosa*, *A. verna*, and *A. phalloides*) (Wolfe et al. 2012). One possible explanation for this result is that the trait of toxin production has been lost in some lineages in sect. *Phalloideae*.

Fig. 3.2 Basidiomata (fruiting bodies) of 12 lethal *Amanita* species growing in China. (**a**) *A. exitialis* (**b**) *A. fuliginea* (**c**) *A. fuligineoides* (**d**) *A. griseorosea* (**e**) *A. olliuscula* (**f**) *A. pallidorosea* (**g**) *A. parviexitialis* (**h**) *A. rimosa* (**i**) *A. subfuliginea* (**j**) *A. subjunquillea* (**k**) *A. subpallidorosea* (**l**) *A. virosa*. Scale bars = 2 cm. (From Cai et al. (2016). Reprinted by permission of Taylor & Francis, LLC (http://www.tandfonline.com))

Table 3.1 Survey of mushrooms for presence of α-amanitin, β-amanitin, phallacidin, and phalloidin

Species	Section	Accession number (Herbarium Rooseveltensis Amanitarum)	Collection location	α-Amanitin	β-Amanitin	Phallacidin	Phalloidin
Amanita phalloides	*Phalloideae*	278–8	Gironde, France	+	+	+	+
Amanita phalloides	*Phalloideae*	520–7	Monmouth County, New Jersey	+	+	+	+
Amanita bisporigera	*Phalloideae*	505–7	Northampton County, Pennsylvania	+	−	+	+
Amanita bisporigera 04	*Phalloideae*	492–10	Warren County, New York	−	−	−	+
Amanita bisporigera 04	*Phalloideae*	327–10	Newfoundland and Labrador, Canada	−	−	−	+
Amanita bisporigera 04	*Phalloideae*	368–8	Newfoundland and Labrador, Canada	−	−	−	+
Amanita bisporigera 04	*Phalloideae*	397–8	Cape May County, New Jersey	−	−	−	+
Amanita bisporigera 04	*Phalloideae*	480–1	Franklin County, New York	−	−	+	−
Amanita bisporigera 05	*Phalloideae*	508–1	Cumberland County, Tennessee	+	+	−	+
Amanita ocreata	*Phalloideae*	041–1	Santa Cruz County, California	+	+	+	−
Amanita ocreata	*Phalloideae*	277–6	Monterey County, California	+	+	+	+
Amanita ocreata	*Phalloideae*	604–5	Contra Costa County, California	+	+	+	+
Amanita ocreata	*Phalloideae*	274–4	Sonoma County, California	+	+	+	+
Amanita ocreata	*Phalloideae*	010–8	Cowlitz County, Washington	+		+	−
Amanita suballiacea	*Phalloideae*	490–1	Middlesex County, Connecticut	−	−	−	+
Amanita suballiacea	*Phalloideae*	539–2	Lee County, Florida	+	−	+	+
Amanita suballiacea	*Phalloideae*	491–7	McComb co, Michigan	+	−	+	+
Amanita suballiacea	*Phalloideae*	524–8	Luzerne County, Pennsylvania	+	−	+	+
Amanita suballiacea	*Phalloideae*	478–6	Clarion County, Pennsylvania	+	+	+	+
Amanita verna	*Phalloideae*	313–8	Landes, France	+	+	+	−
Amanita virosa	*Phalloideae*	310–9	Norway	+	−	+	+

Amanita virosa	*Phalloideae*	291–3	France	+	–	+	+
Amanita arocheae	*Phalloideae*	155–2	Costa Rica	+	–	–	+
Amanita arocheae	*Phalloideae*	155–3	Costa Rica	+	–	–	+
Amanita longitibiale	*Phalloideae*	125–9	Mexico	–	–	–	–
Amanita longitibiale	*Phalloideae*	081–6	Brevard County, Florida	–	–	–	–
Amanita cf. *marmorata*	*Phalloideae*	623–7	New South Wales, Australia	–	–	+	+
Amanita cf. *marmorata*	*Phalloideae*	563–3	New South Wales, Australia	–	–	+	+
Amanita marmorata	*Phalloideae*	044–6	Oahu, Hawaii	–	–	+	–
Amanita mediinox	*Phalloideae*	152–9	Transylvania County, North Carolina	–	–	–	–
Amanita mediinox	*Phalloideae*	376–1	Sevier County, Tennessee	–	–	–	–

All mushrooms were identified and supplied by R. Tulloss (Herbarium Rooseveltensis Amanitarum, www.amanitaceae.org). Analyses were performed by Brandon DuBois in the laboratory of the author. Compound identifications were based on retention time, absorption at 295 nm and 305 nm, and mass ($M+H^+$ of the parent ion). Due to variation in levels of toxins caused by age and postharvest handling and among different tissues of a single mushroom, the results are presented as simply presence (+) or absence (–). Each analysis was performed in duplicate by 50% methanol extraction. Each HPLC injection represented the equivalent of ~3 mg dry weight (Chap. 2). The detection limit was ~0.1 µg

Although overall it appears that the trait of cycloamanide biosynthesis is monophyletic in *Amanita,* a full understanding of the evolution of toxin biosynthesis within *Amanita* will require broader and deeper genome sequencing of more species, both toxin producers and close relatives.

The cycloamanide gene family has been identified by PCR or whole genome sequencing in many species known to make the toxins, including *A. bisporigera, A. phalloides, A. fuliginea, A. fuligineoides, A. pallidorosea, A. verna, A. virosa, A. rimosa, A. ocreata,* and *A. marmorata.* It has never been found in any toxin non-producing species (Hallen et al. 2007; Li et al. 2014b; Pulman et al. 2016; unpublished work from the author's laboratory; unpublished GenBank entries FN555143 and FN555144). The presence of the cycloamanide gene family is therefore the genetic hallmark of the potential for cyclic peptide production in *Amanita* and other fungi (Chap. 4). Understanding the origin, expansion, and diversification of this gene family is the key to understanding the evolution of the biosynthesis of the deadly amatoxins and related cyclic peptides (Chap. 6).

Firm conclusions about the total amatoxin and phallotoxin levels in poisonous mushrooms are difficult due to the large number of methods of estimating toxin concentrations and the many factors that influence toxin levels (such as tissue, mushroom age, handling, and soil type). However, collectively, averaging across studies, a reasonable estimate is that the total content of the major toxic amanitins (α, β, and γ) in the average poisonous *Amanita* mushroom (e.g., *A. phalloides, A. bisporigera,* or *A. exitialis*) is in the range of 0.1–0.5 mg/g fresh weight, corresponding to 1–5 mg/g dry weight (Sgambelluri et al. 2014; Wieland 1986). An average basidiocarp (fruiting body) of *A. phalloides* weighing 100 gm (fresh weight) could therefore contain >20 mg active amatoxins. Considering the LD_{50} for a human adult to be ~0.1 mg/kg (i.e., 7 mg for a 70 kg person), one mushroom could easily contain three times the human LD_{50}.

Amanita phalloides This is the iconic poisonous mushroom, responsible for more deaths than any other (Fig. 3.1a). Its common name is "death cap." It is native throughout Europe, where it forms symbiotic mycorrhizal associations with many tree species. It has been introduced, most likely on tree roots, into Africa, Australia, and North America. It occurs in the eastern United States but is especially widespread and expanding in western North America from California to Vancouver (Wolfe et al. 2010). In California it grows preferentially on the roots of native oaks (*Quercus agrifolia* and others). It can be seasonably very abundant in some areas of its current ranges, including California and Australia (New South Wales and Victoria). *A. phalloides* is readily identified by its greenish-yellow cap, white spores, and the presence of a prominent annulus and volva (see frontispiece). The cap color can vary to pure white in a variant known as *A. phalloides* var. *alba.*

All of the classical chemistry on the *Amanita* cyclic peptide toxins was performed with European specimens of *A. phalloides.* Yocum and Simon (1977), using LH-20 chromatography and TLC, found that American and European specimens of *A. phalloides* have very similar toxin profiles. The toxin profiles of an Italian specimen and a California specimen of *A. phalloides*, determined by LC/MS, were also almost identical (Sgambelluri et al. 2014).

Enjalbert et al. (1996, 1999) compared the toxin profiles of ~27 specimens of *A. phalloides* from different geographical regions and soil types in France. Although there was considerable quantitative variation between samples, none of the specimens lacked phallotoxins or amatoxins altogether. Beutler and Vergeer (1980) surveyed 206 individual basidiocarps of *A. phalloides* collected in Tomales Bay State Park, Marin County, California. All but one of them showed the presence of amatoxins by the Wieland-Meixner test and TLC. This is the largest survey ever performed on a local population of any poisonous mushroom. This important study indicates that toxin-minus "sports" of *A. phalloides* certainly do occur but are rare compared to some other species, such as *A. virosa*, *A. bisporigera*, and *A. verna* (see below). A specimen of *A. phalloides* var. *alba* from Turkey was shown to contain α-amanitin, β-amanitin, and phallacidin based on HPLC retention times compared to standards and 302/291 nm UV absorption ratios (Kaya et al. 2013).

The sequenced genome of a specimen of *A. phalloides* from California contained >30 members of the cycloamanide gene family, including genes for α-amanitin, β-amanitin, phallacidin, and phalloidin (Pulman et al. 2016; Chap. 4). A gene encoding β-amanitin was amplified by PCR from an Italian specimen (GenBank accession number FN555142; S. Epis, Universitá degli Studi di Milano, unpublished results).

Li et al. (2014b) amplified by PCR several cycloamanide genes from a specimen of *A. phalloides* from Italy, including those encoding β-amanitin, cycloamanide B (CylB), and a truncated version of phalloidin (predicted amino acid sequence WLATCP). However, genes for the other cycloamanides including antamanide were not found in the complete genome of *A. phalloides* collected in California (Pulman et al. 2016). Although sequencing errors, incomplete genome coverage, or chemical misidentification are possible explanations for this result, natural genetic variation seems the most plausible explanation for the lack of a perfect correlation between known cyclic peptides and the genes of the cycloamanide family (Pulman et al. 2016).

Based on multiple studies, the production of the major cyclic peptide toxins (amatoxins and phallotoxins) seems to be phenotypically quite stable within *A. phalloides*. Although there are reports of *A. phalloides* specimens that lack one or more of the major toxins, the majority of specimens from both Europe and North America make both types, and Beutler and Vergeer (1980) found production of the major toxins to be stable within a local population. However, this is apparently not true for the minor cyclic peptides, including antamanide, perhaps because they are under weaker natural selection (Chap. 6).

Amanita bisporigera This white mushroom, one of the destroying angels, is native to eastern and midwestern North America (www.mushroomexpert.com/amanita_bisporigera.html) (Fig. 3.1b). Based on morphological characteristics, *A. bisporigera* is likely to be one name applied to multiple cryptic species forming a "species complex" (R. Tulloss, http://www.amanitaceae.org/?Amanita+bisporigera, and

B. Bunyard, personal communications). This conclusion is supported by molecular studies still in progress as of this writing. Names that have been applied to toxic white *Amanita* species in North America include *A. elliptosperma*, *A. magnivelaris*, *A. tenuifolia* (*A. virosiformis*), and *A. suballiacea* (Jenkins 1982; www.amanitaceae.org). One of the destroying angels that grows in eastern North America, and which caused a human poisoning (Yarze and Tulloss 2012), has been given the provisional name *A. amerivirosa* (http://www.amanitaceae.org/?Amanita+amerivirosa). A provisional name for another of the North American destroying angels is *A. sturgeonii* (http://www.amanitaceae.org/?Amanita+sturgeonii). Collectively, *A. bisporigera* and related species have not been thoroughly evaluated at the chemical or genomic level for the diversity of their cyclic peptide toxins, but all appear to have the genetic potential and should be considered highly toxic.

Yocum and Simons (1977) reported the presence of α-amanitin, β-amanitin, γ-amanitin, phalloidin, and phalloin in a specimen of *A. bisporigera* collected in Michigan, USA, using separation by Sephadex LH-20 chromatography and TLC. An analysis of seven specimens of *A. bisporigera* obtained from R. Tulloss collected in Pennsylvania, Tennessee, New Jersey, or Canada and analyzed by the superior method of LC/MS found that most lacked β-amanitin and others contained phalloidin but not α-amanitin, β-amanitin, or phallacidin (Table 3.1). Intriguingly, none of these toxin profiles corresponded exactly with the genomic complement of the cycloamanide family in two other specimens from Michigan (Hallen et al. 2007; Pulman et al. 2016). The genome of the sequenced specimen of *A. bisporigera* contains genes for α-amanitin and phallacidin but not β-amanitin or phalloidin, which some other specimens identified as *A. bisporigera* produce (Table 3.1). As with *A. phalloides* (see above), the discrepancy between genes and toxins could be due to either natural variation or gaps in the genome sequence. Natural variation could occur at either the population level (i.e., intraspecific variation) or between species (i.e., different species in the *A. bisporigera* species complex produce different complements of cycloamanides). There might well be both intra- and interspecific natural variation.

The cycloamanide gene family was originally discovered in a specimen of *A. bisporigera* collected in central Michigan (Hallen et al. 2007; Chap. 4). Later, deeper sequencing of a different specimen collected in the same geographical area showed little overlap in the complement of cycloamanide genes and, based on their ITS sequences (Chap. 1), the two specimens are probably different species (Pulman et al. 2016). This result suggests that there is high diversity in the cycloamanide gene family even within the *A. bisporigera* species complex (Chap. 6).

Amanita marmorata This species (the "marbled death cap") is native to southern Australia but has been exported to South Africa and Hawaii, probably on the roots of horticultural plants (Fig. 3.1c). It forms mycorrhizal associations with trees of the southern hemisphere, including *Eucalyptus*, *Casuarina*, *Araucaria*, and *Melaleuca* (Miller et al. 1988). Hallen et al. (2002) tentatively detected α-amanitin, β-amanitin, phallacidin, and phalloidin, based on HPLC retention time and detection at 295 nm compared to standards, in a South African specimen of *A. reidii*, which might be synonymous with *A. marmorata* (www.amanitaceae.org/?Amanita%20marmorata). Three specimens of *A. marmorata* obtained from R. Tulloss, two of which were from Australia and one from Hawaii, were analyzed by LC/MS in the author's laboratory. All three contained phallotoxins but none contained α- or β-amanitin (Table 3.1).

Several cycloamanide genes were amplified by PCR in a Hawaiian specimen of *A. marmorata* (H. Hallen, unpublished results from the author's lab), including a gene encoding phallacidin (toxin core region: AWLVDCP) and three unknowns encoding putative cycloamanides (core sequences FMFFRYPFP, FMFFRYPLP, and VWGIGCSP). This is consistent with the toxin analyses showing the presence of phallacidin and absence of α-amanitin in *A. marmorata* (Table 3.1). However, while successful PCR amplification is proof that a gene is present, due to possible primer mismatches it cannot be used to prove the absence of a gene whose exact DNA sequence is not known, and therefore *A. marmorata* should be assumed to have the potential to be deadly poisonous.

Amanita ocreata This white destroying angel is native to western North America (Fig. 3.1d). Next to *A. phalloides*, it is the most common cause of amatoxin mushroom poisonings in western North America (Duffy 2008). As *A. phalloides* continues to spread in western North America (it is now becoming common as far north as Vancouver), poisonings from *A. ocreata* will probably decline in proportion. Horgen et al. (1976) and Ammirati et al. (1977) showed the presence of α-amanitin and β-amanitin in *A. ocreata* from California by TLC and RNA polymerase inhibition assays. This was confirmed by Beutler and Der Marderosian (1981) with the Wieland-Meixner test and TLC. Specimens of *A. ocreata* collected in Santa Cruz, Monterey, and Contra Costa counties in northern California and in Cowlitz County, Washington, contained α-amanitin, β-amanitin, phalloidin, and phallacidin as determined by LC/MS (Table 3.1). The gene for phalloidin was amplified by PCR from a California specimen of *A. ocreata* (Hallen et al. 2007).

Amanita virosa The common name of this mushroom is the "European destroying angel" (Fig. 3.1f). It is native to northern Europe and eastern Asia. Bresinsky and Besl (1990) state that *A. virosa* grows in northern Europe, whereas the other European destroying angel, *A. verna* (see below), grows only in the Mediterranean region. A specimen of *A. virosa* collected in Hunan, China was confirmed by ITS sequencing to be the same species as a specimen from Germany, and both formed a clade with *A. subpallidorosea*, an Asian species (Wei et al. 2017).

Fungi identified as *A.virosa* from Europe, the United States, Canada, and Japan have been implicated in human poisonings and shown to contain amatoxins and phallotoxins by multiple authors and multiple methods (Bresinsky and Besl 1990; Fig. 2.3). Faulstich et al. (1974) reported α-amanitin but not β-amanitin or γ-amanitin by LH-20 chromatography and TLC in two specimens collected in Italy and Sweden. Both specimens also contained phallacidin and phalloidin. In another study, some North American specimens of "*A. virosa*" did not contain one or more of the major cyclic peptide toxins (Yocum and Simons 1977). Using the Wieland-Meixner test and TLC, Beutler and Vergeer (1980) found amatoxins in 10 of 17 specimens of "*A. virosa*" collected from unstated locations in North America. Beutler and Der Marderosian (1981) examined multiple specimens of "*A. virosa*" collected in Michigan, New Jersey, Pennsylvania, and Delaware. About half the specimens contained detectable levels of amatoxins and half did not. All specimens contained significant levels of phallotoxins, but the amatoxin-minus specimens had levels of phallotoxins that were on average twice as high as the amatoxin-positive specimens. Two specimens, from Norway and France, obtained from R. Tulloss and analyzed by the author, contained α-amanitin, phallacidin, and phalloidin but not β-amanitin (Table 3.1). The same toxin spectrum was found by Sgambelluri et al. (2014) in a specimen of *A. virosa* from Italy. *A. virosa* was also shown to contain amaninamide (Fig. 2.3). Wei et al. (2017) also found the same three major toxins in a German specimen, but a specimen from China contained α-amanitin and phalloidin but not phallacidin. Yocum and Simons (1977) reported amanin from *A. virosa* in North America, but Buku et al. (1980) argued that they actually detected amaninamide. Amanin was not found in an Italian specimen of *A. virosa* (Fig. 2.3).

A. virosa also makes virotoxins, which are chemically related to the phallotoxins (Chap. 2). Wei et al. (2017) found viroidin but not Ala-viroidin or viroisin in single specimens of *A. virosa* from Germany and China.

At the genetic level, Li et al. (2014b) identified cycloamanide genes by PCR from a specimen of *A. virosa* collected in the Czech Republic. A gene encoding phallacidin was identified in an Italian specimen of *A. virosa* (S. Epis et al., GenBank entry FN555144). There are to date no records of the genes encoding virotoxins, but they are predicted to belong to the cycloamanide family and be identical or nearly identical to the genes encoding phallotoxins found in other peptide toxin-producing *Amanita* species (Chap. 4).

Overall, the collective evidence indicates that *A. virosa* in Europe and the fungus (or fungi) bundled under the same name in North America do make clinically significant levels of the cyclic peptide toxins, but the trait is highly variable, at least in North America. How much of this variability is due to natural genetic variation within a species and how much to misidentification and/or cryptic speciation is still unknown.

Amanita verna Like *A. virosa*, this toxic white mushroom is native to Europe (Fig. 3.1e). Eleven specimens of *A. verna* collected from several locales in Germany and Switzerland and analyzed by high-performance TLC (HPTLC) all contained similar levels of α-, β-, and γ-amanitin (Seeger and Stijve 1979). Average levels of the three amatoxins were 1.55, 1.29, and 0.37 mg/g dry weight, respectively. By LC/MS, a specimen from France contained α-amanitin, β-amanitin, and phallacidin, but not phalloidin (Table 3.1).

Like *A. virosa*, the name *A. verna* has been applied to superficially similar mushrooms in eastern North America. Like *A. bisporigera* and *A. virosa*, North American "*A. verna*" show high variability in toxin profiles. Tyler et al. (1966) reported that four out of ten specimens of *A. verna* collected in North Carolina, New York, Texas, and Tennessee did not contain any detectable α- or β-amanitin. Of four specimens from Delaware analyzed by Yocum and Simons (1977) by LH-20 chromatography and TLC, one contained only β-amanitin, and three contained no amatoxins or phallotoxins. Beutler and Der Marderosian (1981) found no amatoxins in a sample of *A. verna* from New Jersey. Preston et al. (1975, 1982) documented specimens of *A. verna* that lacked amatoxins as assayed by inhibition of RNA polymerase. Faulstich et al. (1974) found α-amanitin, phalloidin, and phallacidin but not β-amanitin in a specimen of *A. verna* from Germany, yet a gene for β-amanitin was identified by PCR in an Italian specimen of *A. verna* (Sara Epis, GenBank CBG76462.1).

In summary, *A. verna*, like *A. virosa*, seems to consistently contain amatoxins and phallotoxins in Europe, but fungi identified by these names are more variable in North America. The probable existence of cryptic species currently lumped under historical epithets such as *bisporigera*, *verna*, and *virosa* might well explain at least some of the observed variability in cyclic peptide toxin distribution.

Amanita exitialis This mushroom (Fig. 3.2a) is a common cause of mushroom poisonings in Asia, and multiple studies have consistently detected high levels of amatoxins and phallotoxins (Deng et al. 2011; Hu et al. 2012; Tang et al. 2016). Analysis of nine specimens of *A. exitialis* from Thailand, which were taxonomically confirmed by ITS sequencing, found α- and β-amanitin in all nine (Parnmen et al. 2016). Members of the cycloamanide gene family have been identified from *A. exitialis* by genomic PCR and RNA-Seq (transcriptomics) (Li et al. 2013). *A. exitialis* makes a cycloamanide called amanexitide, of structure cyclo(VFFPVFSLP); this is the first cycloamanide described from a fungus other than *A. phalloides* (Xue et al. 2011).

Other *Amanita* Species There are many other species of *Amanita* in sect. *Phalloideae* (Fig. 3.2). Most of them probably make amatoxins and/or phallotoxins, but definitive evidence is often lacking or incomplete. Although cyclic peptide production is monophyletic in *Amanita*, the inclusion of species which do not make amatoxins and/or phallotoxins in sect. *Phalloideae* is supported by the molecular phylogenetic analysis of Cai et al. (2014), which places *A. areolata* (=*A. zangii*) in sect. *Phalloideae.* In addition, a chemical analysis in the author's laboratory found that *A. longitibiale* and *A. mediinox*, which are in sect. *Phalloideae*, do not contain amatoxins or phallotoxins (Table 3.1). However, at this point, it is not possible to conclusively state that any species in sect. *Phalloideae* do not make the toxins because toxin production can be variable within a species, and only a small number of specimens of each species have been chemically analyzed to date. As discussed earlier, additional uncertainty is introduced by our incomplete understanding of the taxonomic diversity and phylogenetic relationships among some species of *Amanita* (Pulman et al. 2016). Currently, there are many specimens in the literature described with epithets such as "*Amanita* sp.," "*Amanita* affin.," or "*Amanita* subspecies," indicating that the authors believe them to be distinct from already named species, but they have not yet been formally named (e.g., Cai et al. 2014; Parnmen et al. 2016). Combining molecular identification by multi-gene sequencing with chemical toxin analysis is an encouraging trend that should reduce confusion in determining taxonomic distribution of the peptide toxins (Parnmen et al. 2016; Pulman et al. 2016; Sgambelluri et al. 2014; Tang et al. 2016; Wei et al. 2017).

Another important approach for resolving the relationship between taxonomy and toxin production within *Amanita* and also other genera is a molecular analysis of the cycloamanide gene family (Chap. 4). Possession of the gene family is necessary but not sufficient for a fungus to be able to make the cyclic peptide toxins. For example, a particular species or specimen might lack the toxins due to a mutation outside the cycloamanide core sequences. The *Amanita* species comprising the toxin-nonproducing basal lineage of sect. *Phalloideae* have not yet been analyzed for the presence of the cycloamanide gene family, nor have *A. longitibiale* and *A. mediinox* (Table 3.1). Therefore, whether any species of *Amanita* sect. *Phalloideae* completely lacks the genetic potential to make the cyclic peptide toxins remains an open question until such time as genomic sequences are obtained from these fungi and the presence or absence of the cycloamanide family is experimentally established.

In regard to toxin and related molecular genetic analyses in other species of *Amanita*, the cyclic peptide toxins have been detected in *A. rimosa*, *A. fuligineoides*, *A. fuliginea*, *A. subjunquillea* var. *alba*, *A. pallidorosea*, *A. rimosa*, and *A. subpallidorosea* from China (Tang et al. 2016; Wei et al. 2017). Amatoxins and phallotoxins were found in specimens of *A. arocheae* from Costa Rica and *A. suballiacea* from North America by LC/MS (Table 3.1). Based on HPLC retention time, Vargas et al. (2011) reported amatoxins including α-amanitin in a number of species of *Amanita* from Colombia, including species such as *A. muscaria* that are not in sect. *Phalloideae* and lack the cycloamanide gene family; these results are thus not in agreement with most other studies. Mullersman and Preston (1982) used a novel assay (protection of deoxyribonuclease I activity from inhibition by actin) to measure phallotoxins in several species of *Amanita*. They detected high levels (>8 mg/g dry weight) in *A. suballiacea*, as expected, and low levels (2–5 μg/g) in *A. mutabilis*, *A. rubescens*, and *A. hygroscopica*, only the last of which is in sect. *Phalloideae*. These levels are ~1000 times lower than found in fungi such as *A. phalloides*, and furthermore *A. hygroscopica* was probably misidentified by the same laboratory (see below). Considering also that Hallen et al. (2002) found no phallotoxins in *A. rubescens* from South Africa, these results must be considered tentative until they have been replicated.

A. subpallidorosea, which is closely related to the European *A. virosa*, produces the virotoxins Ala-viroidin, viroisin, and viroidin (Wei et al. 2017). *A. suballiacea* collected in Florida has been reported to make virotoxins and in addition a virotoxin isomer known as alloviroidin, in which Leu #3 has the (2*S*,4*S*) configuration instead of (2*S*,4*R*) (Little et al. 1986).

Specimens of two new species of *Amanita*, *A. ballerina* and *A. brunneitoxicaria*, were described from Thailand (Thongbai et al. 2017). On molecular and morphological criteria, *A. brunneitoxicaria* was placed in sect. *Phalloideae*, but the taxonomic affinity of *A. ballerina* was uncertain. The authors concluded that *A. ballerina* did not produce amatoxins or phallotoxins whereas *A. brunneitoxicaria* did. Cycloamanide C (CylC; Chap. 2) was also claimed to be detected in *A. brunneitoxicaria*. However, identification of the toxins and CylC was based only on HPLC retention time and UV absorption, and not even these essential data were shown for CylC. The number of specimens of *A. ballerina* analyzed was not given, so it is premature to conclude that the cyclic peptide toxins are completely absent from the whole species. Critically, the mass spectral data of the unknowns was not presented, and therefore the presence or absence of the cyclic peptide toxins in these two new species remains uncertain.

Genes in the cycloamanide family have been amplified by PCR from *A. fuliginea*, *A. fuligineoides*, *A. pallidorosea*, and *A. rimosa* (Cai et al. 2014).

3.2.1.1 Summary of the Distribution of the Cyclic Peptide Toxins in *Amanita*

From all of the published and unpublished studies across the entire genus, the following general conclusions can be drawn: (1) production of the cyclic peptide toxins is restricted to sect. *Phalloideae*; (2) α-amanitin is the toxic peptide that is most consistently present, although many species also contain β-amanitin, phalloidin, and phallacidin; and (3) *A. phalloides* is the most consistently positive for the presence of amatoxins and phallotoxins.

A. verna, *A. virosa*, and *A. bisporigera* are more variable than *A. phalloides* in composition and presence/absence of particular cyclic peptide toxins. Some of this might be due to confusion in identification and the weak state of our taxonomic understanding of the destroying angels in North America. On the other hand, the overall variability seems too high to be accounted for simply by misidentifications and therefore probably represents genuine natural variation, at least in part. A possible contributing factor to the observed high level of natural variation is genetic instability – the toxin structural genes (i.e., members of the cycloamanide family) and/or other essential genes in the pathway might be lost at relatively high frequency, for example, due to unequal crossing over among members of the gene family. Instability of genes involved in secondary metabolism is well-documented in ascomycetes, although little is known about this phenomenon in basidiomycetes (e.g., Pitkin et al. 2000; Walton 2000; Wight et al. 2013).

3.2.2 The Genus Galerina

The Wikipedia entry for *Galerina* provides an excellent description of this genus of "little brown mushrooms" (https://en.wikipedia.org/wiki/Galerina) (Fig. 3.3d). The evidence that some species of *Galerina* make amatoxins is substantiated by multiple analyses of mushrooms in North America and Europe, extraction of amatoxins from

Fig. 3.3 (**a**) *Lepiota subincarnata* (synonym *L. josserandii*) from Vancouver Island, British Columbia, Canada. (Photo credit: Oluna and Adolf Ceska (mushroomobserver.org/258712). Used by permission.) (**b**) *Lepiota subincarnata* from Italy. (Photo credit: Giovanni Consiglio, Associazione Micologica Bresadola, Italy. Used by permission.) (**c**) *Lepiota subincarnata* from Linn County, Oregon, USA. (Photo credit: Britney Wharton mushroomobserver.org/81447.) (**d**) *Galerina marginata* from Italy. (Photo credit: Giovanni Consiglio, Associazione Micologica Bresadola, Italy. Used by permission)

axenic cultures, multiple documented cases of human poisonings, and full genome sequencing. Human poisonings caused by *Galerina* have been reported in Finland, Japan, the United States, and China (Bresinsky and Besl 1990; Enjalbert et al. 2004; Kaneko et al. 2001; Tyler and Smith 1963; Tyler et al. 1963; Yin and Yang 1993). *Galerina* species have also killed dogs (Mungenast 2008). Being a saprobic fungus rather than obligately mycorrhizal, species of *Galerina* grow in culture faster than any species of *Amanita*, albeit still slowly. The cultured mycelium contains 0.5–1 mg α-amanitin per gm dry weight (the toxin is not secreted into the medium) (Benedict et al. 1966; Luo et al. 2012). *G. marginata* is transformable, i.e., genetically modifiable with foreign DNA (Luo et al. 2014). Luo et al. (2015) used cultures of *G. marginata* to produce highly enriched ^{15}N-labeled α-amanitin.

Like *Amanita*, the trait of toxin biosynthesis is not universal in the genus *Galerina*, being restricted to subgenus *Naucoriopsis*. Species described as making amatoxins include *G. marginata*, *G. autumnalis*, *G. unicolor*, and *G. venenata*, all of which have been conflated into a single species, *G. marginata* (Gulden et al. 2001, 2005). Recent molecular phylogenetics combined with chemical analyses performed by Brandon Landry from the laboratory of Mary Berbee (University of British Columbia) support a monophyletic origin of amanitin biosynthesis in *Galerina*. It is noteworthy that amanitin could be detected in 20-year-old herbarium specimens.

Measured concentrations of amatoxins in *G. marginata* are comparable to toxin-producing species of *Amanita*. Representative examples of published values are 0.6 mg/g dry weight (α-amanitin from cultured mycelium; Sgambelluri et al. 2014); 0.7–2.1 mg/g dry weight, assuming 88% moisture content (α, β, and γ from fruiting bodies; Enjalbert et al. 2004); and 0.8–2.2 mg/g dry weight (α, β, and γ from fruiting bodies; Wieland 1986).

Some isolates of *Galerina* apparently contain β-amanitin, whereas others have only α-amanitin. Tyler and Smith (1963) and Tyler et al. (1963) reported α- and β-amanitin by TLC in specimens of *G. venenata*, *G. marginata*, and *G. autumnalis* collected in the Pacific Northwest of America. Specimens of *G. marginata* from France have been reported to make β-amanitin as well as α-amanitin, based on HPLC retention time and UV absorption (Enjalbert et al. 1992, 2004). Over 27 specimens were analyzed, and some specimens even contained more β-amanitin than α-amanitin. β-Amanitin was detected by Johnson et al. (1979) by TLC in several specimens of *G. marginata* collected in Ohio, North America. On the other hand, Faulstich et al. (1974) reported only a trace of β-amanitin in a specimen of *G. marginata* from Switzerland by LH-20 chromatography and TLC, and a specimen of *G. marginata* from Bronx, New York, involved in a human poisoning did not contain any detectable β-amanitin by LC/MS (unpublished results from the author's lab).

The complete genome of strain CBS 339.88 of *G. marginata* has been sequenced (Luo et al. 2012; Riley et al. 2014; genome.jgi.doe.gov/Galma1/Galma1.home.html). This strain, which is monokaryotic (i.e., it has a single copy of its genetic complement and is therefore functionally haploid), is no longer available from the original source, the Centraalbureau voor Schimmelcultures (CBS-KNAW Fungal Biodiversity Centre, Utrecht). The geographical origin of this strain was not recorded before it was deaccessioned, but it most likely came from France like all of the other *Galerina* accessions in the CBS-KNAW collection. The genome of *G. marginata* strain CBS 339.88 does not contain a gene for β-amanitin (i.e., encoding IWGIGCDP), although some extracts of this particular strain grown in culture do occasionally show the presence of β-amanitin (Luo et al. 2012; Sgambelluri et al. 2014). These low levels could conceivably arise by artifactual deamidation of α-amanitin during aging of the mushrooms/mycelium or during the extraction procedure. The levels of β-amanitin seen in some European specimens seem to be too high to be due to artifactual deamidation of α-amanitin, which leads to the prediction that some specimens of *G. marginata* have a cycloamanide gene specifically encoding β-amanitin, like *A. phalloides* (Pulman et al. 2016).

In conclusion, α-amanitin is consistently present in *G. marginata* (and in its synonymous and sister species in section *Naucoriopsis*) from everywhere in the world, and β-amanitin is present in some but not all isolates of *G. marginata* from both Europe and North America. γ-Amanitin, which is a post-translational variant of α-amanitin, is present in some specimens of *G. marginata* (Sgambelluri et al. 2014) (Fig. 2.3). Phallotoxins have never been reported from any species of *Galerina*, and the genome of *G. marginata* CBS 339.88 does not contain any phallotoxin or any other cycloamanide genes (Riley et al. 2014).

The genes for the amanitin precursor peptide (*GmAMA1*) and the prolyl oligopeptidase (*GmPOPB*) that processes and cyclizes α-amanitin have been isolated from *G. marginata* (Luo et al. 2012) (Chap. 4). Southern blotting indicated that *G. badipes*, which is also in subgenus *Naucoriopsis* but is a distinct species from *G. marginata*, also has the *AMA1* and *POPB* genes, but *G. hybrida* (which is in subgenus *Tubariopsis* and does not make α-amanitin) lacks both *AMA1* and *POPB* (Gulden et al. 2005; Luo et al. 2012). This result is consistent with the trait of toxin biosynthesis being monophyletic in *Galerina* like it is in *Amanita*.

G. marginata has the distinction of being one of the few macrofungi that grows in Antarctica (antarcticsun.usap.gov/science/contenthandler.cfm?id=2723). It was collected growing on moss and identified by Dr. Robert Blanchette (University of Minnesota). Similar mushrooms were collected in 1962 on the South Orkney Islands and named *G. antarctica*; to date it is not known if *G. antarctica* is a bonafide distinct species. Dr. Hong Luo and Brandon DuBois analyzed a sample of the antarctic *G. marginata* obtained from Dr. Blanchette. Separation was by reverse-phase HPLC with detection by UV absorption at 295 and 305 nm, and identification was based on the 305/295 nm ratio and retention times relative to standards. α-Amanitin was present in the two samples tested (unpublished results from the author's lab). The presence or absence of β-amanitin could not be established due to limiting amounts of material. It is not known if *G. marginata* is native to Antarctica or whether it was introduced by humans on construction wood. If native, it is intriguing to speculate what selective advantage the toxin might confer in this extreme habitat.

3.2.3 The Genus Lepiota

Lepiota is a large cosmopolitan genus of saprobic fungi (en.wikipedia.org/wiki/Lepiota) (Fig. 3.3a, b, c). Considering their generally small size and obscurity, toxic species of *Lepiota* have been responsible for a surprising number of human and canine poisonings. Cases have been reported from Italy, Spain, France, the United States, Tunisia, and Turkey (Ben Khelil et al. 2010; Enjalbert et al. 2002; Furia et al. 1982; Kervégant et al. 2013; Kose et al. 2015; Meunier et al. 1994; Mottram et al. 2010; Varvenne et al. 2015). Among 27 cases of *Lepiota* poisoning followed in one study in Turkey, 14 patients died of liver failure (Paydas et al. 1990).

Like *Amanita* and *Galerina*, cyclic peptide production is not universal in this genus. Epithets for amatoxin-producing species include *L. brunneoincarnata*, *L. subincarnata*, *L. josserandii*, *L. brunneolilacea*, *L. castanea*, *L. helveola*, and *L. cristata*, found in Europe, North Africa, and North America (Fig. 3.3). The genome of the nontoxicogenic relative *Macrolepiota fuliginosa* has been sequenced (http://genome.jgi.doe.gov/Macfu1/Macfu1.home.html). It is expected that the taxonomy of this genus will undergo significant revision in the future, and so some names might be abandoned and/or synonymized.

α-Amanitin has been detected many times in specimens of *Lepiota* by TLC and the Wieland-Meixner test. By TLC, amatoxins were detected in 10 of 36 European *Lepiota* species (Gérault and Girre 1975). The sole analysis of any species of *Lepiota* using LC/MS found α-amanitin levels of 0.7–4.3 mg/g dry weight in *L. josserandii* and *L. brunneoincarnata* collected in Italy (Sgambelluri et al. 2014). Two specimens of *L. brunneoincarnata* contained β-amanitin, amanin, and amaninamide, and two specimens of *L. josserandii* did not contain these three amatoxins but did contain γ-amanitin (Fig. 2.3). No amatoxins or phallotoxins were observed in Italian specimens of *L. clypeolaria*, *L. cristata*, *L. magnispora*, or *L. echinacea* (Sgambelluri et al. 2014). All species were confirmed by ITS sequencing. Two specimens of *L. josserandii* collected on Vancouver Island, British Columbia, and specimens of *L. subincarnata* from California and Washington state analyzed by LC/MS contained only α-amanitin and not γ-amanitin, β-amanitin, amanin, amaninamide, or any phallotoxins. This chemical profile was consistent with the presence of a gene encoding α-amanitin, but not genes for β-amanitin or any phallotoxins, in the genome of the same specimens of *L. subincarnata* (Fig. 4.13; unpublished results from the author's laboratory).

Like *Galerina*, no *Lepiota* species has ever been reported to make phallotoxins (Sgambelluri et al. 2014). However, the genome of an α-amanitin-containing specimen of *L. subincarnata* from Washington state encoded five predicted cycloamanide genes in addition to the gene for α-amanitin (Fig. 4.13; unpublished results from the author's laboratory). Two of the predicted cycloamanides were detectable by LC/MS in mushroom extracts. Thus, the diversity of the cycloamanide gene family in *Lepiota* falls between *Galerina* (which has only a gene for α-amanitin) and *Amanita* (several species of which have >30 cycloamanide genes) (Chap. 4).

3.2.4 The Genus Conocybe

Conocybe is a large genus of agarics in the family *Bolbitiaceae* (Fig. 3.4). Mice injected intraperitoneally with a crude ethanolic extract of *C. filaris* collected in Seattle, Washington, died with symptoms and autopsy observations consistent with the known effects of α-amanitin (Brady et al. 1975). Extracts of *C. filaris* were analyzed by TLC and shown to contain a material having the same R_f and color (violet-lavender with cinnamaldehyde-HCl; Fig. 2.2) as authentic α-amanitin (Brady et al. 1975). Based on this single report, *C. filaris* (also known as *Pholiotina filaris*) is widely included on all lists of amatoxin-containing poisonous mushrooms. However, the collective evidence that *C. filaris* contains α-amanitin and is therefore potentially lethal to humans is not well substantiated. First, the injection method of toxicity testing used by Brady et al. (1975) cannot distinguish between amatoxins and other toxic compounds that might have been in the tested mushrooms, including

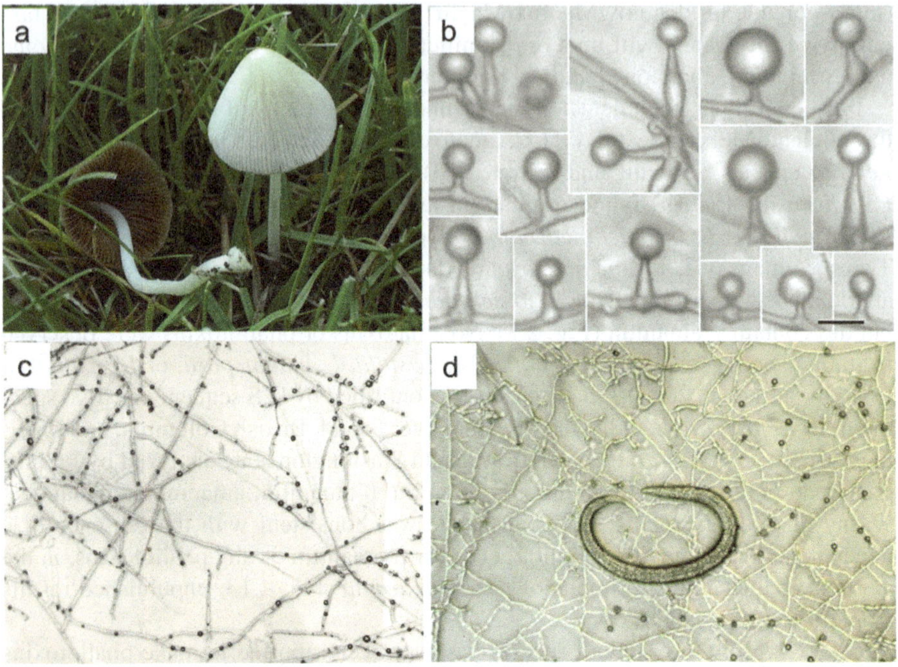

Fig. 3.4 (**a**) *Conocybe apala* growing in a lawn in Michigan. (**b**) Droplets forming on mycelium of *C. apala* growing in culture. These droplets contain unidentified materials toxic to nematodes, although *C. apala* is not nematophagous. Other agarics also kill without consuming nematodes, i.e., the toxins are "protective antifeedants" (Hutchinson et al. 1996) (**c**) Mycelial droplets on *C. albipes*. (**d**) Nematode killed by contact with mycelial droplets. (Photo credits: **a** and **b**, Hong Luo, Kunming Institute of Botany, Chinese Academy of Sciences; **c** and **d**, George Barron, University of Guelph, used by permission, https://atrium.lib.uoguelph.ca/xmlui/handle/10214/6184, and http://www.uoguelph.ca/~gbarron/2008/conocybe.htm)

phallotoxins. Second, neither before nor in the 40+ years since this one report have there been any cases in the peer-reviewed scientific literature of human or animal poisonings by any species of *Conocybe* that are consistent with amanitin poisoning. This is especially significant considering that some species of *Conocybe* contain psilocybin, a powerful human attractant. Third, there have been no further adequately documented chemical analyses showing the presence of amanitin in any species of *Conocybe*, and instead there are several negative reports. For example, Bresinsky and Besl (1990) were not able to detect amatoxins in any German species of *Pholiotina*.

Duffy (2008) relates a secondhand account of a nonfatal poisoning due to *C. filaris* in San Francisco in the early 1960s, but no clinical details were provided, and therefore it is unknown whether the symptoms were consistent with amatoxin exposure. Duffy (2008) also states "Not all specimens of *Conocybe* (*Pholiotina*) *filaris* contained amatoxins on the basis of Meixner testing performed by the Mycological

Society of San Francisco (MSSF) and by the Los Angeles Mycological Society (LAMS)" but does not say what percentage *were* positive. In any case, by modern standards, the Wieland-Meixner test alone cannot be considered an adequate positive or negative indicator for amatoxins.

Another *Conocybe* species, *C. apala* (synonyms *C. badipes* and *C. lactea*) was reported to contain phallotoxins (Hallen et al. 2003) (Fig. 3.4). Like the evidence that *C. filaris* makes amatoxins, the evidence that *C. apala* makes phalloidin is based on a single study. Phalloidin was detected by LC/MS, giving high confidence of identification of the compound. However, the levels were very low, ~3 ng/g fresh weight, equivalent to ~30 ng/g dry weight, or ~100,000-fold lower than typically seen in species such as *A. phalloides*. This raises the reasonable concern that the phalloidin detected in *C. apala* was the result of accidental contamination of glassware or the HPLC instrumentation (see below). Furthermore, phalloidin was not detectable in a large percentage (39%) of the specimens of *C. apala* tested (Hallen et al. 2003). Additional analyses of several specimens of *C. apala* collected in Michigan failed to detect any amatoxins or phallotoxins (unpublished results from the author's lab).

A prolyl oligopeptidase (POP) capable of cleaving a phalloidin precursor peptide to release the linear mature peptide of phalloidin was purified from *C. apala* (Luo et al. 2009). This result could be interpreted to signify that because *C. apala* possesses the biochemical machinery to process the phalloidin precursor peptide, it probably makes phalloidin. However, Luo et al. (2009) could not exclude the possibility that the POP that they isolated was the "housekeeping" POP (i.e., POPA) and not the POP (POPB) dedicated to peptide toxin biosynthesis that was later identified in *Amanita* and *Galerina* (Chap. 4).

In summary, it is very dubious whether any species of *Conocybe* makes phallotoxins or amatoxins. One way this question could be resolved would be to sequence and then search the genome of one or more *Conocybe* species for any cycloamanide genes. Although cyclic peptide toxin biosynthesis does not appear to be a general trait of this genus, it can never be excluded that some rare subspecies or specimens do make the peptide toxins.

3.2.5 Do Edible Mushrooms Make Cyclic Peptide Toxins?

Using radioimmunoassay and inhibition of RNA polymerase II (pol II) as bioassays, Faulstich and Cochet-Meilhac (1976) reported the presence of amatoxins in several *Amanita* species outside sect. *Phalloideae* (i.e., *A. muscaria*, *A. rubescens*, *A. citrina*, and *A. pantherina*) and in several popular wild mushrooms, including *Agaricus sylvaticus* (pinewood mushroom), *Boletus edulis* (cep, bolete, or porcini), and *Cantharellus cibarius* (chanterelle). The levels of amatoxins were very low, 0.2–16 ng/g fresh weight, equivalent to 10,000- to 50,000-fold lower than found in *A. phalloides* using the same methods. However, when repeated in a laboratory where amatoxin-containing mushrooms had not previously been analyzed, the authors detected no amatoxins in the same mushrooms tested earlier (cited as personal communication in Wieland 1986). Therefore, in at least this study, it appears that the apparent presence of amatoxins was an artifact caused by contamination of the laboratory glassware or the analytical equipment.

Using inhibition of pol II (Chap. 5) to quantitate amatoxins, Preston et al. (1982) reported α-amanitin-equivalent levels of ~2–12 μg/g dry weight in various mushrooms outside sect. *Phalloideae*, including *A. alliacea* (sect. *Lepidella*), *A. brunnescens* (sect. *Validae*), *A. komarekensis* (sect. *Amanita*), and *A. mutabilis* (sect. *Lepidella*). Some other species of *Amanita* outside sect. *Phalloideae* had levels of α-amanitin <1 μg/g. These levels are far below those found in well-established toxin-producing mushrooms. Furthermore, the authors could not exclude that the amatoxin-positive mushrooms contained inhibitors of pol II other than the amatoxins. Johnson and Preston (1979) reported that they were able to induce rapid growth of *A. hygroscopica*, which is in sect. *Phalloideae* and makes amatoxins, on defined medium in culture. This is highly unusual behavior for any species of *Amanita*. The cultured strain was deposited in the American Type Culture Collection (ATCC) with accession number 26843. This strain was obtained from ATCC and confirmed to grow quite well in culture. However, on the basis of the sequence of its ITS region, ATCC 26843 is not a species of *Amanita* but instead an unidentified species in the *Polyporales*, a taxon that does not make amatoxins (H. Heather-Adams, unpublished results from the author's laboratory).

Molecular analysis of the cycloamanide gene family does not support the production of the cyclic peptide toxins by any fungus outside *Amanita* sect. *Phalloideae*, *Galerina*, and *Lepiota*, including any edible fungi. By Southern blotting and whole genome sequencing, the *AMA1* and *PHA1* genes as well as any recognizable members of the cycloamanide superfamily are absent from *Amanita* species outside sect. *Phalloideae* and from all other agarics that have been sequenced to date including *A. muscaria* and *A. thiersii* (Hallen et al. 2007; Kohler et al. 2015).

3.2.6 Summary of the Distribution of the Cyclic Peptide Toxins Across All Taxa

Amatoxins are unequivocally produced by a subset of species in the genus *Amanita* sect. *Phalloideae*, by species in *Galerina* subgenus *Naucoriopsis*, and by certain species of *Lepiota*. Within each group, the trait of amatoxin biosynthesis is probably monophyletic, i.e., it evolved once. However, the evolutionary trajectory that resulted in this trait occurring in three unrelated genera of mushrooms is open to speculation, as discussed further in Chap. 6. For the phallotoxins and virotoxins, only mushrooms in *Amanita* sect. *Phalloideae* unequivocally produce them. The evidence for production of amatoxins or phallotoxins by any species of *Conocybe*, by species of *Amanita* outside sect. *Phalloideae*, or by any edible mushroom is very weak.

3.3 Factors that Affect Quantitative Toxin Levels

Enjalbert et al. (1996, 1999) showed an effect of soil type on levels of amatoxins and phallotoxins in *A. phalloides* in France. Garcia et al. (2015) also found that site-specific environmental factors influenced amatoxin levels in *A. phalloides* in Portugal.

Several studies have examined the influence of developmental stage on toxin concentrations. Mycelium growing in culture can make amatoxins, as shown for both *Galerina marginata*, which is a saprobic fungus, and *A. exitialis*, which is mycorrhizal (Benedict et al. 1966; Luo et al. 2015; Zhang et al. 2005). It is not known if the toxins are present in mycelium growing in soil, on and around tree roots, but this could be ecologically significant (Chap. 6).

Preston et al. (1982) found only small changes in levels of the toxins per g dry weight during a developmental time course of basidiocarps of *A. suballiacea*. Phallotoxin levels in *A. phalloides* varied modestly in different tissues as a function of developmental stage but with no consistent pattern (Enjalbert et al. 1989). In contrast, Hu et al. (2012) found that amatoxin levels in *A. exitialis* were mainly stable from early through late development but fell by two-thirds in old, wilted basidiocarps. If this same pattern is valid for *A. phalloides* and other toxin-producing mushrooms, it might be sensible when harvesting mushrooms for the toxins to avoid collecting old mushrooms.

Amatoxins and phallotoxins have been detected in dried herbarium specimens up to 20 years old, consistent with their renowned stability to cooking and the mammalian digestive tract (Hallen et al. 2002; Brandon Landry, personal communication).

3.4 Tissue Distribution of the Cyclic Peptide Toxins

Most studies on the distribution of the amatoxins and phallotoxins within mushrooms have been at the tissue level, i.e., cap, gills, stipe, volva, annulus, etc. These studies have documented quantitative differences among the tissues, but without doubt the toxins can occur at toxicologically significant levels in all parts of a mushroom. The conclusion from multiple studies on tissue distribution in *A. phalloides* growing in Europe is that the amatoxins are higher in the cap, gills, and stipe and lower in the volva and spores, whereas phallotoxins show the opposite pattern (Enjalbert et al. 1996, 1999; Garcia et al. 2015; Kaya et al. 2013; McKnight et al. 2010). Yilmaz et al. (2014) found higher levels of amatoxins and phallotoxins in the pileus, gills, and stipe of *A. verna* collected in Turkey compared to the spores and volva. Hu et al. (2012) found that levels of α-amanitin and β-amanitin in *A. exitialis* were also highest in the gills followed by the cap, stipe, and finally annulus; spores had the lowest levels. Preston et al. (1982) found a gradient of amatoxins in *A. suballiacea* from the volva through the stipe to the cap and very low levels in the spores. Toxicologically, these studies indicate that no parts of amatoxin-producing mushrooms are safe to eat. The biological significance of these collective results is less clear. There is no apparent ecological rationale for the observed differences in relative levels and proportions of the cyclic peptides in different tissues and at different developmental stages.

Li et al. (2014a) took the study of tissue distribution to a new level by quantifying expression of *AMA1* (the gene encoding the α-amanitin precursor peptide) by quantitative reverse transcriptase PCR (qRT-PCR). In *A. exitialis*, *AMA1* RNA was expressed throughout the mushroom but at somewhat higher levels in the pileus than the cap or volva. Expression in all tissues declined with age of the basidiocarp; at later developmental stages, expression was on average reduced about fivefold. These results are in reasonable agreement with measurements of the toxin levels themselves. The demonstration that toxin gene expression and toxin levels are correlated developmentally and histologically is significant because it argues against several interesting speculative scenarios, for example, that the toxins are made only at the earliest stages of basidiocarp development or are made in one part of the mushroom and transported to the other tissues.

3.5 Toxin Biosynthesis and Accumulation at the Cellular Level

Mushroom tissues are not homogeneous; they are composed of many different types of cells. Mushroom gills, for example, are composed of trama, hymenium, sterile and fertile basidia, and cystidia. To date there has been only a single study on the distribution of the cyclic peptide toxins at the cellular level. Luo et al. (2010) showed by immunolocalization and confocal microscopy that α-amanitin is localized throughout the tissues of the basidiocarp, consistent with the chemical analyses discussed above, but that not all cells within any one tissue type contain it. For example, in gills, α-amanitin is present in a subpopulation of cells in the hymenium (i.e., the spore-bearing surface of the gills (Fig. 3.5)). By phase-contrast light microscopy, no distinction could be made between those hymenial cells that contained α-amanitin and those that did not. It could be excluded that the differential staining of cells was due to differential penetration of the antibody, because all cells stained equally well with an anti-actin antibody (Fig. 3.5). Some of the amanitin-containing cells have sterigmata (i.e., projections to which basidiospores are attached), whereas others might be sterile cells or immature basidia (Fig. 3.5). Although some immunostaining was observed in the internal gill tissues (the trama), it was consistently weaker (Fig. 3.5).

α-Amanitin is also present throughout the pileus (cap) of *A. bisporigera*, but similar to the situation in the gills, it is not present in all cells. Furthermore, the individual areas of signal do not correspond neatly to the boundaries of individual cells (Luo et al. 2010). This suggests that amanitin is localized in subcellular structures, perhaps specialized storage vacuoles. A similar irregular and apparent subcellular pattern of amanitin distribution was seen in the stipe and in the universal veil (Luo et al. 2010). The amanitin biosynthetic enzyme, POPB, co-localizes strongly with α-amanitin itself in *A. bisporigera* (Chap. 4).

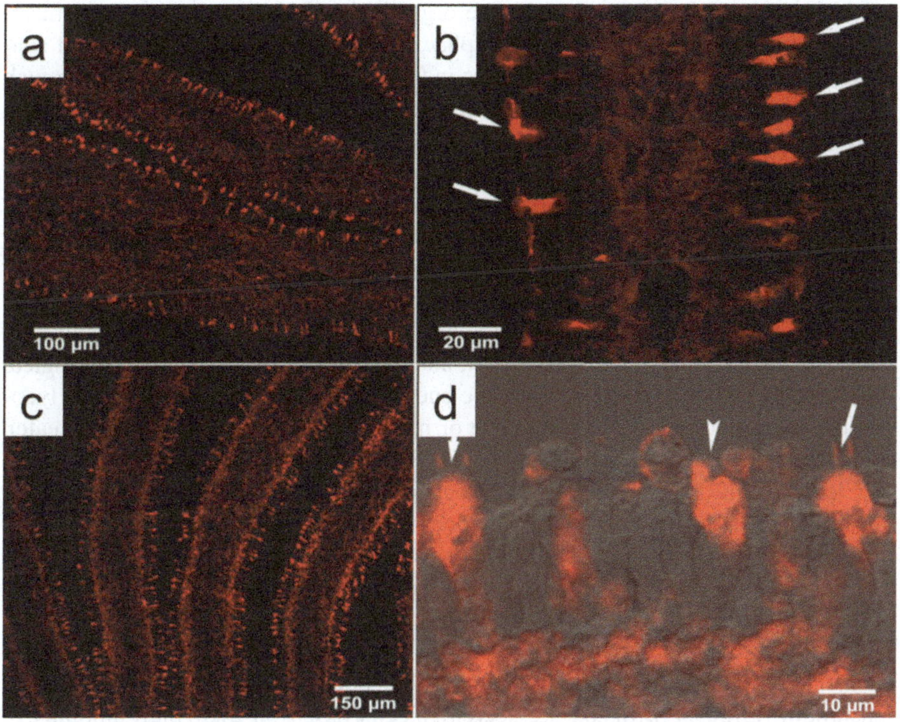

Fig. 3.5 Immunolocalization of α-amanitin in basidiocarps (fruiting bodies) of *A. bisporigera* by confocal laser scanning microscopy. (**a**) Low magnification view showing cross section of two lamellae (gills) and part of a third. (**b**) Higher magnification of a single lamella. Arrows point to cells in the hymenium containing α-amanitin. (**c**) Lamellae in a different fruiting body. (**d**) Higher magnification of a lamella from **c**, in which the confocal image is superimposed on a differential interference contrast (DIC) image (Reprinted from Luo et al. (2010), with permission of the American Society for Microbiology)

3.6 The Relationship Between Specialized Cellular Structures and Natural Product Accumulation in Agarics

Many agarics make specialized structures and types of hyphal cells that have been postulated to be sites of secondary metabolite biosynthesis and/or accumulation. These cell types can be delineated microscopically by differential cytological stains and by their unique morphologies. *Cystidia* are specialized cells found in many agarics. They are often unique to particular taxonomic groups and hence useful for mushroom identification and classification. It seems compelling to consider that the two traits – specialized chemistry and specialized structures – are functionally related. Yet, this relationship has been almost completely unmapped in agarics.

In his outstanding and highly original monograph on the cytology of the agarics and related fungi, Clémençon (2004) documented many unique cell types in mushrooms and showed that they are chemically distinguishable by histological stains for particular chemical classes. Despite the organizational chaos of the typical mushroom ground tissue (the intertwined hyphae have been compared to a plate of spaghetti), many micrographs and diagrams in Clémençon's monograph reveal a much greater degree of chemical specialization among different cells of an agaric basidiocarp than is visible by conventional light microscopy. Clémençon (2004) furthermore documents numerous examples of chemicals of particular classes that accumulate nonuniformly in cells in other parts of mushrooms such as the cap and stipe.

Most histological stains are not very informative about chemical structures or biological function. In addition to cyclic peptides, the staining compounds described by Clémençon (2004) could belong to any of a number of chemical classes, including terpenoids (Schmidt-Dannert 2015) and compounds that could act as physical barriers, such as gums, mucilages, latex, and polyphenolics. In regard to biological function, the staining materials could be energy storage compounds such as glycogen or defense compounds against one or more groups of organisms that threaten basidiocarps, such as nematodes, molluscs, or insects.

Clémençon (2004) refers to multiple subtypes of "secretory hyphae" and "excretory hyphae." However, there is no evidence that these cells actively secrete anything outside of themselves, but rather they appear to accumulate substances that could then be released defensively when damaged, for example, by wounding or herbivory. A functional connection between secretory hyphae and secondary metabolites was postulated by researchers already in the 1800s and early 1900s (cited in Clémençon 2004). Gull and Newsam (1975) hypothesized that the cystidia of the mushroom *Agrocybe praecox* are secretory cells based on their high amounts of smooth endoplasmic reticulum.

Besides the easily identified secretory cells in genera such as *Lactarius* that ooze milky latex, other types of secretory cells can be differentiated from normal surrounding cells by reaction with a large variety of cytological stains such as Sudan III, iron salts, potassium permanganate, silver ammine, vanillin/sulfuric acid, and ruthenium red. Although these types of stains do not indicate with any degree of precision the chemical nature of the staining material, which can range from anionic polysaccharides to oils, they do offer tantalizing clues that agarics contain previously unrecognized types of chemically specialized cells.

Cystidia are of particular interest from the point of view of natural product accumulation. Cystidia are typically found interspersed with the basidia (where the basidiospores, i.e., products of meiosis, are borne) on the spore-forming surfaces (hymenia) of the gills of fruiting bodies. Although cystidia are useful for mushroom identification, their functions are unknown. Based on their peculiar morphologies and staining characteristics, they could act as storage cells for defensive chemicals. Their close proximity to critical reproductive structures (the basidia) supports such a function. Buller (1909) hypothesized that cystidia have a defensive function. Cystidia are also found in the mycelium of some fungi (Clémençon 2004).

The potential significance of cystidia as sites of natural product storage can be illustrated with a few examples. A taxonomically useful hallmark of species of *Galerina* in the subgenus *Naucoriopsis* (which produce α-amanitin) is the elaboration of distinctive cystidia (Fig. 3.6). The large size and translucent appearance of the *Galerina* cystidia are consistent with a storage role.

Fig. 3.6 Cystidium in a lamella (gill) of *Galerina*. Cystidia are often translucent and protrude from the surrounding tissue (Photo credit: Dan Molter, http://mushroomobserver.org/observer/show_observation/27740 (Creative Commons))

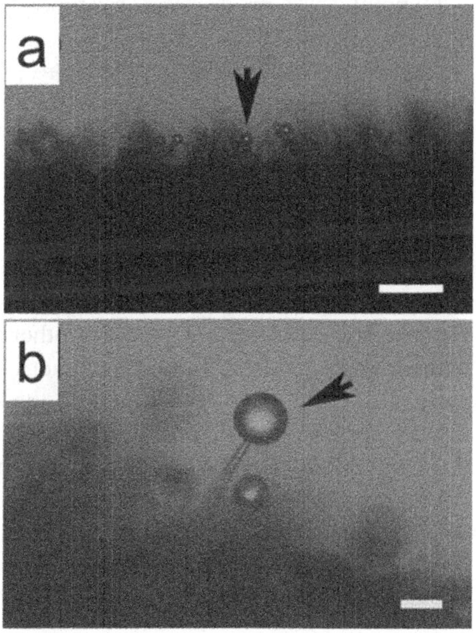

Fig. 3.7 Cystidia that kill springtails (*Collembola*, in the subphylum Hexapoda, phylum Arthropoda). (**a**) Cystidia on the surface of a pileus (cap) of *Russula bella*. (**b**) Single cystidium at higher magnification. Arrow indicates a droplet secreted by the cystidium (Photo credit: T. Nakamori, Yokohama National University. Reprinted from Nakamori and Suzuki (2007) with permission of Elsevier)

Cystidia in the hymenium, stipes, and caps of *Russula bella* were shown experimentally to protect this agaric against herbivory by springtails (Hexapoda) (Nakamori and Suzuki 2007). The tips of these cystidia exude droplets of ~10 μm diameter, whose chemical composition is still unknown (Fig. 3.7). The droplets are not surrounded by a membrane and are easily washed off. *Conocybe lactea* in culture produces secretory cells that form droplets 5–7.5 μm in diameter (Fig. 3.4). The droplets are toxic to nematodes, but the nature of the nematicidal compound, given the trivial name conocybin, is unknown (Hutchinson et al. 1996). Conocybin could plausibly be the same as phalloidin, given that the latter compound was presumptively identified in *C. apala* (Hallen et al. 2003). A specimen of *Inocybe* involved in

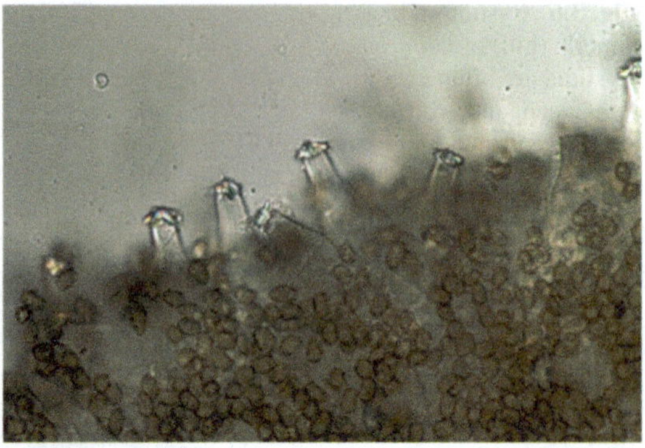

Fig. 3.8 Cystidia on the lamellae (gills) of a species of *Inocybe* that poisoned a dog. Each brown basidiospore is ~7 μm in length. Photo credit: Kathie Hodge, Cornell University, used with permission. www.flickr.com/photos/cornellfungi

the fatal poisoning of a dog has prominent transparent cystidia embedded in its hymenium, perhaps serving as repositories for the unknown canicidal toxin (Fig. 3.8).

Some species of the agaric *Pleurotus* produce specialized structures, called toxocysts, that contain nematicidal compounds. The toxocyst bursts when a nematode touches it, releasing the toxic compound. The fungal mycelium then penetrates and devours the worm (Truong et al. 2007). A nematicidal toxin from *P. ostreatus*, the edible oyster mushroom, was identified as *trans*-2-decenedioic acid, but it is unknown if this same compound is present in toxocysts (Kwok et al. 1992).

In all of these examples, the relationship, if any, between the specialized structures and the accumulation of biologically active compounds is poorly understood. In those few cases where cystidia or other specialized structures have been shown to contain compounds with biological activity, the compounds have not been chemically purified or identified. In light of the fact that plants accumulate defensive secondary metabolites in specialized structures such as trichomes (Wheeler and Krimmel 2015), and that cystidia often appear to contain something that is too transparent to be cytoplasm, it seems plausible that cystidia are repositories for defensive secondary metabolites and that their primary function is to contribute to the survival of the mushrooms in which they are found by protecting against predators such as nematodes, insects, and slugs.

References

Ammirati JF, Thiers HD, Horgen PA (1977) Amatoxin-containing mushrooms: *Amanita ocreata* and *A. phalloides* in California. Mycologia 69:1095–1108

Ben Khelil M, Zhioua M, Bakir O, Allouche M, Gloulou F, Banasr A, Haouet S, Hedhili A, Hamdoun M (2010) Intoxication mortelle par *Lepiota brunnioincarnata*: à propos de 4 cas. Ann Biol Clin (Paris) 68:561–567. https://doi.org/10.1684/abc.2010.0467

Benedict RG, Tyler VE Jr, Brady LR, Weber LJ (1966) Fermentative production of amanita toxins by a strain of *Galerina marginata*. J Bacteriol 91:1380–1381

Beutler JA, Der Marderosian AH (1981) Chemical variation in *Amanita*. J Nat Prod (Lloydia) 44:422–431

Beutler JA, Vergeer PP (1980) Amatoxins in American mushrooms: evaluation of the Meixner test. Mycologia 72:1142–1149

Brady LR, Benedict RG, Tyler VE, Stuntz DE, Malone MH (1975) Identification of *Conocybe filaris* as a toxic basidiomycete. Lloydia 38:172–173

Bresinsky A, Besl H (1990) A colour atlas of poisonous fungi. Wolfe Publishing, Regensburg

Brown DW, Butchko RA, Busman M, Proctor RH (2012) Identification of gene clusters associated with fusaric acid, fusarin, and perithecial pigment production in *Fusarium verticillioides*. Fungal Genet Biol 49:521–532. https://doi.org/10.1016/j.fgb.2012.05.010

Buku A, Wieland T, Bodenmüller H, Faulstich H (1980) Amaninamide, a new toxin of *Amanita virosa* mushrooms. Experientia 36:33–34

Buller AHR (1909) Researches on fungi. Longmans, Green, London

Cai Q, Tulloss RE, Tang LP, Tolgor B, Zhang P, Chen ZH, Yang ZL (2014) Multi-locus phylogeny of lethal amanitas: implications for species diversity and historical biography. BMC Evol Biol 14:143. https://doi.org/10.1186/1471-2148-14-143

Cai Q, Cui YY, Yang ZL (2016) Lethal *Amanita* in China. Mycologia 108:993–1009. https://doi.org/10.3852/16-008

Clémençon H (2004) Cytology and plectology of the hymenomycetes. J. Cramer, Berlin

Deng WQ, Li TH, Xi PG, Gan LX, Xiao ZD, Jiang ZD (2011) Peptide toxin components of *Amanita exitialis* basidiocarps. Mycologia 103:946–949. https://doi.org/10.3852/10-319

Duffy TJ (2008) Toxic fungi of western North America. Available on line at http://www.mykoweb.com/TFWNA/index.html. (retrieved December, 2017)

Enjalbert F, Cassanas G, Andary C (1989) Variation in amounts of main phallotoxins in *Amanita phalloides*. Mycologia 81:266–271

Enjalbert F, Gallion C, Jehl F, Monteil H, Faulstich H (1992) Simultaneous assay for amatoxins and phallotoxins in *Amanita phalloides* Fr. by high-performance liquid chromatography. J Chromatogr 598:227–236

Enjalbert F, Cassanas G, Guinchard C, Chaumont JP (1996) Toxin composition of *Amanita phalloides* tissues in relation to the collection site. Mycologia 88:909–921

Enjalbert F, Cassanas G, Salhi SL, Guinchard C, Chaumont JP (1999) Distribution of the amatoxins and phallotoxins in *Amanita phalloides*. Influence of the tissues and the collection site. CR Acad Sci III 322:855–862

Enjalbert F, Rapior S, Nouguier-Soulé J, Guillon S, Amouroux N, Cabot C (2002) Treatment of amatoxin poisoning: 20-year retrospective analysis. J Toxicol Clin Toxicol 40:715–757

Enjalbert F, Cassanas G, Rapior S, Renault C, Chaumont JP (2004) Amatoxins in wood-rotting *Galerina marginata*. Mycologia 96:720–729. https://doi.org/10.2307/3762106

Faulstich H, Cochet-Meilhac M (1976) Amatoxins in edible mushrooms. FEBS Lett 64:73–75

Faulstich H, Georgopoulos D, Bloching M, Wieland T (1974) Analysis of the toxins of amanitin-containing mushrooms. Z Naturforsch 29c:86–88

Ford WW (1906) The toxicological constitution of *Amanita phalloides*. J Exp Med 8:437–450

Ford WW (1909) The distribution of poisons in mushrooms. Science 30:97–108. https://doi.org/10.1126/science.30.760.97

Furia A, Bernicchia A, Alberton F (1982) Intossicazione parafalloidea da *Lepiota brunneoincarnata*. Micol Ital 11:29–34

Garcia J, Oliveira A, de Pinho PG, Freitas V, Carvalho A, Baptista P, Pereira E, de Lourdes BM, Carvalho F (2015) Determination of amatoxins and phallotoxins in *Amanita phalloides* mushrooms from northeastern Portugal by HPLC-DAD-MS. Mycologia 107:679–687. https://doi.org/10.3852/14-253

Gérault A, Girre L (1975) Recherches toxicologiques sur le genre *Lepiota* Fries. CR Acad Sci, Sér D 280:2841–2843

Gulden G, Dunham S, Stockman J (2001) DNA studies in the *Galerina marginata* complex. Mycol Res 105:432–440. https://doi.org/10.1017/S0953756201003707

Gulden G, Stensrud O, Shalchian-Tabrizi K, Kauserud H (2005) *Galerina* Earle: a polyphyletic genus in the consortium of dark-spored agarics. Mycologia 97:823–827. https://doi.org/10.3852/mycologia.97.4.823

Gull K, Newsam RJ (1975) Ultrastructural organization of cystidia in the basidiomycete *Agrocybe praecox*. J Gen Microbiol 91:74–78

Hallen HE, Adams GC, Eicker A (2002) Amatoxins and phallotoxins in indigenous and introduced South African *Amanita* species. S Afr J Bot 68:322–326

Hallen HE, Watling R, Adams GC (2003) Taxonomy and toxicity of *Conocybe lactea* and related species. Mycol Res 107:969–979

Hallen HE, Luo H, Scott-Craig JS, Walton JD (2007) Gene family encoding the major toxins of lethal *Amanita* mushrooms. Proc Natl Acad Sci U S A 104:19097–19101. https://doi.org/10.1073/pnas.0707340104

Horgen PA, Ammirati JF, Thiers HD (1976) Occurence of amatoxins in *Amanita ocreata*. Lloydia 39:368–371

Hu J, Zhang P, Zeng J, Chen Z (2012) Determination of amatoxins in different tissues and development stages of *Amanita exitialis*. J Sci Food Agric 92:2664–2667. https://doi.org/10.1002/jsfa.5685

Hutchinson LJ, Madzia SE, Barron GL (1996) The presence and antifeedant function of toxin-producing secretory cells on hyphae of the lawn-inhabiting agaric *Conocybe lactea*. Can J Bot 74:431–434

Jenkins DT (1982) A study of *Amanita* types. IV Taxa described by G F Atkinson. Mycotaxon 14:237–246

Johnson BC, Preston JF (1979) Unique amatoxin resistance of RNA synthesis in isolated nuclei from *Amanita* species accumulating amatoxins. Arch Microbiol 122:161–167

Johnson BEC, Preston JF, Kimbrough JW (1979) Quantitation of amanitins in *Galerina autumnalis*. Mycologia 68:1248–1252. https://doi.org/10.2307/3758759

Kaneko H, Tomomasa T, Inoue Y, Kunimoto F, Fukusato T, Muraoka S, Gonmori K, Matsumoto T, Morikawa A (2001) Amatoxin poisoning from ingestion of Japanese *Galerina* mushrooms. J Toxicol Clin Toxicol 39:413–416

Kaya E, Yilmaz I, Sinirlioglu ZA, Karahan S, Bayram R, Yaykasli KO, Colakoglu S, Saritas A, Severoglu Z (2013) Amanitin and phallotoxin concentration in *Amanita phalloides* var. *alba* mushroom. Toxicon 76:225–233

Kervégant M, de Haro L, Patat AM, Pons C, Thomachot L, Minodier P (2013) Phalloides syndrome poisoning after ingestion of *Lepiota* mushrooms. Wilderness Environ Med 24:170–172. https://doi.org/10.1016/j.wem.2012.11.002

Kohler A, Kuo A, Nagy LG, Morin E, Barry KW, Buscot F, Canback B, Choi C, Cichocki N, Clum A, Colpaert J, Copeland A, Costa MD, Dore J, Floudas D, Gay G, Girlanda M, Henrissat B, Herrmann S, Hess J, Hogberg N, Johansson T, Khouja HR, LaButti K, Lahrmann U, Levasseur A, Lindquist EA, Lipzen A, Marmeisse R, Martino E, Murat C, Ngan CY, Nehls U, Plett JM, Pringle A, Ohm RA, Perotto S, Peter M, Riley R, Rineau F, Ruytinx J, Salamov A, Shah F, Sun H, Tarkka M, Tritt A, Veneault-Fourrey C, Zuccaro A, Tunlid A, Grigoriev IV, Hibbett DS, Martin F (2015) Convergent losses of decay mechanisms and rapid turnover of symbiosis genes in mycorrhizal mutualists. Nat Genet 47:410–415. https://doi.org/10.1038/ng.3223

Kose M, Yilmaz I, Akata I, Kaya E, Guler K (2015) A case study: rare *Lepiota brunneoincarnata* poisoning. Wilderness Environ Med 26:350–354. https://doi.org/10.1016/j.wem.2014.12.025

Kwok OCH, Plattner R, Weisleder D, Wicklow DT (1992) A nematicidal toxin from *Pleurotus ostreatus* NRRL 3526. J Chem Ecol 18:127–136. https://doi.org/10.1007/BF00993748

Li P, Deng WQ, Li TH, Song B, Shen YH (2013) Illumina-based de novo transcriptome sequencing and analysis of *Amanita exitialis* basidiocarps. Gene 532:63–71. https://doi.org/10.1016/j.gene.2013.09.014

Li P, Deng W-Q, Li T-H (2014a) Molecular cloning of α-amanitin and characterization of its expression pattern in different parts and development stages of *Amanita exitialis* fruitbody. Mycol Progress 13:1011–1016

Li P, Deng W, Li T (2014b) The molecular diversity of toxin gene families in lethal *Amanita* mushrooms. Toxicon 83:59–68. https://doi.org/10.1016/j.toxicon.2014.02.020

Little MC, Preston JF III, Jackson C, Bonetti S, King RW, Taylor LCE (1986) Alloviroidin, the naturally occurring toxic isomer of the cyclopeptide viroidin. Biochemistry 25:2867–2872

Luo H, Hallen-Adams HE, Walton JD (2009) Processing of the phalloidin proprotein by prolyl oligopeptidase from the mushroom *Conocybe albipes*. J Biol Chem 284:18070–18077. https://doi.org/10.1074/jbc.M109.006460

Luo H, Hallen-Adams HE, Scott-Craig JS, Walton JD (2010) Co-localization of amanitin and a candidate toxin-processing prolyl oligopeptidase in *Amanita* basidiocarps. Eukaryot Cell 9:1891–1900. https://doi.org/10.1128/EC.00161-10

Luo H, Hallen-Adams HE, Scott-Craig JS, Walton JD (2012) Ribosomal biosynthesis of α-amanitin in *Galerina marginata*. Fungal Genet Biol 49:123–129. https://doi.org/10.1016/j.fgb.2011.12.005

Luo H, Hong SY, Sgambelluri RM, Angelos E, Li X, Walton JD (2014) Peptide macrocyclization catalyzed by a prolyl oligopeptidase involved in α-amanitin biosynthesis. Chem Biol 21:1610–1617. https://doi.org/10.1016/j.chembiol.2014.10.015

Luo H, DuBois B, Sgambelluri RM, Angelos ER, Li X, Holmes D, Walton JD (2015) Production of ^{15}N-labeled α-amanitin in *Galerina marginata*. Toxicon 103:60–64. https://doi.org/10.1016/j.toxicon.2015.06.018

McKnight TA, McKnight KB, Skeels MC (2010) Amatoxin and phallotoxin concentration in *Amanita bisporigera* spores. Mycologia 102:763–765. https://doi.org/10.3852/09-131

Meunier B, Messner M, Bardaxoglou E, Spiliopoulos G, Terblanche J, Launois B (1994) Liver transplantation for severe *Lepiota helveola* poisoning. Liver 14:158–160. https://doi.org/10.1111/j.1600-0676.1994.tb00066.x

Miller OK Jr, Hemmes DE, Wong G (1988) *Amanita marmorata* subsp. *myrtacearum*, a new subspecies in *Amanita* section *Phalloideae* from Hawaii. Mycologia 88:140–145

Mottram AR, Lazio MP, Bryant SM (2010) Lepiota subincarnata J.E. Lange induced fulminant hepatic failure presenting with pancreatitis. J Med Toxicol 6:155–157. https://doi.org/10.1007/s13181-010-0062-1

Mullersman JE, Preston JF (1982) A microassay for phallotoxins: quantification of phallotoxins in *Amanita* species. Anal Biochem 119:266–273

Mungenast T (2008) Eulogy for a lost dog. https://blog.mycology.cornell.edu/2008/10/09/eulogy-for-a-lost-dog/. Retrieved 24 Oct 2017

Nakamori T, Suzuki A (2007) Defensive role of cystidia against *Collembola* in the basidiomycetes *Russula bella* and *Strobilurus ohshimae*. Mycol Res 111:1345–1351. https://doi.org/10.1016/j.mycres.2007.08.013

Parnmen S, Sikaphan S, Leudang S, Boonpratuang T, Rangsiruji A, Naksuwankul K (2016) Molecular identification of poisonous mushrooms using nuclear ITS region and peptide toxins: a retrospective study on fatal cases in Thailand. J Toxicol Sci 41:65–76. https://doi.org/10.2131/jts.41.65

Paydas S, Kocak R, Erturk F, Erken E, Zaksu HS, Gurcay A (1990) Poisoning due to amatoxin-containing *Lepiota* species. Br J Clin Pract 44:450–453

Pitkin JW, Nikolskaya A, Ahn JH, Walton JD (2000) Reduced virulence caused by meiotic instability of the *TOX2* chromosome of the maize pathogen *Cochliobolus carbonum*. Mol Plant-Microbe Interact 13:80–87. https://doi.org/10.1094/MPMI.2000.13.1.80

Preston JF, Stark HJ, Kimbrough JW (1975) Quantitation of amanitins in *Amanita verna* with calf thymus RNA polymerase B. Lloydia 38:153–161

Preston JF, Johnson BEC, Little M, Romeo T, Stark HJ, Mullersman JE (1982) Investigations on the function of amatoxins in *Amanita* species: a case for amatoxins as potential regulators of transcription. In: Kleinkauf H, von Döhren H (eds) Peptide antibiotics: biosynthesis and functions. Walter de Gruyter, Berlin, pp 399–426

Pulman JA, Childs KL, Sgambelluri RM, Walton JD (2016) Expansion and diversification of the MSDIN family of cyclic peptide genes in the poisonous agarics *Amanita phalloides* and *A. bisporigera*. BMC Genomics 17:1038. https://doi.org/10.1186/s12864-016-3378-7

Riley R, Salamov AA, Brown DW, Nagy LG, Floudas D, Held BW, Levasseur A, Lombard V, Morin E, Otillar R, Lindquist EA, Sun H, LaButti KM, Schmutz J, Jabbour D, Luo H, Baker SE, Pisabarro AG, Walton JD, Blanchette RA, Henrissat B, Martin F, Cullen D, Hibbett DS, Grigoriev IV (2014) Extensive sampling of basidiomycete genomes demonstrates inadequacy of the white-rot/brown-rot paradigm for wood decay fungi. Proc Natl Acad Sci U S A 111:9923–9928. https://doi.org/10.1073/pnas.1400592111

Schmidt-Dannert C (2015) Biosynthesis of terpenoid natural products in fungi. Adv Biochem Eng Biotechnol 148:19–61. https://doi.org/10.1007/10_2014_283

Seeger R, Stijve T (1979) Amanitin content and toxicity of *Amantia verna* bull. Z Naturforsch 34C:330–333

Sgambelluri RM, Epis S, Sassera D, Luo H, Angelos ER, Walton JD (2014) Profiling of amatoxins and phallotoxins in the genus *Lepiota* by liquid chromatography combined with UV absorbance and mass spectrometry. Toxins 6:2336–2347. https://doi.org/10.3390/toxins6082336

Tang S, Zhou Q, He Z, Luo T, Zhang P, Cai Q, Yang Z, Chen J, Chen Z (2016) Cyclopeptide toxins of lethal amanitas: compositions, distribution and phylogenetic implication. Toxicon 120:78–88. https://doi.org/10.1016/j.toxicon.2016.07.018

Thongbai B, Miller SL, Stadler M, Wittstein K, Hyde KD, Lumyong S, Raspé O (2017) Study of three interesting *Amanita* species from Thailand: morphology, multiple-gene phylogeny and toxin analysis. PLoS One 12:e0182131. https://doi.org/10.1371/journal.pone.0182131

Truong BN, Okazaki K, Fukiharu T, Takeuchi Y, Futai K, Le XT, Suzuki A (2007) Characterization of the nematocidal toxocyst in *Pleurotus* subgen. *Coremiopleurotus*. Mycoscience 48:222–230. https://doi.org/10.1007/s10267-007-0358-4

Tyler VE Jr, Smith AH (1963) Chromatographic detection of amanita toxins in *Galerina venenata*. Mycologia 55:358–359. https://doi.org/10.2307/3756526

Tyler VE Jr, Brady LR, Benedict RG, Khana JM, Malone MH (1963) Chromatographic and pharmacologic evaluation of some toxic *Galerina* species. Lloydia 26:154–157

Tyler VE Jr, Benedict RG, Brady LR, Robbers JE (1966) Occurrence of *Amanita* toxins in American collections of deadly amanitas. J Pharm Sci 55:590–593

Vargas N, Bernal A, Sarria V, Franco-Molano A, Restrepo S (2011) Amatoxin and phallotoxin composition in species of the genus *Amanita* in Colombia: a taxonomic perspective. Toxicon 58:583–590. https://doi.org/10.1016/j.toxicon.2011.09.005

Varvenne D, Retornaz K, Metge P, De Haro L, Minodier P (2015) Amatoxin-containing mushroom (*Lepiota brunneoincarnata*) familial poisoning. Pediatr Emerg Care 31:277–288. https://doi.org/10.1097/PEC.0000000000000399

Walton JD (2000) Horizontal gene transfer and the evolution of secondary metabolite gene clusters in fungi: an hypothesis. Fungal Genet Biol 30:167–171. https://doi.org/10.1006/fgbi.2000.1224

Wei J, Wu J, Chen J, Wu B, He Z, Zhang P, Li H, Sun C, Liu C, Chen Z, Xie J (2017) Determination of cyclopeptide toxins in *Amanita subpallidorosea* and *Amanita virosa* by high-performance liquid chromatography coupled with high-resolution mass spectrometry. Toxicon 133:26–32. https://doi.org/10.1016/j.toxicon.2017.04.012

Wheeler AG Jr, Krimmel BA (2015) Mirid (hemiptera: heteroptera) specialists of sticky plants: adaptations, interactions, and ecological implications. Annu Rev Entomol 60:393–414. https://doi.org/10.1146/annurev-ento-010814-020932

Wieland T (1986) Peptides of poisonous *Amanita* mushrooms. Springer, New York

Wight WD, Labuda R, Walton JD (2013) Conservation of the genes for HC-toxin biosynthesis in *Alternaria jesenskae*. BMC Microbiol 13:165. https://doi.org/10.1186/1471-2180-13-165

Wolfe BE, Richard F, Cross HB, Pringle A (2010) Distribution and abundance of the introduced ectomycorrhizal fungus *Amanita phalloides* in North America. New Phytol 185:803–816. https://doi.org/10.1111/j.1469-8137.2009.03097.x

Wolfe BE, Tulloss RE, Pringle A (2012) The irreversible loss of a decomposition pathway marks the single origin of an ectomycorrhizal symbiosis. PLoS One 7:e39597. https://doi.org/10.1371/journal.pone.0039597

Xue JH, Wu P, Chi YL, Xu LX, Wei XY (2011) Cyclopeptides from *Amanita exitialis*. Nat Prod Bioprospect 1:52–66

Yarze JC, Tulloss RE (2012) Acute liver injury and renal failure due to poisonous mushroom ingestion. Am J Gastroenterol 107:790–791. https://doi.org/10.1038/ajg.2012.29

Yilmaz I, Kaya E, Sinirlioglu ZA, Bayram R, Surmen MG, Colakoglu S (2014) Clinical importance of toxin concentration in *Amanita verna* mushroom. Toxicon 87:68–75. https://doi.org/10.1016/j.toxicon.2014.05.019

Yin W, Yang ZR (1993) A clinical analysis of twelve patients with *Galerina autumnalis* poisoning. Zhonghua Nei Ke Za Zhi 32:812–812

Yocum RR, Simons DM (1977) Amatoxins and phallotoxins in *Amanita* species of the northeastern United States. Lloydia 40:178–190

Zhang P, Chen Z, Hu J, Wei B, Zhang Z, Hu W (2005) Production and characterization of Amanitin toxins from a pure culture of *Amanita exitialis*. FEMS Microbiol Lett 252:223–228. https://doi.org/10.1016/j.femsle.2005.08.049

Chapter 4
Biosynthesis of the *Amanita* Cyclic Peptide Toxins

Cyclic peptides are not very widespread in nature... Besides their stability to enzymatic hydrolysis, they acclaimed attention by their non-ribosomal biosynthesis... It is very probable, but not yet established, that the Amanita peptides will be manufactured in an analogous way. Unfortunately, the biosynthesis of the Amanita peptides cannot as yet be investigated.

(Wieland 1986)

To a scientist in 1986, the prospects of ever identifying the enzymatic pathway, much less the genes, involved in the biosynthesis of the *Amanita* peptide toxins must have seemed remote. The organisms are intractable to culture, the molecules are uniquely complex, gene cloning was in its infancy, and high-throughput genomic sequencing was not yet invented. Several indirect approaches to gain at least some insight into the pathway were attempted in the author's and other's laboratories in the first decade of the twenty-first century, including feeding radiolabeled amino acids to mushroom slices, in vitro assays for nonribosomal peptide synthetases (NRPSs) using the ATP/PP_i exchange assay (Walton 1987), DNA hybridization with heterologous NRPS probes, and polymerase chain reaction (PCR) based on conserved motifs in NRPSs. All these attempts were unsuccessful.

4.1 Identification of the Genes for the *Amanita* Cyclic Peptides

The development of "next-generation" (post-Sanger) DNA sequencing dropped the price of sequencing more than 1000-fold over a span of a few years, to the point that it became feasible to sequence an entire genome, at least ones as small as the average fungus (40–80 MB), in order to find a single gene. This approach seemed

J. Walton, *The Cyclic Peptide Toxins of Amanita and Other Poisonous Mushrooms*, https://doi.org/10.1007/978-3-319-76822-9_4

particularly attractive in the case of the *Amanita* biosynthetic genes because it had been widely assumed since the 1980s that the *Amanita* toxins were biosynthesized by NRPSs, like all other fungal cyclic peptides known at that time (Walton et al. 2004). Considering that activation of each amino acid in a nonribosomal peptide requires its own module of ~140 kDa, an eight-module NRPS for α-amanitin would require ~30 kb of protein-encoding DNA, and a seven-module NRPS for phalloidin would require ~27 kb. Thus, relatively shallow coverage of a ~50 MB genome would have a statistically high probability of yielding an identifiable fragment of one of the hypothetical NRPSs for amatoxins or phallotoxins, and with a fragment in hand, one could simply "walk" along the chromosome to find the entire gene.

4.1.1 Biosynthesis of α-Amanitin in the Genus Amanita

In a search for the putative NRPS(s) involved in amatoxin and phallotoxin biosynthesis, *Amanita bisporigera* was sequenced to ~2× coverage by first-generation 454 pyrosequencing (Hallen et al. 2007). Surprisingly, the genome of this fungus contains no NRPS genes at all, in contrast to some fungi that have more than 25 (von Döhren 2009; http://genome.jgi.doe.gov/pages/sm-clusters-summary.jsf). Based on the many basidiomycete genomes now available (over 330 as of this writing), *A. bisporigera* is not exceptional within this phylum in its general lack of NRPS genes (Riley et al. 2014). Fungal species in the phylum Ascomycota (ascomycetes) have many more NRPS and other secondary metabolite genes than any fungi in the phylum Basidiomycota, including the class Agaricomycetes (which includes most mushrooms; Chap. 1). Furthermore, most of the putative agaric NRPS genes tabulated at JGI (http://genome.jgi.doe.gov/pages/sm-clusters-summary.jsf) are probably not true NRPSs but rather single-module members of the adenylating superfamily.

Instead of an eight-module NRPS, Hallen et al. (2007) found nucleotide sequences that could directly encode α-amanitin and phallacidin, i.e., nucleotide sequences encoding IWGIGCNP and AWLVDCP, respectively, in single-letter amino acid code. Sequencing of the surrounding genomic region and of corresponding cDNAs revealed that the pathway to α-amanitin begins with a ribosomally synthesized 35-amino acid precursor peptide and that phallacidin begins with a 34-amino acid precursor peptide. The 8 or 7 amino acids (the core peptides) that are found in the mature toxins of α-amanitin or phallacidin, respectively, are located in the central regions of the precursor peptides, flanked upstream by 10 amino acids (the leader peptide) and downstream by 17 amino acids (the follower peptide) (Arnison et al. 2013) (Fig. 4.1). The gene for α-amanitin in *A. bisporigera* is called *AMA1* (or sometimes when necessary to avoid confusion, *AbAMA1*) and the gene for phallacidin is *PHA1* (*AbPHA1*). Genes for the same toxins in other fungi are prefaced with a two-letter code for the source organism, e.g., *GmAMA1* designates the gene for α-amanitin in *G. marginata* and *ApAMA1* for the α-amanitin gene in *A. phalloides* (see below).

a

```
                                                                  M   S
  a ccc aac tcc cat tcg aac cta act cca aga cct cta aac ctc aca atc cca atg tct

 D   I   N   A   T   R   L   P   I   W   G   I   G   C   N   P   C   I   G   D
gac atc aat gct acc cgt ctt ccc atc tgg ggt atc ggt tgc aac ccg tgc atc ggt gac

 D   V   T   T   L   L   T   R   G   E   A   L   C   *
gac gtc act aca ctc ctc act cgt ggc gag gc^c ctt tgt taa attccccatccatttgtccg
ctgctatgacacgaag^tagtgggcgatacaagttgtggacgttatcaggcttggccgttgagcctgcatcggaa^ac
aacttatgttccttctttttctgttttcatttgttaaaatacagaacccatgtcgatgatctgtgttgtagtcaatat
aaagttgtactgtgtttcttgtcaaaaaaaaaaaaaaaaaaaa
```

b

```
                           M   S   D   I   N   A   T   R   L   P   A   W
ga cct ctg ctc taa atc aca atg tct gac atc aat gcc acc cgt ctt ccc gct tgg

 L   V   D   C   P   C   V   G   D   D   V   N   R   L   L   T   R   G   E
ctt gta gat tgc cca tgc gtc ggt gac gat gtc aac cgt ctc ctc act cgt ggc gag

 S   L   C   *
ag^c ctt tgg taa atgtctcatccactagtcaag^gcaagttgttgacaatgtcaggcttgcgggtttg
agcctgcatcggaa^acgactcacgttctttctcattctttctgattctcatttgtaaacatataaacccacgt
aaatgatccgttgtgctatggaatgcaatatacttgtgaaaaaaaaaaaaaaaaaaaaaa
```

Fig. 4.1 cDNA sequences and conceptual translations for (**a**) *AMA1* and (**b**) *PHA1* from *Amanita bisporigera*. The GenBank accession numbers are EU196139.2 and EU196140.1 for α-amanitin and EU196142.1 and EU196143.1 for phallacidin. The carats (^) indicate the positions of the three introns in each gene. The toxin (core) regions are underlined. * indicates stop codon. (From Hallen et al. 2007; copyright (2007) National Academy of Sciences, used with permission)

In *A. bisporigera*, *AMA1* and *PHA1* both contain three introns. The first intron in *AMA1* is 58 bp (base pairs) and interrupts the fourth to the last codon (counting the stop codon as one of the four) (Fig. 4.1). The other two introns, of 53 and 59 bp, interrupt the 3′ untranslated region (3′-UTR). *PHA1* also has three introns of 57, 70, and 51 bp, with the intron interrupting the coding region being in the same position (Hallen et al. 2007; Fig. 4.1). Due to the exon-interrupting introns, in both *AMA1* and *PHA1* the nucleotides encoding the last three amino acids of the precursor peptides are not contiguous in the genome with the rest of the precursor peptide-coding amino acids. However, due to high conservation of the follower peptides and of the intron positions and lengths not only in other species of *Amanita* but also in *Galerina* and *Lepiota* (see below), these last three amino acids are usually readily deduced bioinformatically (Pulman et al. 2016).

The presence of introns and polyadenylation of the cDNAs for *AMA1* and *PHA1* indicate that these are eukaryotic genes. Therefore, the genes and the trait of cyclic peptide toxin biosynthesis cannot be attributed to a prokaryotic symbiont of *A. bisporigera*. This is an important point insofar as many other ribosomally encoded peptides (RiPPs), and other natural products, are often found to actually be produced by bacterial symbionts of plants, fungi, and animals (e.g., Donia et al. 2006; Partida-Martinez and Hertweck 2005).

There are nine characterized naturally occurring amatoxins (Fig. 2.6). The core amino acid sequences can be encoded by just two genes, represented by α- and β-amanitin. α-Amanitin and its derivatives have the amino acid sequence IWGIGCNP, whereas β-amanitin and its derivatives have the sequence IWGIGCDP. The sole difference between these two (Asn or Asp at position #7) is probably not due to posttranslational modification (i.e., deamidation of Asn to Asp or amidation of Asp to Asn) because genes encoding the amino acid sequences of both α- and β-amanitin are present in the genome of *A. phalloides* and other fungi. The other amatoxins (γ-amanitin, amanin, etc.) differ from α- and β-amanitin in their patterns of hydroxylations, which are posttranslational modifications not encoded in the DNA.

There are seven known phallotoxins (Fig. 2.8). The core amino acid sequences of the phallotoxins can be encoded by two genes, represented by phalloidin (AWLATCP) and phallacidin (AWLVDCP). Genes encoding both of these are found in the genomes of *A. bisporigera*, *A. phalloides*, and/or other species in sect. *Phalloideae*. The other phallotoxins are the result of different posttranslational modifications.

Despite the fact that α-amanitin and phallacidin have only three amino acids in common (Trp, Cys, and Pro), *AMA1* and *PHA1* are highly similar outside their core regions, with 86% amino acid identity and 85% nucleotide identity (Fig. 4.2).

By DNA (Southern) blotting, the *AMA1* and *PHA1* genes are present only in known cyclic peptide toxin-producing species of *Amanita* in sect. *Phalloideae*

a

```
              M  S  D  I  N  A  T  R  L  P  I  W  G  I  G  C  N  P
amanitin     ATGTCTGACATCAATGCTACCCGTCTCCCCATCTGGGGTATCGGTTGCAACCCG
             ||||||||||||||||| |||||||| |||   |||  | | |  |||   || 
phallacidin  ATGTCTGACATCAATGCCACCCGTCTTCCCGCTTGGCTTGTAGACTGC---CCA
              M  S  D  I  N  A  T  R  L  P  A  W  L  V  D  C     P

              C  I  G  D  D  V  T  T  L  L  T  R  G  E  A  L  C  *
             TGCATCGGTGACGACGTCACTACTCTCCTCACTCGTGGCGATGCCCTTTGTTAA
             ||| |||||||||| ||||    ||||||||||||||||||   || ||| |||
             TGCGTCGGTGACGATGTCAACCGTCTCCTCACTCGTGGCGAGAGCCCTTGGTAA
              C  V  G  D  D  V  N  R  L  L  T  R  G  E  S  L  C  *
```

b

```
MSDINATRLPIWGIGCNPCIGDDVTTLLTRGEALC     [α-amanitin]
MSDINATRLP W   C PC GDDV  LLTRGE LC     [consensus]
MSDINATRLPAWLVDC-PCVGDDVNRLLTRGESLC     [phallacidin]
```

Fig. 4.2 (**a**) Alignments of the coding nucleotide and amino acid sequences of *AMA1* for α-amanitin and *PHA1* for phallacidin from *A. bisporigera*. This is an emended version of Fig. 4 from Hallen et al. (2007), which was erroneously missing six nucleotides and two amino acids. The gap in the phallacidin core region is to compensate for one less amino acid in phallacidin compared to α-amanitin. (**b**) Alignment of *AMA1* and *PHA1* precursor peptides. (Emended from Hallen et al. 2007. Copyright (2007) National Academy of Sciences, used with permission)

(Fig. 4.3; Chap. 3). Based on their whole-genome sequences, neither *A. muscaria* nor *A. thiersii* have the genetic potential to make any cycloamanide including amatoxins and phallotoxins (Kohler et al. 2015; Wolfe et al. 2012). Genes with the potential to encode amatoxins, phallotoxins, or any other cycloamanide are not present in any other fungal genomes in the JGI (genome.jgi.doe.gov/programs/fungi/index.jsf) or NCBI (www.ncbi.nlm.nih.gov/) databases (as of March, 2018). Collectively, the genetic results strongly support the conclusions of Chap. 3 that the cyclic peptide toxins are present only in sect. *Phalloideae* of *Amanita* (and some species of *Galerina* and *Lepiota*; see below).

4.1.2 Biosynthesis of α-Amanitin in Galerina marginata

The gene for α-amanitin in *G. marginata*, *GmAMA1*, was identified by whole-genome sequencing of a monokaryotic isolate (Luo et al. 2014; Riley et al. 2014). *G. marginata* has two genes for α-amanitin, *GmAMA1-1* on scaffold 42 and

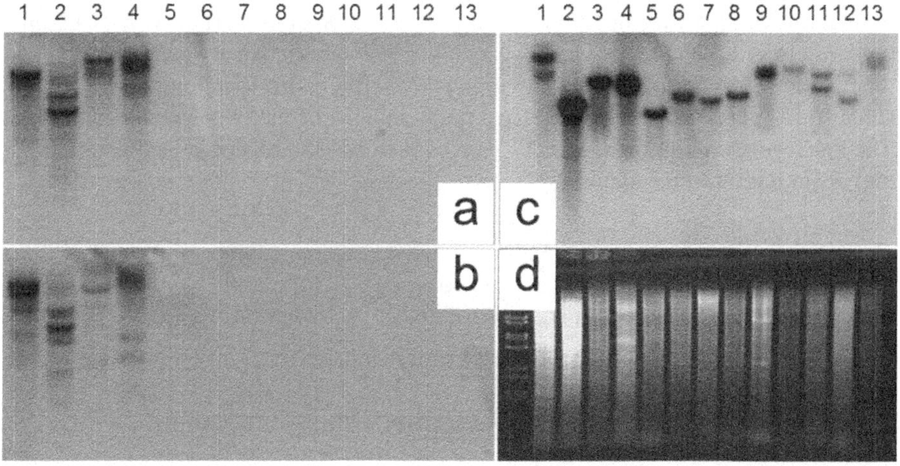

Fig. 4.3 DNA (Southern) blots of different species of *Amanita*. (**a**) Probed with *AMA1*, (**b**) probed with *PHA1*, (**c**) probed with the *A. bisporigera* tubulin gene. (**d**) Ethidium-stained gel showing DNA loading. Species and provenances were lane 1, *A.* aff. *suballiacea* (Ingham County, Michigan); lane 2, *A. bisporigera* (Ingham County); lane 3, *A. phalloides* (Alameda County, California); lane 4, *A. ocreata* (Sonoma County, California); lane 5, *A. novinupta* (Sonoma County); lane 6, *A. franchetii* (Mendocino County, California); lane 7, *A. porphyria* (Sonoma County); lane 8, a second isolate of *A. franchetii* (Sonoma County); lane 9, *A. muscaria* (Monterey County, California); lane 10, *A. gemmata* (Mendocino County); lane 11, *A. hemibapha* (Mendocino County); lane 12, *A. velosa* (Napa County, California); and lane 13, *Amanita* section *Vaginatae* (Mendocino County). Mushrooms represent sections *Phalloideae* (1–4), *Validae* (5–8), *Amanita* (9 and 10), *Caesareae* (11), and *Vaginatae* (12 and 13). (Emended from Hallen et al. 2007 in which panel labels **b** and **c** were switched during production. Copyright (2007) National Academy of Sciences, used with permission)

GmAMA1-2 on scaffold 1. The former has two introns and the latter has three. In both genes, the first intron is in the same location as the first intron in *AbAMA1*, i.e., interrupting the fourth from the last codon, and the other introns are in the 3′ untranslated regions. *GmAMA1-1* but not *GmAMA1-2* is expressed at the RNA level (Hong Luo, unpublished results from the author's laboratory). Identical to *AbAMA1* in *A. bisporigera*, *GmAMA1* also encodes a precursor peptide of 35 amino acids with a 10-amino acid leader and a 17-amino acid follower peptide. The last six amino acids of the leader (amino acids 5 through 10) are identical in *G. marginata* and *A. bisporigera*, but otherwise there is little conservation between the genes from the two fungi outside the toxin-encoding region itself (Fig. 4.4). However, the amino acids that are conserved between the two precursor peptides in the follower peptide are probably significant, and both follower peptides are predicted to form α-helices (see below). Species in the genus *Galerina* that do not produce α-amanitin lack sequences that hybridize to *GmAMA1* and *GmPOPB* (encoding the first processing enzyme; see below), which explains why they do not biosynthesize α-amanitin (Luo et al. 2012).

An as-yet unresolved paradox is the existence of β-amanitin and its encoding gene in *G. marginata*. The sequenced genome of *G. marginata* CBS 339.88 does not have a gene encoding β-amanitin, yet this strain sometimes shows a low level of β-amanitin when grown in culture (Luo et al. 2012; Sgambelluri et al. 2014). Some wild specimens of *G. marginata* have levels of β-amanitin that are as high, or higher, than α-amanitin (Chap. 3). This situation is most likely due to natural variation, i.e., some specimens of *G. marginata* have the β-amanitin gene, whereas others do not. The low levels sometimes seen in *G. marginata* CBS 339.88 could be due to artifactual deamidation of α-amanitin.

Fig. 4.4 Pair-wise alignments of α-amanitin precursor peptides (AMA1) from *Lepiota subincarnata* (Ls), *Amanita bisporigera* (Ab), and *Galerina marginata* (Gm). (Data taken from Hallen et al. 2007; Luo et al. 2012; and unpublished results from the author's laboratory. Amanitin core regions are *underlined* and in red)

G. marginata has been a useful model system for elucidation of the pathway of α-amanitin because it grows in culture, is transformable, and its genome has a three-member cluster of known or putative pathway genes (Luo et al. 2014; Riley et al. 2014; see below).

4.1.3 Biosynthesis of α-Amanitin in Lepiota subincarnata

In addition to *Amanita* and *Galerina*, α-amanitin is made by some species of *Lepiota*, such as *L. brunneoincarnata*, *L. castanea*, and *L. subincarnata* (*L. josserandii*) (Sgambelluri et al. 2014; Chap. 3). Most species in the genus *Lepiota* make neither amatoxins nor phallotoxins. Two specimens of *L. subincarnata* (as confirmed by ITS sequencing) collected in Washington state were sequenced to ~25× coverage, and sequences encoding the α-amanitin core region, IWGIGCNP, were identified (unpublished results from the author's lab). The *LsAMA1* gene encodes a 34-amino acid precursor peptide of similar structure to the *AMA1* genes of *Galerina* and *Amanita* (Fig. 4.4). The predicted LsAMA1 precursor peptide is one amino acid shorter than those of *Galerina* and *Amanita* due to one fewer amino acid in the leader peptide (shown as a gap in Fig. 4.4). *L. subincarnata* does not make phallotoxins nor does its genome contain any phallotoxin-encoding genes. However, *L. subincarnata* does have genes encoding novel cycloamanides, i.e., unmodified monocyclic peptides (see below).

4.2 Proteolytic Processing and Cyclization of the Toxin Precursor Peptides by Prolyl Oligopeptidase B (POPB)

Biosynthetically, at least several additional reactions are necessary to convert the linear ribosomal precursor peptides to the mature cyclic amatoxins and phallotoxins. These include proteolysis (to remove the leader and follower), cyclization of the core peptides, tryptathionine formation (i.e., cross-linking of Trp and Cys), hydroxylation at up to four positions (in the case of the amatoxins) or five positions (in the phallotoxins) positions, epimerization of amino acid #5 (in the phallotoxins only), and oxidation of the Cys sulfur (amatoxins only). These reactions do not necessarily occur in that order.

Multiple lines of evidence point to the initial posttranslational processing steps, proteolysis and cyclization, being catalyzed by a member of the prolyl oligopeptidase (POP) family of serine proteases (family S9A, EC 3.4.21.26). First, the two Pro residues flanking the core regions are strongly conserved in the genes for α-amanitin and phallacidin and generally in the cycloamanide family (Pulman et al. 2016; Figs. 4.2, 4.4, 4.12, and 4.13). The Pro at the C-terminal end of the core peptide remains in the mature toxin, and the N-terminal Pro is discarded with the leader peptide. The conservation of these flanking Pro residues suggests the involvement of a Pro-specific endoprotease. Second, of the known Pro-specific endoproteases, only POP is an endo-acting enzyme active exclusively on small peptides (Szeltner and Polgár 2008). Third, toxin-producing fungi whose genomes have been sequenced (i.e., *A. phalloides*, *A. bisporigera*, and *G. marginata*) have two POP genes, whereas other agarics (with only a few exceptions, some of which are probably artifacts of poor genome assembly) have only one. Phylogenetically, the POPB enzymes of *A. bisporigera*, *A. phalloides*, and *G. marginata* cluster with each other, whereas their POPAs cluster with each other and with the solitary POPs found in toxin-nonproducing fungi such as *A. muscaria* and *A. thiersii* (Pulman et al. 2016; Fig. 4.5). POPA can be considered to be the "housekeeping" POP of all agarics (albeit its function is unknown), whereas POPB can be assigned a specific role in cycloamanide biosynthesis. Fourth, pure POPB obtained by expression in yeast catalyzes conversion of the 35mer precursor peptide of α-amanitin to cyclo(IWGIGCNP) at high efficiency (Luo et al. 2014). POPB is unusual among RiPP-processing proteases in catalyzing two reactions in the cycloamanide pathway: proteolysis at the upstream Pro (after amino acid #10 in the precursor peptide) to remove the leader sequence, and transpeptidation (transamidation) at the downstream Pro (amino acid #18) to form the homodetic macrocycle (Fig. 4.6). The two steps are nonprocessive as evidenced by transient accumulation of the 25mer linear intermediate (Fig. 4.6). This catalytic path has been confirmed (Czekster and Naismith 2017). A fifth line of evidence in support of POPB being involved in cycloamanide biosynthesis is that POPB and α-amanitin itself are co-localized in mushroom tissues (Fig. 4.7; Luo et al. 2010).

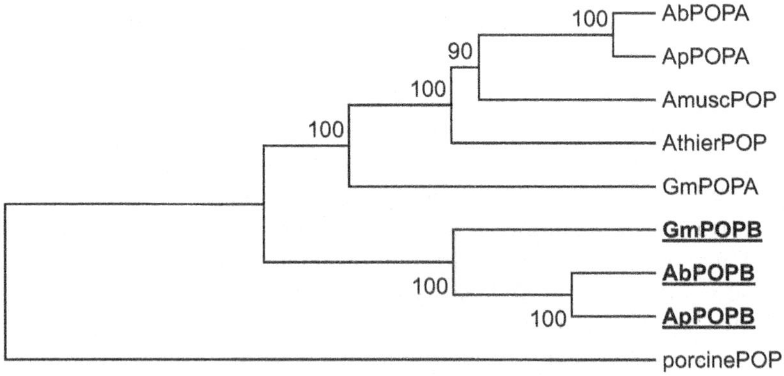

Fig. 4.5 Phylogenetic tree of POP proteins from species of *Amanita*, *Galerina marginata*, and pig (*Sus scrofa*). *Ap*, *A. phalloides*; *Ab*, *A. bisporigera*; *Amusc*, *A. muscaria*; *Athier*, *A. thiersii*; *Gm*, *G. marginata*. GenBank accession numbers of AbPOPA, AbPOPB, GmPOPA, and GmPOPB are ADN19204, ADN19205, AEX26937, and AEX26938, respectively. Department of Energy Joint Genome Institute (JGI) identifiers for AmuscPOP and AthierPOP are Amamu1l74086 and Amath1l193040, respectively. POPB proteins are in bold and underlined. (From Pulman et al. 2016, Creative Commons Attribution License)

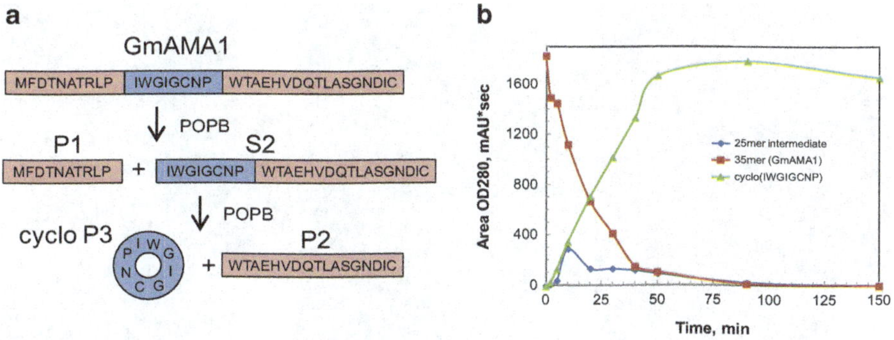

Fig. 4.6 Protease/macrocyclase activities of POPB. (**a**) Reactions catalyzed by POPB. (**b**) Time course of the POPB reaction showing consumption of 35mer GmAMA1 propeptide (red), transient accumulation of 25mer intermediate S2 (blue), and accumulation of cyclic product P3 (green). (Reprinted with permission from Luo et al. 2014)

4.2.1 Precursor Peptide Requirements for POPB Processing

The features of the precursor peptide that are important for POPB processing have been partially elucidated. The structural requirements actually involve two different sets of criteria, one for the initial hydrolysis that removes the 10-amino acid leader and the second for the cyclization reaction and simultaneous removal of the follower peptide.

The length and composition of the 10-amino acid leader are important for the first hydrolysis reaction (Luo et al. 2014). However, the leader is not necessary for cyclization because the 25mer (i.e., the precursor peptide lacking the leader) is cyclized as efficiently as the full-length 35mer precursor peptide (Czekster and Naismith 2017; Luo et al. 2014). Considering the diversity present in the natural cycloamanide family and the demonstrated versatility of POPB (Sgambelluri et al. 2018; see below), the core region is expected to have few direct interactions with POPB. This is consistent with the X-ray crystal structure of the structurally related peptide macrocyclase PCY1 bound to its precursor peptide (Chekan et al. 2017) and with the structure of GmPOPB resolved to 2.4 Å (Czekster et al. 2017; see below).

GmAMA1 and AbAMA1 are both 35 amino acids in length, and the *Lepiota subincarnata* AMA1 precursor peptide (LsAMA1) is only 1 amino acid shorter (Fig. 4.4). Although the primary amino acid sequences of the three precursor peptides outside the core region are not well-conserved, all three have the same six amino acids immediately upstream of the core region (i.e., Asn-Ala-Thr-Arg-Leu-Pro or NATRLP in single-letter code; Fig. 4.4). It is therefore likely that the pentapeptide sequence NATRL is important for initial recognition and subsequent hydrolytic cleavage by POPB, which thereby releases the 25mer (i.e., amino acids 11–35). Because the 25mer lacking the leader peptide is efficiently cyclized, the

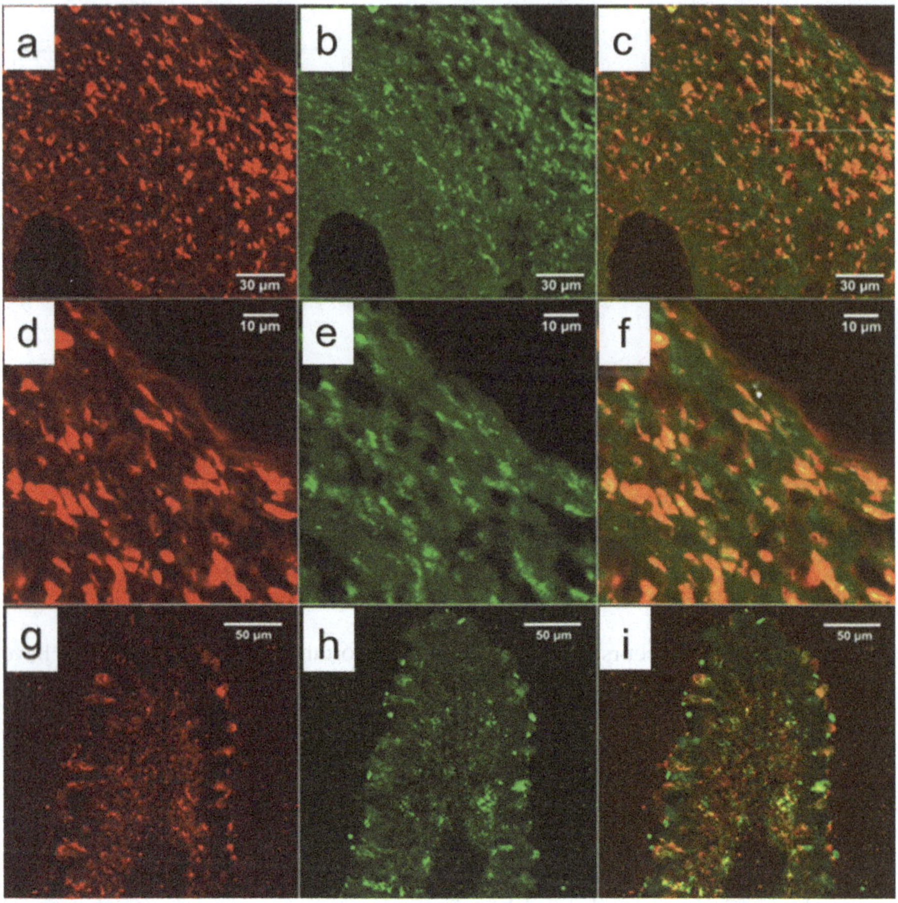

Fig. 4.7 Co-localization of α-amanitin and prolyl oligopeptidase B (POPB) in *A. bisporigera*. Panels **a**, **d**, and **g** show immunolocalization of α-amanitin in red. Panels **b**, **e**, and **h** show immunolocalization of POPB in green. Panels **c**, **f**, and **i** show co-localization in orange. Panels **a**, **b**, and **c** show low magnification of the pileus (cap). Panels **d**, **e**, and **f** show higher magnification of the same tissue. Panels **g**, **h**, and **i** show a single lamella. (Reprinted from Luo et al. 2010, with permission of the American Society for Microbiology)

pentapeptide motif cannot be necessary for macrocyclization (Luo et al. 2014). The Ala and Thr residues in the NATRL motif are probably less important than the Asn, Arg, and Leu residues because they are less well-conserved in the cycloamanide superfamily (Pulman et al. 2016; see below). Despite the conservation of NATRL in the known cycloamanides, Czekster et al. (2017) found no evidence for binding of the leader decapeptide (MFTDNATRLP) to GmPOPB, and in the co-crystal structure, the first nine residues make few contacts with the enzyme. If the leader is not important for binding to GmPOPB, its strong conservation in all three genera of amanitin-producing fungi indicates that it has some other function, perhaps to determine the overall three-dimensional shape of the 35-amino acid precursor peptide or to target the precursor peptide to a subcellular compartment (Chap. 3).

Amino acid sequence conservation in the AMA1 follower peptides is low between *Amanita*, *Galerina*, and *Lepiota* (Fig. 4.4). However, Val, Leu, Gly, (Leu/Ile), and Cys are positionally conserved in all three, and all have predicted α-helix secondary structure (shown for GmAMA1 and AbAMA1 in Fig. 4.8). The conserved Leu, Cys, and Val form one face of the α-helices. The α-helix in AbAMA1 is amphipathic and GmAMA1 to a lesser extent (Fig. 4.8). Preliminary studies introducing Gly substitutions in order to weaken the α-helix support a role for this secondary structure motif in processing of the α-amanitin precursor peptide by GmPOPB (Luo et al. 2014). In other RiPPs, such as microcin B17, patellamide, and carnobacteriocin B2, the leader peptides have α-helical and/or amphipathic properties that are important for processing (Houssen et al. 2010; Roy et al. 1998; Sprules et al. 2004).

The terminal Cys of the cycloamanide precursor peptides, which is highly conserved in AMA1 from all three fungi (Fig. 4.4), is important for cyclization, because

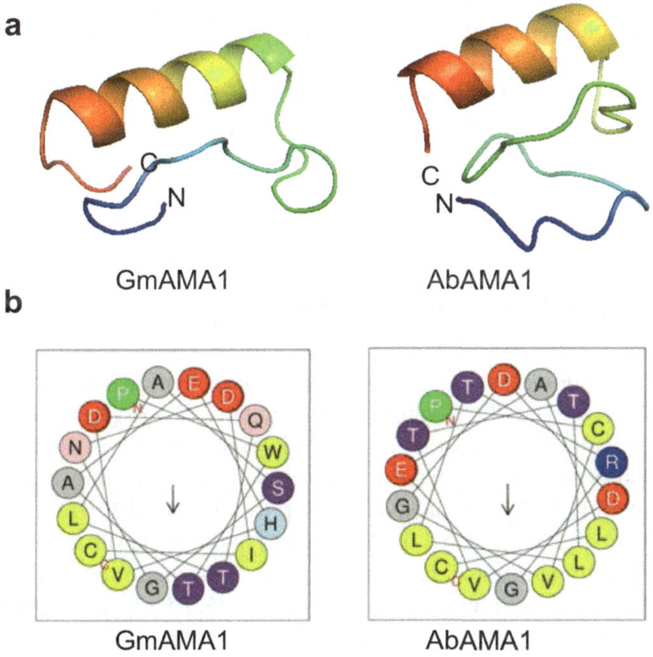

Fig. 4.8 (**a**) Predicted secondary structures of the α-amanitin precursor peptides from *G. marginata* (GmAMA1) and *A. bisporigera* (AbAMA1). From Luo et al. (2014) (**b**) Helical wheel depiction, showing amphipathic nature of the follower peptides of AbAMA1 and GmAMA1. Nonpolar residues are in yellow, polar residues are in purple, and charged residues are in red. In (**b**), only the 18 amino acids of the C-termini follower peptides are shown (starting with the dipeptide Pro-Cys)

when it is changed to Ala, cyclization does not occur (Luo et al. 2014). Similarly, the carboxylate terminal amino acid, Val, was shown to be important for macrocyclization of the cyclotide segetalin A1 by the S9A protease PCY1 (Barber et al. 2013; Chekan et al. 2017; see below).

In regard to the structural requirements of the core peptide, POPB is highly tolerant. There are >100 known natural members of the cycloamanide gene family from multiple species in *Amanita*, *Galerina*, and *Lepiota*, ranging from six (e.g., CylA) to ten (e.g., antamanide), amino acids, with a mode of eight (Pulman et al. 2016). There is amino acid bias in the natural cycloamanide core peptides with a preference for Pro and hydrophobic amino acids (Pulman et al. 2016).

Studies with engineered amino acid replacements in the core sequence of α-amanitin indicate that different types of amino acids (e.g., large/small, hydrophilic/hydrophobic) are tolerated in all positions although small amino acids (Gly or Ala) are preferred at positions #3 and #5 (Sgambelluri et al. 2018). In vitro, GmPOPB cyclizes core peptides of 8–16 amino acids and can tolerate 4-hydroxy-Pro (Hyp) at position #8, 5-hydroxy-Trp at #2 and to some extent N-methyl-Ala or β-Ala at #3 (Sgambelluri et al. 2018). Although POPB in vitro does not cyclize the heptapeptide phallacidin precursor AWLATCP, it can cyclize it to some extent if L-Thr is replaced with D-Thr, the isomer found in native phallacidin. This suggests that epimerization occurs before cyclization in phallotoxin and virotoxin biosynthesis (Sgambelluri et al. 2018).

Mechanistically, POPB is a classic serine protease. The active site triad is composed of His698, Ser577, and Asp661 (Czekster et al. 2017; Luo et al. 2014). Trp619 is critical for stacking with the Pro of the substrate where peptide bond cleavage occurs.

4.2.2 Structural Features of POPB

POPs have two domains, a seven-bladed β-propeller and a catalytic domain (Figs. 4.9 and 4.10). The POPA and POPB enzymes of *G. marginata* and *A. bisporigera* align very well with each other and with porcine POP (Luo et al. 2014). All four of these mushroom enzymes have the conserved catalytic triad typical of serine proteases in family S9A and the stacking Trp residue (W619 in GmPOPB) (Fülöp et al. 2001; Li et al. 2010). PCY1 has this same Trp (W603) even though it transamidates preferentially at Ala residues instead of Pro (Barber et al. 2013). The POPB enzymes of *G. marginata* and *A. bisporigera* are 75.5% identical, the two POPAs are 65.7% identical, and the average identity between the POPAs and the POPBs of the two fungi is 57.6% (Luo et al. 2014). The amino acid homology among the four POPs is roughly distributed throughout the proteins, as are the ~60 amino acid differences between the POPAs and the POPBs. GmPOPB is 36% identical to PCY1 (Barber et al. 2013) and 37% identical to porcine POP (Fülöp et al. 1998). Based on alignment with porcine POP, the catalytic domain of GmPOPB is predicted to comprise amino acids 1–80 and 451–730.

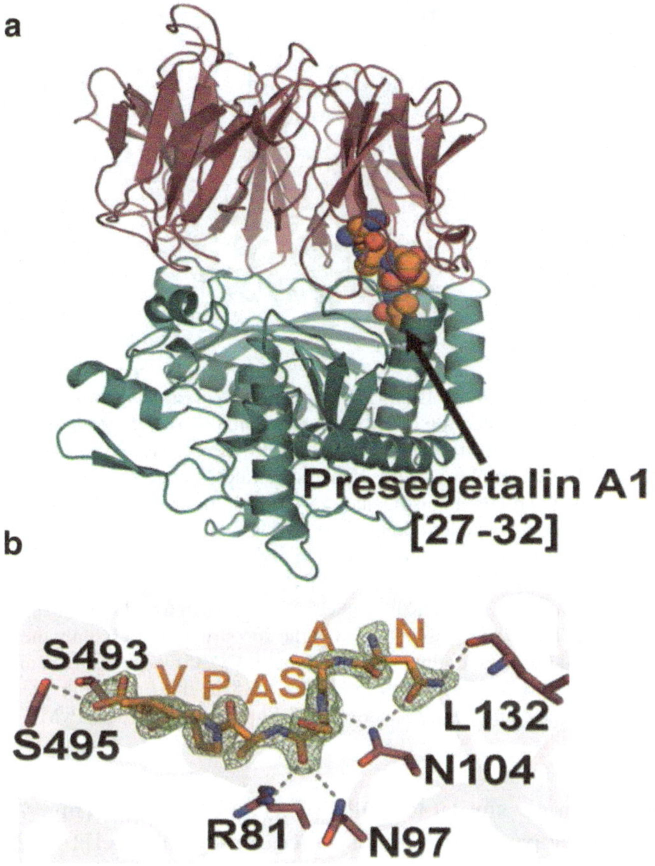

Fig. 4.9 (**a**) Crystal structure of prolyl oligopeptidase PCY1, which catalyzes macrocyclization of the cyclotide segetalin A1 bound to its substrate, presegetalin A1[14–32]. All POP enzymes, including PCY1 and POPB, have two domains, a β-propeller (shown in red), which controls access to the active site, and an α/β hydrolase catalytic domain (shown in green). (**b**) Close-up view of substrate binding. The crystal structure could resolve only the terminal six amino acids of the follower peptide (Val-Pro-Ala-Ser-Ala-Asn; i.e., presegetalin A [27–32]). (From Chekan et al. 2017, used with permission)

The amino acid alignments of the mushroom POPs do not indicate any major indels or unique domains that are likely to account for the unique properties of POPB, unlike the macrocyclase PatG that has two α-helices that distinguish it from other subtilisin-like proteases (Agarwal et al. 2012; Koehnke et al. 2012). However, one of the regions of relatively high divergence between the mushroom POPA and POPB enzymes corresponds to loop A, which along with loop B has been implicated in substrate entry, size selection, and specificity (Szeltner et al. 2013). This divergence is due to a six-amino acid indel and a region of poor overall conservation (amino acids ~220 to 237 in GmPOPB and AbPOPB) (Luo et al. 2014).

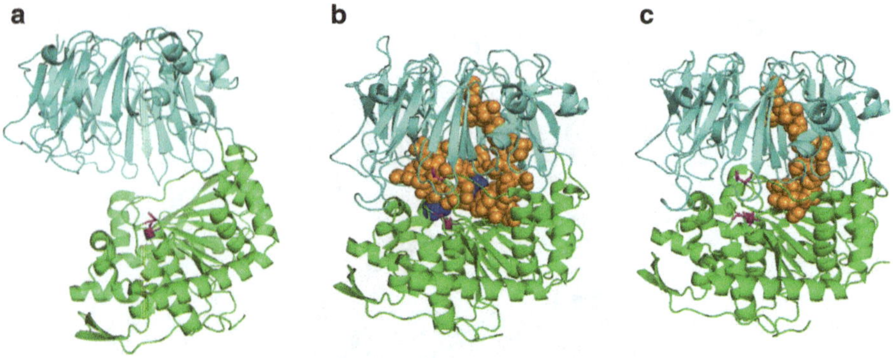

Fig. 4.10 Three views of GmPOPB from *Galerina marginata*. (**a**) Apo enzyme, i.e., in absence of substrate. The 22 Å gap between the ends of the two domains reduces to 4 Å when the enzyme folds up around the bound substrate. (**b**) GmPOPB complexed with 35mer substrate (first catalytic step). (**c**) GmPOPB complexed with 25mer substrate (second catalytic step). Protein Data Bank accession numbers are 5N4F, 5N4C, and 5N4B for **a**, **b**, and **c**, respectively (Czekster et al. 2017). In each figure, the β-propeller domain is shown in light blue on top, the catalytic domain in green on the bottom, the three active site residues in magenta, and the substrate in orange. In **b** the two Pro residues that are the sites of hydrolysis and cyclization are shown in dark blue; Pro10 is on the left and Pro18 on the right. The active site Ser577 residues in **a** and **b** had been mutated to Ala. The active site residue His698 was not resolved in the apo structure. In **b**, only the last 33 residues of the precursor peptide (corresponding to Asp3 through Cys35) were resolved and in **c**, only the last 17 residues (corresponding to Trp19 through Cys35 in the full-length precursor peptide). (Figures are based on the results of Czekster et al. 2017 as visualized with PyMol 1.5.0.5)

PCY1 is structurally similar to POPB (Chekan et al. 2017) (Fig. 4.9). In PCY1, the target precursor peptide (a 19mer called presegetalin A1[14–32]) binds to a hydrophobic pocket enriched in aromatic residues. Based on the crystal structure, the His in the catalytic site PCY1 has been proposed to provide two functions. One is to activate the Ser nucleophile as in other serine proteases, and the other is to deprotonate the N-terminal α-amine of the core peptide precursor, thus facilitating transamidation over hydrolysis. Strong binding of the follower peptide could maintain PCY1 in a closed state and thereby exclude water solvent from the active site. In contrast to strong binding by the follower peptide, there was little evidence in the crystal structure for interactions between the enzyme and the core amino acids to be cyclized, thus accounting for substrate promiscuity. A similar scenario is likely to occur in the case of GmPOPB and the AMA1 precursor peptide, where it has been shown that the follower peptide has conserved secondary structure and whose release after catalysis is the overall rate-limiting step (Czekster and Naismith 2017; Luo et al. 2014).

Chekan et al. (2017) noted that some putative non-cyclizing POPs have a Gly residue next to the active site His695, which is missing in PCY1. They hypothesized that this Gly forces displacement of the His away from the catalytic Ser, thereby promoting hydrolysis over transamidation. In support of this hypothesis, PCY1 with an insertion of a Gly residue catalyzed some degree of hydrolysis to the linear form rather than only macrocyclization. However, this Gly is present in all of the known fungal POPs, including the characterized macrocyclase GmPOPB and its orthologs in *A. bisporigera* and *A. phalloides*, so this particular Gly residue cannot play the same role in the mushroom POPB macrocyclases as it is hypothesized to do in PCY1.

Czekster and Naismith (2017) performed detailed kinetic studies on GmPOPB. They confirmed the conclusions of Luo et al. (2014) that it is "among the fastest macrocyclases described to date" and that it catalyzes two, nonprocessive reactions, peptide bond cleavage (hydrolysis) and peptide bond formation (cyclization). They concluded that release of the bound follower peptide after macrocyclization is the rate-limiting step. Czekster and Naismith (2017) found some differences in kinetic properties compared to those reported by Luo et al. (2014), probably due to differences in assay conditions. For example, Czekster et al. (2017) performed their enzyme assays at room temperature rather than 37 °C, added a protease inhibitor cocktail during purification that contained inhibitors of serine proteases (to which class POPB belongs), and used a different protein assay of unknown comparability.

Czekster et al. (2017) crystallized GmPOPB and determined its structures by X-ray crystallography in both the open (apo) form, i.e., when not bound to substrate and when bound to the 25mer and the 35mer substrates (Fig. 4.10). (To remind the reader, the 35mer is the native precursor peptide. The 25mer is the product of the initial hydrolysis step, after removal of the first ten amino acids. The first 8 amino acids of the 25mer comprise the core octapeptide which is cleaved from the C-terminal 17 amino acids and simultaneously cyclized to form cyclo[IWGIGCNP]; see Fig. 4.6). The structures with bound substrate (both the 35mer and the 25mer) were obtained using proteins with the three catalytic triad residues (Ser577, Asp661, and His698) mutated to inactive residues. The rationale for this is that the substrates can still bind as in the wild type but become trapped within the enzyme, enabling their visualization in the crystal structure. In the apo form, the two domains (β-propeller and catalytic) are separated by 22 Å (Fig. 4.10a). In the bound conformation, the enzyme undergoes a large conformational change so that the two domains become only 4 Å apart (Fig. 4.10b, c).

The structure of GmPOPB changes very little whether it is binding the 25mer or the 35mer (Fig. 4.10). However, the substrate structure does change significantly. In particular, the six amino acids immediately following the core region (i.e., Trp-Thr-Ala-Glu-His-Val), which Czekster et al. (2017) call the *linker region*, have a significantly different orientation in the two bound substrates. Czekster et al. (2017) proposed that this conformational change in the linker is necessary for transpeptidation but that the shift from one conformation to the other cannot occur while the substrate is still bound following the initial hydrolysis reaction. That is, after hydrolysis of the leader peptide the 25mer becomes kinetically trapped, and transpeptidation cannot occur until the 25mer is released and rebound with a different linker conformation. This is consistent with earlier results showing that GmPOPB is nonprocessive (Luo et al. 2014).

The first nine amino acids of the 35mer (the leader peptide) make few contacts with the enzyme (Czekster et al. 2017). This is somewhat surprising because Luo et al. (2014) showed that the leader peptide has compositional and length requirements (see above). Less surprising is that the N-terminal residues of the 25mer (i.e., the core domain) are disordered in the crystal structure, indicating few or no contacts between the amino acids that will end up in the final cyclic product and the enzyme. This is consistent with the ability of GmPOPB to cyclize peptides that differ widely in length and composition (Sgambelluri et al. 2018). The crystal structure of POPB bound to the 25mer did not resolve the eight-amino acid core region (Czekster et al. 2017), so insights into the mechanism of transpeptidation were limited. Also, the structure gave no indications why the terminal Cys of the precursor peptide should be important for cyclization activity (Luo et al. 2014).

Czekster et al. (2017) made site-specific changes to five of the amino acids that distinguish housekeeping POPs such as GmPOPA from macrocyclizing POPs such as POPB. Residues R663 and W695 were thereby shown to be important for peptide bond hydrolysis and macrocyclization. The mutations affected both hydrolysis of the 35mer and cyclization of the 25mer. The different POPs show overall low amino acid identity (e.g., GmPOPA and GmPOPB are only 58% identical), and therefore more work will be needed to understand fully how a conventional proteolytic POP evolved into a macrocyclase.

4.2.3 POPB in Relation to Other POPs

POP activities have been characterized in plants, protozoans, bacteria, and animals. Genes encoding POPs are present in all branches of life (Venäläinen et al. 2004). Interestingly, although POP genes and POP activity are widespread in the basidiomycetes (phylum Basidiomycota), one major taxonomic group that appears to largely lack POP genes is the ascomycetes (phylum Ascomycota), the sister group to the Basidiomycota (Chap. 1).

In humans, POP has been implicated in processes as diverse as amnesia, neuroprotection, cognition, memory, depression, diabetes, and T-cell activation (Brandt et al. 2007; Jalkanen et al. 2014; Nagatsu 2017; Peltonen et al. 2011; Szeltner and Polgár 2008). POPs are probably involved in processing linear neuropeptides in the mammalian brain (Aertgeerts et al. 2004). POP is required for virulence of trypanosomes (Bastos et al. 2013). POP is being investigated as a treatment for celiac disease, which is characterized by an immune response to Pro-rich oligopeptides from cereal gluten (Kaukinen and Lindfors 2015; Szeltner and Polgár 2008).

As discussed above, all POP enzymes have the same bipartite structure consisting of a seven-bladed β-propeller domain and a catalytic domain with an α/β hydrolase fold (Figs. 4.9 and 4.10). The β-propeller is thought to gate the enzyme to prevent larger peptides and proteins from accessing the active site. As a general rule (but this has not been experimentally tested in all cases), POPs cleave only peptides smaller than ~30 amino acids, can utilize the artificial chromogenic substrate Z-Gly-Pro-*p*-nitroanilide (Z-Gly-Pro-*p*NA), and are inhibited by Z-Pro-prolinal (ZPP) (Luo et al. 2009, 2014). POPA of *G. marginata* (GmPOPA), the housekeeping POP of this fungus, behaves in these regards like a canonical POP. GmPOPB, on the other hand, hydrolyzes larger substrates (at least up to 37 amino acids), utilizes the artificial substrate only poorly, and, most importantly, catalyzes macrocyclization (Luo et al. 2014).

In addition to GmPOPB and PCY1, several crystal structures of mammalian, protist, and bacterial POP and related enzymes have been solved (Canning et al. 2013; Fülöp et al. 1998; Li et al. 2010; McCluskey et al. 2010; Shan et al. 2005). The rationale for why normal POPs cannot hydrolyze peptides larger than 30 amino acids is that such peptides are too big to allow the two domains to close completely. However, this limitation is not applicable to GmPOPB, which can act on larger peptides (up to 37 residues in the case of a cyclic decapeptide such as antamanide).

Water is a co-substrate for the hydrolysis reaction catalyzed by POPB, but water must be excluded during the cyclization step to prevent competition for the transpeptidation reaction. The structure of another two-domain protease, dipeptidyl peptidase III (DPP III), might be a reasonable model by which to explain this, albeit speculatively. Elucidation of the structures of the free and bound forms of DPP III to its peptide substrate indicated a large ligand-induced movement of the two domains relative to each other, causing the ligand to be completely buried when bound (Bezerra et al. 2012). Upon closure, a large amount of water was released from the hydrophobic binding cleft, providing an entropic driving force for ligand binding and lowering the local water concentration. If POPB follows a similar trend, then this exclusion of water from the binding site would promote transpeptidation over hydrolysis.

POP enzymes have been shown to be involved in the biosynthesis of other RiPPs. A POP catalyzes the removal of the leader peptide from the precursor of class III lanthipeptide in the bacterium *Kribbella flavida* (Völler et al. 2013). Another POP, PCY1, acts as a macrocyclase in cyclotide biosynthesis (see above).

A prolyl oligopeptidase is probably also involved in processing and peptide macrocyclization of omphalotin A, which comes from another agaric, *Omphalotus olearius* (the jack-o'-lantern mushroom). Omphalotin A is a cyclic dodecapeptide with nine N-methylated amino acids. The precursor peptide of omphalotin A is part of the C-terminus of an N-methyltransferase known as OphMA and is immediately preceded by a Pro residue. A gene for a predicted POP (OphP) is clustered with the gene for OphMA. OphMA auto-methylates the precursor peptide while it is still attached to itself, and then OphP is proposed to catalyze cleavage and cyclization to produce the mature N-methylated cyclic dodecapeptide (Ramm et al. 2017; van der Velden et al. 2017). The name *borosin* has been suggested for this family of N-methylated cyclic peptides.

4.2.4 POPB in Relation to Other Peptide Macrocyclases

Peptide macrocyclases have been described from bacteria and plants. Prominent examples include the asparaginyl endoprotease AEP for the cyclotide kalata B1 from plants, PatG for biosynthesis of patellamide from the symbiotic cyanobacterium *Prochloron didemni*, PCY1 for biosynthesis of the orbitide segetalin A1 from the plant *Vaccaria hispanica*, and butelase 1 also for plant cyclotides (Agarwal et al. 2012; Barber et al. 2013; Chekan et al. 2017; Koehnke et al. 2012; Nguyen et al. 2014, 2015; Saska et al. 2007).

One common theme of all known peptide macrocyclization reactions is that the precursors include leader and/or follower amino acids that are required for, and removed during, cyclization (Oman and van der Donk 2010). A second common theme is that the precursor peptides require two proteolytic/transpeptidation reactions in going from the primary translation products to the final cyclic peptide product. In the case of segetalins and patellamides, two separate enzymes catalyze the two reactions. For example, removal of the leader peptide during segetalin biosynthesis is catalyzed by OLP1 (protease oligopeptidase 1), and then PCY1 removes the follower peptide and simultaneously catalyzes macrocyclization (Barber et al. 2013). In contrast, the same enzyme, POPB, catalyzes both reactions in the case of the cycloamanides.

4.3 Steps in *Amanita* Cyclic Peptide Biosynthesis After Cyclization

Although studies on POPB indicate that leader removal and cyclization are the first steps after translation of the precursor peptide, the order of the subsequent biosynthetic steps is currently unknown.

Tryptathionine Formation This chemical linkage between Trp (amino acid #2) and Cys (amino acid #6) is a unique characteristic of the amatoxins and phallotoxins. The mechanism, the enzyme, and the gene responsible for tryptathionine biosynthesis are not known as of this writing. An intriguing gene candidate to encode the hypothetical "tryptathionine synthase" is the predicted flavoprotein monooxygenase (FMO) that clusters in the genome of *G. marginata* between GmPOPB and GmAMA1 (Fig. 4.11) (genome.jgi-psf.org/Galma1/Galma1.home.html). Genes of secondary metabolite pathways are frequently, although not universally, clustered in fungi (Rep and Kistler 2010; Seo et al. 2001; Walton 2000; Wight et al. 2009). Although this fact alone makes GmFMO1 a candidate for involvement in amatoxin biosynthesis, it does not indicate its precise role.

FMOs are a large and diverse family of enzymes with multiple activities, including regioselective hydroxylation and enantioselective sulfoxidation (Ceccoli et al. 2014; Dijkman et al. 2013; Hansen et al. 2007; Moroni et al. 1995). Several lines of evidence suggest that GmFMO1 might catalyze tryptathionine biosynthesis by activating Trp by hydroxylation at the 3-carbon. First, the sequence of GmFMO1 is similar to a bacterial FMO (PDB 2XVE) that catalyzes 3-hydroxylation of indole (Choi et al. 2003; Cho et al. 2011). Second, the Savige-Fontana synthetic route to tryptathionine involves a 3-hydroxy-Trp intermediate (May and Perrin 2007; Chap. 2). Subsequent spontaneous attack by the sulfhydryl

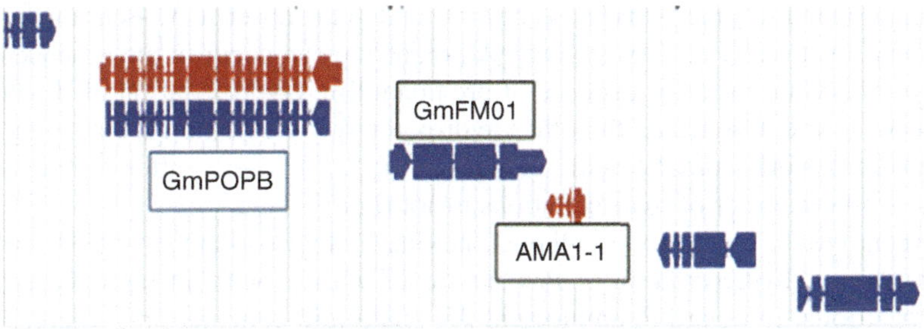

Fig. 4.11 Gene cluster in *G. marginata* containing *GmPOPB*, *GmFMO1*, and *GmAMA1*. Approximate coordinates of the region are scaffold 42:161000-168000. *GmAMA1-1* encodes the 35mer precursor core peptide that is processed by the product of *GmPOPB* to produce the unmodified cyclic octapeptide, cyclo(IWGIGCNP) (see Fig. 4.7). The *G. marginata* genome has a second copy of *AMA1* (*GmAMA1-2*) at scaffold 1:692000, but this copy is not expressed at the mRNA level. The function of *GmFMO1*, predicted to encode a flavoprotein monooxygenase, is unknown as of this writing. Gene models shown in blue are predicted and those in red have been experimentally verified. (The figure was based on the *G. marginata* web site at the DOE Joint Genome Institute (http://genome.jgi.doe.gov/Galma1/Galma1.home.html; Riley et al. 2014)

of Cys on the oxidatively activated indole of Trp yields tryptathionine (Fig. 2.12). In the hypothetical pathway, GmFMO1 would be a "cryptic hydroxylase," i.e., the hydroxyl group added by the FMO is subsequently removed and therefore not present in the final product. May and Perrin (2008) showed that the Savige-Fontana reaction can be used to synthesize intraannular bicyclic peptides including Pro^8-Ile^1-*S*-deoxo-amaninamide (i.e., α-amanitin lacking the sulfoxide and the hydroxyl groups on Pro, Ile, and Trp).

The involvement of FMO in tryptathionine biosynthesis is also consistent with the proposed mechanism of a key step in fumiquinazoline biosynthesis in the fungus *Aspergillus fumigatus*. An FMO (the product of the *FqzB* gene) in this pathway catalyzes epoxidation of Trp, which resolves into a 3-hydroxy-Trp and then spontaneously to the 2,3-diol in the absence of other enzymes and substrates (Tsunematsu et al. 2013). It seems plausible that in proximity to Cys, the 3-hydroxy-Trp could resolve into tryptathionine instead of the diol.

Hydroxylations The amatoxins have one to four hydroxylations and the phallotoxins one to five. In the amatoxins, two of the hydroxylations occur on the same amino acid (Ile #1), whereas in the phallotoxins up to three hydroxylations occur on Leu #3 (Figs. 2.6 and 2.8).

All of the hydroxylations are on aliphatic (primary or secondary) carbons except the one on Trp in the amatoxins, which is an aromatic hydroxylation. Among all the known amatoxins and phallotoxin hydroxylation sites, Pro is the most common (Figs. 2.6 and 2.8), but hydroxylation at Pro can come with (e.g., α-amanitin and β-amanitin) or without (e.g., amanin and amaninamide) hydroxylation of Trp. Hydroxylation of the 5-carbon of Ile (δ position) is the least common hydroxylation modification in the amatoxins and is never found in the absence of 4-carbon hydroxylation (γ position) (Fig. 2.6). Hydroxylation at Trp can occur in the absence of the other hydroxylations, as in the case of proamanullin. Hydroxylation of Ile and Trp can occur in the absence of the other hydroxylations, as in the case of θ-amanitin (unpublished work from the author's lab; see Fig. 2.6). These patterns indicate that the biosynthetic order of hydroxylations is not rigidly fixed.

In regard to the phallotoxins, all phallotoxins are hydroxylated at C4 of Leu. Like the amatoxins, hydroxylation of Pro can occur with or without hydroxylations at other positions. In regard to amino acid #5, phalloidin contains Thr, whereas phallacidin contains 3-hydroxy-Asp. The corresponding genes encode Thr or Asp (Hallen et al. 2007; Pulman et al. 2016), respectively, i.e., the hydroxyl group of Thr is genetically encoded, whereas the hydroxyl group of Asp is a posttranslational modification. Phallacidin must therefore undergo an extra hydroxylation step that phalloidin does not.

The hydroxylations of the amatoxins and phallotoxins could plausibly be catalyzed by cytochrome P450 monooxygenases (CYP450), α-ketoglutarate-dependent dioxygenases, or flavoprotein monooxygenases (FMOs) (Haefelé et al. 1997; Hara and Kino 2009; Hibi et al. 2013). All known Pro hydroxylations, in plants, animals, and bacteria, are catalyzed by dioxygenases, including the enzymes that produce hydroxy-Pro (Hyp) for collagen, plant Hyp-rich glycoproteins, cone snail toxins, and fungal nonribosomal peptides (Hara and Kino 2009; Houwaart et al. 2014). The stereochemical isomer of the majority of Hyp in nature is *trans*-4-Hyp (2*S*,4*R*), which is also the isomer of Hyp found in the amatoxins. In contrast, the Hyp in phallotoxins has the *cis* configuration (2*S*,4*S*) (Chap. 2). To date, there are no candidate dioxygenase genes for Pro hydroxylation in any species of *Amanita*, *Galerina*, or *Lepiota*. It is possible that the 4-hydroxylation of Pro is catalyzed by different types of enzymes in these three amanitin-producing fungi. Furthermore, because there is apparently no evidence that the stereochemistry of the 4-Pro hydroxylation is critical for biological activity, it is not beyond the realm of the possible that α-amanitin from *Galerina* or *Lepiota* contains *cis*-Hyp instead of *trans*-Hyp as found in α-amanitin from *Amanita*. If this were true, it would argue strongly for convergent evolution rather than horizontal gene transfer of the amanitin biosynthetic trait among the three fungi (Chap. 6).

The enzyme(s) that hydroxylate Ile #1 in the amatoxins and Leu #3 in the phallotoxins are likewise unknown as of this writing. Hydroxylated derivatives of Ile are found elsewhere in nature. 4-Hydroxy-Ile from fenugreek has antidiabetic and anti-obesity activity (Jetté et al. 2009; Maurya et al. 2014). Its stereochemistry (2*S*,3*R*,4*S*) is the same as 4-hydroxy-Ile and 4,5-dihydroxy-Ile in the amatoxins (Chap. 2). All known Leu and Ile hydroxylations in plants and bacteria are catalyzed by dioxygenases (Haefelé et al. 1997; Hibi et al. 2013; Smirnov et al. 2013).

CYP450s catalyze many biosynthetic hydroxylation reactions (Meunier et al. 2004; Podust and Sherman 2012). Some fungal CYP450s catalyze multiple, vicinal hydroxylations, such as those occur in both the amatoxins (Ile #1) and the phallotoxins (Leu #3) (Seo et al. 2001; Tokai et al. 2007; Tudzynski et al. 2002). Like most fungi, the genomes of *Amanita bisporigera*, *A. phalloides*, and *Galerina marginata* contain many CYP450 genes (Pulman et al. 2016; Riley et al. 2014). At least four CYP450 genes are near *GmPOPB* and *GmAMA1* in the genome of *G. marginata.* Targeted mutation of one, *CYP450-29,* caused disappearance of α-amanitin and accumulation of a new product, which was purified and characterized as θ-amanitin (H. Luo, unpublished work from the author's lab). θ-Amanitin is a previously undescribed member of the amatoxin family, being α-amanitin lacking the hydroxyl groups on Pro #8 and the 5-carbon of Ile #1 (Fig. 2.6). Therefore, CYP450-29 catalyzes either Pro hydroxylation, Ile hydroxylation, or both.

No amatoxins are known having the 5-carbon hydroxyl on Ile #1 without the 4-carbon hydroxyl (Fig. 2.6). 5-Hydroxy-Ile does, however, occur in one of the orbitides, which are plant RiPPs (Morita et al. 1994).

Epimerization at Amino Acid #5 The phallotoxins but not the amatoxins contain one amino acid with the D configuration. This occurs in both the phalloidin and the phallacidin subfamilies, which contain D-Thr or hydroxy-D-Asp, respectively, at position #5. Nothing is currently known about the gene or enzyme catalyzing epimerization, nor when in the biosynthetic pathway it occurs. However, Sgambelluri et al. (2018) showed that a 25mer precursor peptide corresponding to phallacidin (core sequence AWLATCP) with D-Thr pre-introduced was partially cyclized, whereas the all-L peptide was not. This suggests that the epimerization might occur before macrocyclization. Ribosomally synthesized peptides that contain D-amino acids have been described from multiple sources, including spider venoms, cone snail toxins, platypus venom, and amphibian skin peptides. In the cases that have been studied, the epimerization reaction occurs posttranslationally and is catalyzed by a specific peptide epimerase (also known as peptide isomerase) (Bansal et al. 2008; Buczek et al. 2005; Jilek et al. 2005; Ollivaux et al. 2014; Torres et al. 2006). No clear orthologs of any known peptide epimerases are present in the genomes of *A. bisporigera* or *A. phalloides.*

Mechanistically, hydroxy-D-Asp could conceivably arise via a dehydro intermediate, i.e., by dehydrogenation of L-Asp to 2,3-dehydro-Asp followed by stereospecific hydration. This would be the reverse of what happens in the biosynthesis of some lantibiotics such as lacticin 3147, in which L-Ser is dehydrated to dehydro-Ala and then hydrogenated to D-Ala (Suda et al. 2012). However, this pathway is probably not the mechanism by which D-Thr is biosynthesized in the case of the neutral phallotoxins such as phalloidin, because the genomes of *A. bisporigera* and *A. phalloides* contain cycloamanide family members that directly encode Thr (codon ACN) (Pulman et al. 2016).

Sulfoxidation of Cys The sulfur atom in the Cys residue is oxidized to the sulfoxide in the amatoxins but not the phallotoxins. This reaction is probably enzymatically catalyzed because only the (*R*)-isomer is found in native amatoxins, and the (*S*)-sulfoxide is inactive (Wieland 1986). Flavoprotein monooxygenases (FMOs) can catalyze stereoselective sulfoxidation (Jensen et al. 2012; Moroni et al. 1995). A putative FMO gene, whose possible role in tryptathionine biosynthesis was discussed above, is clustered with *GmPOPB* and *GmAMA1* in *G. marginata* (Fig. 4.11).

The closest ortholog of known function to GmFMO1 in GenBank is an FMO from *Arabidopsis* that sulfoxidates methionine (Hansen et al. 2007). Therefore, GmFMO1 might catalyze the sulfoxidation step in amatoxin biosynthesis.

4.4 Biosynthesis of the Cycloamanides

4.4.1 Cycloamanides in A. bisporigera *and* A. phalloides

The genomes of *A. bisporigera* and *A. phalloides* each contain more than 30 genes related to *AMA1* and *PHA1*. All encode conserved 10-amino acid leader and 17-amino acid follower peptides flanking variable core regions (Hallen et al. 2007; Pulman et al. 2016). These sequences can be identified either by PCR amplification using conserved regions of the gene family (i.e., the leaders and followers) or bioinformatically from partial or complete genome sequences. Originally this family was named the "MSDIN" family for the first five highly, but not universally, conserved amino acids of the precursor peptides (Hallen et al. 2007; Zhou et al. 2013). In light of the subsequent discovery that some genes in this family in *Amanita* and especially in *Galerina* and *Lepiota* do not start with the canonical MSDIN, and the demonstration that the amatoxins, phallotoxins, and classic cycloamanides share the same generalized biosynthetic pathway, this gene family is now collectively referred to as the *cycloamanide* family (Sgambelluri et al. 2018). Of the ~30 cycloamanide genes in *A. bisporigera* and *A. phalloides*, only three (encoding α-amanitin, phallacidin, and an unknown) are common among the two fungi (Pulman et al. 2016).

Alignment of the predicted precursor peptides of known cycloamanide family members reveals a structure composed of a conserved leader peptide (almost always 10 amino acids in length), a variable core region of 6 to 10 amino acids, and a conserved follower peptide of 17 amino acids (Fig. 4.12). For cycloamanide genes for which there are no corresponding cDNAs, the ultimate two amino acids (usually Leu-Cys) of the precursor peptides must be deduced bioinformatically because there is an intron (fortunately highly conserved in position and length), interrupting the fourth from the last codon (Pulman et al. 2016).

In regard to how many of the new cycloamanide sequences are expressed, in *A. bisporigera* at least 18 of its 33 cycloamanide gene family members are expressed at the RNA level (Pulman et al. 2016). Other expressed cycloamanide genes have been identified by reverse transcriptase PCR (RT-PCR) (Li et al. 2014; and unpublished results from the author's lab). At least two of the previously undescribed cycloamanide genes, encoding CylE and CylF, are expressed at the chemical level (Pulman et al. 2016; Chap. 2).

In the pre-genomic era, five cycloamanides including antamanide had been isolated from *A. phalloides* (Wieland 1986). In the sequenced genome of *A. phalloides*, genes for the four previously known amatoxins and phallotoxins were found, as well as cycloamanide B (CylB), but not genes for cycloamanides A, C, and D, or antamanide. The most probable explanation for this is natural variation, i.e., different

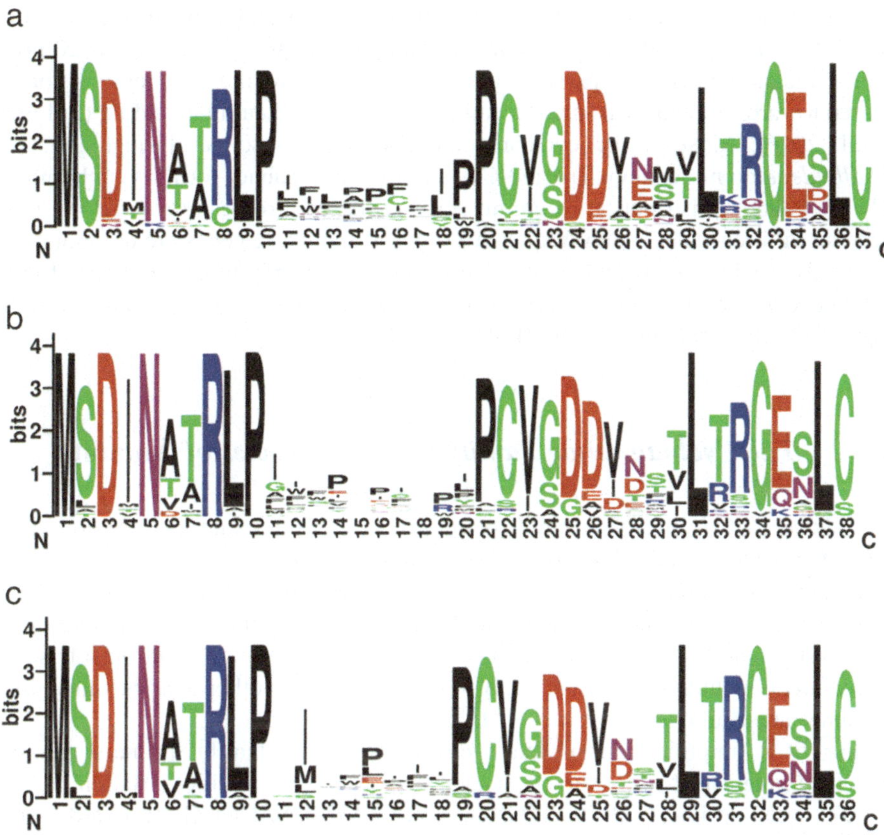

Fig. 4.12 WebLogo alignment of cycloamanide precursor peptide sequences from the genomes of (**a**) *A. phalloides* and (**b**) *A. bisporigera*. (**c**) Alignment of predicted precursor peptides from the transcriptome of *A. bisporigera*. Xs have been added to the core regions having fewer than ten amino acids to make all of the core sequences the same length. (From Pulman et al. 2016, Creative Commons License)

specimens of *A. phalloides* vary naturally in their complement of cycloamanide genes and cyclic peptide natural products. This is consistent with the hypothesis that the cycloamanide family is rapidly diverging evolutionarily (Chap. 6).

More natural variation was uncovered when the genome sequences of two specimens of *A. bisporigera* were compared. One genome contained at least ten cycloamanide genes not found in the other (Pulman et al. 2016). Comparison of the internal transcribed spacer (ITS) sequences (the most widely accepted molecular taxonomic marker for fungi; see Chap. 1) of the two specimens indicated that although morphologically similar and collected in the same geographical area, the two mushrooms were actually distinct species within the *A. bisporigera* species complex (Pulman et al. 2016). This indicates that the cycloamanide family is highly divergent even among closely related taxa.

The amino acid sequences of the core regions in the extended cycloamanide family of *A. bisporigera* and *A. phalloides* are highly diverse (Fig. 4.12). All 20 proteinogenic amino acids are present at least once, but the amino acid composition is biased toward Pro and hydrophobic amino acids such as Leu, Phe, and Ile (Pulman et al. 2016). Most of the cycloamanide core regions found throughout sect. *Phalloideae* do not contain Trp and Cys and therefore cannot form tryptathionine. The nonrandom distribution of amino acids in the core regions (even taking into account differences in codon frequency) suggests that these genes are not mutating randomly. Rather, it suggests that these genes are not only mostly translated into actual cyclic peptides, but furthermore that these peptides confer a selective advantage on the fungi that make them (Pulman et al. 2016; Chap. 6).

4.4.2 The Cycloamanide Family in Other Species of Amanita

In addition to *A. bisporigera* and *A. phalloides*, genes for amatoxins, phallotoxins, and other cycloamanides have been identified in *A. exitialis*, *A. ocreata*, *A. marmorata*, and several other species in *Amanita* sect. *Phalloideae* (Hallen et al. 2007; Li et al. 2014; Pulman et al. 2016; and unpublished work from the author's laboratory). The cycloamanide gene family has not been found in any organism outside sect. *Phalloideae* including *A. muscaria*, *A. thiersii*, and many other agarics whose full genomes are available.

For PCR amplification of the cycloamanide gene family, primers can be designed against the conserved leader and follower peptides. Hallen et al. (2007) used primers 5′-ATGTCNGAYATYAAYGCNACNCG (forward) and 5′-AAGGSYCTCGCCACGAGTGAGGAGWSKRKTGAC (reverse) to amplify cycloamanide genes from *A. phalloides* and *A. ocreata*. (W indicates A or T, S indicates C or G, K indicates G or T, R indicates A or G, and Y indicates T or C). These primers were also used to amplify cycloamanide genes from *A. marmorata*, including core sequences corresponding to phallacidin (core sequence AWLVDCP) and three unknowns (core sequences FMFFRYPFPC, VWGIGCSP, and FMFFRYPLP) (unpublished results from the author's laboratory). It is interesting to note that the second of these has the potential to contain tryptathionine.

Li et al. (2014) used primers 5′-ATGTCNGAYATYAAYGCNACNCG and 5′-CCAAGCCTRAYAWRGTCMA (M indicates A or C) to amplify cycloamanide genes from *A. exitialis*, *A. fuliginea*, *A. fuligineoides*, *A. pallidorosea*, *A. phalloides*, and *A. rimosa*. All of these species except *A. phalloides* are native to Asia. All of the *Amanita* species analyzed by PCR by Li et al. (2014) contained the genes for α-amanitin and phallacidin and most contained the gene for β-amanitin. Many unknown sequences were also found, and every species had cycloamanide genes not found in the other species. For example, none of the 13 novel cycloamanide sequences amplified by PCR by Hallen et al. (2007) from *A. bisporigera* were found in the six Asian species of *Amanita* (Li et al. 2014). Only one unknown from the full genome of *A. bisporigera* was present in any of the Asian *Amanita* specimens (Pulman et al. 2016).

To date, we can generalize that all toxic cyclic peptide-producing species of *Amanita* have a family of related sequences, the cycloamanide family. We can also generalize that there is little overlap between species, even between closely related ones within the *A. bisporigera* species complex. However, only for *A. phalloides* and *A. bisporigera* do we have a picture of the complete complement within a species, and even for these two species, we only know the complement for single specimens. There are probably more cycloamanide sequences in the greater population of each species, based on the observation that the genome of the sequenced specimen of *A. phalloides* (GenBank accession MEHY00000000.1) does not contain the genes for some cyclic peptides that had previously been chemically identified in this fungus, namely, cycloamanides A, C, and D and antamanide (Wieland 1986). Differences in the toxin profiles among specimens within a species are well-documented (Chap. 3), and this is likely due, at least in some cases, to differences in their gene complements.

In regard to the prevalence of the different cycloamanide genes, the gene for α-amanitin appears to be universal or nearly so within *Amanita* sect. *Phalloideae* and β-amanitin and phallacidin are common, but most of the other cycloamanide family members are unique to a particular species. Exceptions to this last rule include core sequence AWLAECP, which is found in *A. rimosa* and *A. bisporigera*, and ISDPTAYP, which is found in *A. bisporigera* and *A. phalloides*.

RNA expression of α-amanitin, β-amanitin, phallacidin, amanexitide, and 11 unknown cycloamanide genes was detected in *A. exitialis*, although its total genomic complement of cycloamanide genes is not yet known (Li et al. 2013). This number of expressed cycloamanide genes is similar to *A. bisporigera* (Pulman et al. 2016).

Nothing specific is known yet about the biosynthesis of the virotoxins made by *A. virosa*, beyond that they can be predicted to be encoded by members of the cycloamanide gene family. All of the virotoxins could be encoded by two genes, assuming that the D-Ser and sulfated Trp are derived from Cys and Trp, respectively (Chap. 2). Ala^4-viroidin and Ala^4-desoxoviroidin could be biosynthetically derived from the same precursor peptide as phalloidin (AWLATCP), whereas the other viroidins and viroisins would require a different gene encoding a hybrid between phalloidin and phallacidin (core sequence AWLVTCP). The gene encoding AWLATCP (sufficient to encode the phalloidins and the Ala^4-virotoxins) is present in the genome of *A. phalloides*, but a gene that could encode AWLVTCP is not (Pulman et al. 2016). Desoxoviroidin is also made by *A. exitialis* collected in China, but a gene encoding AWLVTCP was not found in the transcriptome of *A. exitialis* (Deng et al. 2011; Li et al. 2013).

Faulstich et al. (1980) proposed a biochemical pathway by which the virotoxins could arise from the Cys-Trp moiety (tryptathionine) found in the phallotoxins. In this scheme, methylation of the sulfur leads, through β-elimination, to dehydro-Ala that is then stereospecifically re-reduced to form D-Ser (Wieland 1986). This scenario implies the existence of a sulfur methylase (and perforce the encoding gene) unique to the virotoxin pathway, but the subsequent steps could occur spontaneously. A possible explanation for why *A. phalloides* does not make virotoxins is that it lacks the hypothetical sulfur methylase gene.

4.4.3 Cycloamanide Genes in Galerina *and* Lepiota

Based on its chemical analysis and its full genome sequence, *G. marginata* does not make any cycloamanides beyond α-amanitin and perhaps β-amanitin (Chap. 3). By chemical analysis, species of *Lepiota* (*L. brunneoincarnata* and *L. josserandii*) make amatoxins but not phallotoxins (Sgambelluri et al. 2014). Based on the full genome sequence of *L. subincarnata*, this fungus has the genetic capacity to make five cycloamanides in addition to α-amanitin (Fig. 4.13; unpublished results from the author's laboratory). The five novel cycloamanides, named LsCyl1 through LsCyl5, are all predicted to be nonapeptides. Like the *Amanita* cycloamanide family (Fig. 4.12), the leader and follower peptides of the *Lepiota* cycloamanides show

a

Name	Core sequence	Elemental formula (cyclic)	Mono-isotopic mass (cyclic)	Notes
LsAMA1	IWGIGCNP	$C_{39}H_{54}N_{10}O_{14}S$	918.4	α-amanitin; detected in mushroom extracts
LsCyl1	LVAAVGIPP	$C_{40}H_{67}N_9O_9$	817.5	detected in mushroom extracts
LsCyl2	LFFPLPIPP	$C_{56}H_{79}N_9O_9$	1021.6	not detected
LsCyl3	IALVLSLFP	$C_{49}H_{79}N_9O_{10}$	953.6	detected in mushroom extracts
LsCyl4	FVLLLIVPP	$C_{53}H_{85}N_9O_9$	991.6	detected in mushroom extracts
LsCyl5	IPILAHLGP	$C_{45}H_{73}N_{11}O_9$	911.6	not detected

b

Fig. 4.13 (**a**) Cycloamanide gene family in *Lepiota subincarnata* based on its whole-genome sequence. The formula and mass of amanitin (LsAMA1) include its native posttranslational modifications. The other compounds (LsCyl1-5) are assumed to be unmodified homodetic cyclic peptides (unpublished results from the author's laboratory) (**b**) WebLogo alignment of the *L. subincarnata* cycloamanide precursor peptides

much stronger conservation than the core peptides (Fig. 4.13). The leader peptides of the five *Lepiota* cycloamanide precursor peptides are ten amino acids in length, like all the cycloamanides in *Amanita* and *Galerina*. In contrast, the leader peptide of the gene for amanitin, LsAMA1, is one amino acid shorter (Figs. 4.4 and 4.13).

Based on their masses, three of the predicted cycloamanides in *L. subincarnata* were detected by LC/MS in extracts of mushrooms as unmodified monocyclic peptides (unpublished results from the author's laboratory). Thus, in size the cycloamanide family of *Lepiota* (6 genes) lies between *Galerina* (1 gene) and *Amanita* (31–33 genes).

4.5 Biosynthesis of the *Amanita* Toxins Compared to Other RiPPs

4.5.1 RiPPs in Bacteria and Plants

Although ribosomally synthesized and posttranslationally modified peptides (RiPPs) had been described from bacteria, plants, and animals, the *Amanita* cyclic peptides were the first RiPPs discovered in the Kingdom Fungi. RiPPs from bacteria, animals, and plants include the lanthipeptides, cyanobactins, thiopeptides, bottromycins, microcins, lasso peptides, microviridins, cyclotides, orbitides, and cone snail toxins (Arnison et al. 2013; Ortega and van der Donk 2016). A common feature of all RiPPs is that they are initially biosynthesized as larger precursor peptides which are then processed. Common posttranslational processing steps include removal of leader and/or follower peptides; cyclization by head-to-tail peptide bond formation, internal peptide bond formation, and/or disulfide bonds; and sometimes more drastic modifications such as heterocyclization, lanthionine formation, amino acid epimerization, methylation, and prenylation. A few of the known RiPPs most relevant to an understanding of the biosynthesis of the *Amanita* toxins are discussed here. For a more complete coverage of the diversity of known RiPPs, see Arnison et al. (2013).

Cyanobactins such as patellamides contain heterocyclized or isoprenylated amino acids. Like the *Amanita* cyclic peptide toxins, they can be quite small (6–20 amino acids). Cyanobactins have been isolated from cyanobacterial symbionts of ascidians (tunicates) or free-living cyanobacteria (Czekster et al. 2016; Donia et al. 2006, 2011; Sivonen et al. 2010). The linear precursor peptides of the cyanobactins contain a leader and typically multiple head-to-tail copies of the core region. In this regard they are similar to the ustiloxins from fungi and some cyclotides and orbitides from plants (see below). The core regions are excised and cyclized by two proteases and further modified by a series of "tailoring" enzymes.

Cone snails produce small peptides of 10–30 amino acids called cone snail toxins, conotoxins, or conopeptides, and many of them have neurotoxic activities. Hundreds of cone snail toxins have been described from some of the more than 10,000 species of venomous marine snails. Like the *Amanita* toxins, the cone snail

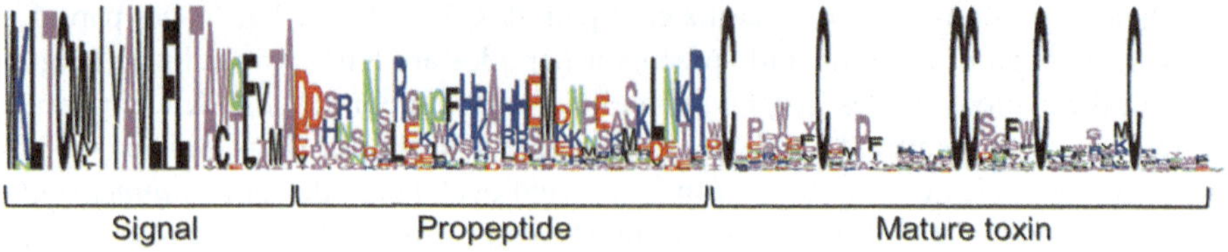

Fig. 4.14 WebLogo showing conservation and variability in different regions of the cone snail toxin precursor peptides. Compare to Figs. 4.12 (*Amanita* cycloamanides) and 4.13 (*Lepiota* cycloamanides). (Reprinted from Conticello et al. 2000 with permission from Elsevier)

toxins exist as gene families encoding precursor peptides composed of variable core regions (i.e., the amino acids present in the mature toxins) and conserved regions of 70–120 amino acids (Buczek et al. 2005; Kaas et al. 2012; Woodward et al. 1990) (Fig. 4.14). Unlike the *Amanita* toxins, the conotoxins are processed from the carboxyl termini of the precursor peptides by a protease that cuts at basic amino acids (Arg or Lys), and most conotoxins are cyclized by disulfide bonds and not by N- to C-terminal (head-to-tail) peptide bonds. Other posttranslational modifications include C-terminal amidation, cyclization of N-terminal Gln, carboxylation, α-carbon epimerization, hydroxylation, and sulfation. All the conotoxin precursor peptides have signal peptides to direct secretion into the venom duct (Fig. 4.14), where processing to the mature toxins occurs. In contrast, the *Amanita* toxins are not secreted and their precursor peptides lack predicted signal peptides (Hallen et al. 2007; Luo et al. 2012; Zhang et al. 2005). The signal peptides of the conotoxins are the most highly conserved regions (Fig. 4.14). The core regions from which the mature toxins are derived are the least conserved, except for the critical Cys residues that are essential for disulfide formation (Fig. 14.4). Like the amatoxins, some conotoxins contain the *trans* isomer of Hyp (Lopez-Vera et al. 2008; Stone and Gray 1982). The enzyme catalyzing Pro hydroxylation in the cone snails is apparently unknown but is presumed to be a dioxygenase like the enzyme that hydroxylates Pro in collagen (Gorres and Raines 2010; Gorson et al. 2015). Another conotoxin contains 4-hydroxy-D-Val (Pisarewicz et al. 2005), reminiscent of some phallotoxins that contain 3-hydroxy-D-Asp (Chap. 2).

The cyclotides, such as kalata B1 and SFTI-1, are cyclic peptides of 14–37 amino acids made by plants (Craik et al. 2010; Mulvenna et al. 2006). They are head-to-tail (homodetic) cyclic peptides but also have one or more internal disulfide bonds (Fig. 4.15). Precursor peptides for cyclotides can contain more than one copy of the core peptide, a situation also found in some cyanobactins and orbitides. An asparagine endopeptidase is involved in removing the C-terminal peptide from the precursor of kalata B1 and cyclization to the mature cyclic 29mer (Harris et al. 2015; Saska et al. 2007). Another asparagine/aspartate endopeptidase, butelase 1, macrocyclizes cyclotides including the kalata B1 precursor peptide to which His-Val has been added to the C-terminus (Nguyen et al. 2014).

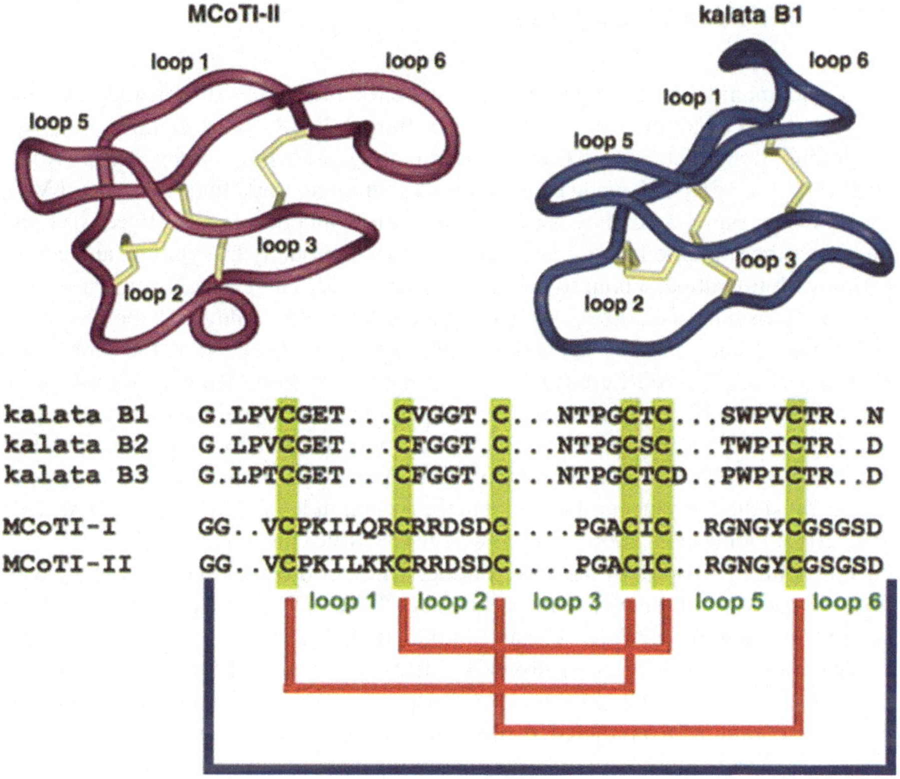

Fig. 4.15 Primary and tertiary structure of cyclotides from the plants *Oldenlandia affinis* (kalatas B1-B3) and *Momordica cochinchinensis* (MCoTI and II). Red and blue connectors indicate disulfide and backbone bonds, respectively. (From Jagadish and Camarero 2010. ©2010 Wiley Periodicals, Inc)

Orbitides are also homodetic cyclic peptides from plants, but they differ from cyclotides in lacking any intramolecular disulfide bonds. Orbitides range in size from 5 to 12 amino acids. Examples include cyclolinopeptides and segetalins. Segetalin macrocyclization is catalyzed by a serine protease related to POPB (see above). Cyclolinopeptides F and G (which are octapeptides) contain methionine sulfoxide (Matsumoto et al. 2001), like some of the cycloamanides from *A. phalloides* (Chap. 2). Cyclolinopeptides D, F, and G are the products of a single gene in flax, *Linum usitatissimum* (Arnison et al. 2013). One orbitide, yunnanin B, a cyclic heptapeptide from *Stellaria yunnanensis*, contains 5-hydroxy-Ile (Morita et al. 1994). This is of interest because some amatoxins contain 4-hydroxy-Ile or 4,5-dihydroxy-Ile (Chap. 2).

4.5.2 RiPPs in Other Fungi

The cycloamanides including α-amanitin were the first RiPPs discovered from any fungus; the second known example of a fungal RiPP came to light in 2014. Ustiloxins, toxic metabolites from the rice pathogen *Ustilaginoidea virens* (Koiso et al. 1994), are unusual cyclic pentapeptides containing norvaline, Tyr, Ala (or Val), Ile, and Gly (Fig. 4.16). They are cyclized through an ether bond between Tyr and Ile and further modified by N-methylation, sulfoxidation, hydroxylation, and conjugation with norvaline, a nonproteinogenic amino acid. The gene cluster involved in ustiloxin biosynthesis has been elegantly characterized in *A. flavus* (Umemura et al. 2014). One of the genes, *ustA*, encodes a precursor peptide encoding the core amino acid sequence (Tyr-Ala-Ile-Gly), revealing ustiloxins to be RiPPs. The precursor peptide encoded by *ustA* contains a signal peptide and 16 tandemly arrayed copies of a 13–18 amino acid peptide each of which contains one copy of the core peptide, Tyr-Ala-Ile-Gly (YAIG). The presence of multiple copies of the RiPP core sequence in one translation product is also found in the patellamides and lipopeptides (Arnison et al. 2013; Donia et al. 2006). The core gene for ustiloxins A and B in *U. virens* also encodes multiple copies of the core peptide. Three of the eight repeats encode ustiloxin B (YAIG) and three encode ustiloxin A (YVIG). The gene cluster in *U. virens* is very similar to the *A. flavus* cluster (Tsukui et al. 2015).

There are 15 genes in the ustiloxin B cluster in *A. flavus*; the other genes encode transporters, kex-like endoproteases, transcription factors, and a number of tailoring enzymes including a predicted methyltransferase, cysteine desulfurase, and two FMOs (Umemura et al. 2014). A C6-type transcription factor and a kex-like protease in the cluster were shown to be necessary for production of ustiloxin B, but it is not yet clear if the kex protease is responsible for releasing the core peptides from the precursor peptide (Yoshimi et al. 2016). Several other genes in the cluster, including a possible cyclase, have also been shown by targeted knockout to be involved in ustiloxin biosynthesis. Using conserved gene sequences in the *A. flavus* ustiloxin

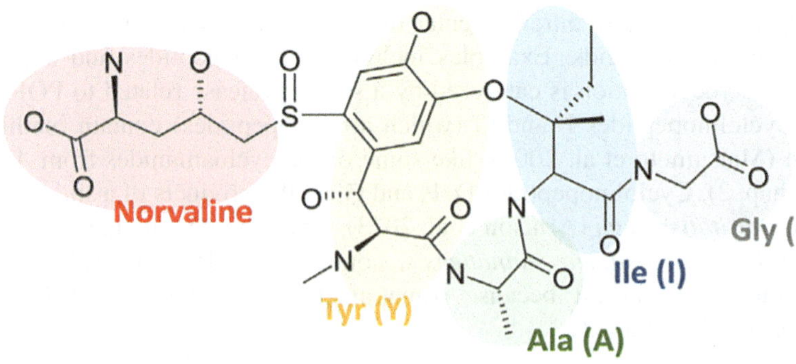

Fig. 4.16 Structure of ustiloxin B. (Reprinted from Umemura et al. 2014 with permission from Elsevier, copyright 2014)

cluster as queries, putative RiPP genes and associated gene clusters were found to be widespread in the genus *Aspergillus* and in other fungi (Nagano et al. 2016).

The phomopsins are another example of fungal cyclic peptides recently shown to be RiPPs (Ding et al. 2016). These are hexapeptide mycotoxins originally isolated from the forage plant lupin (*Lupinus* species) infected with the ascomycete *Phomopsis leptostromiformis*. Plants containing phomopsins cause liver disease in sheep and cattle in Australia and South Africa. In containing an ether-linked Tyr residue, the ustiloxins are chemically related to the ustiloxins. They differ in having a three amino acid "tail" attached to the ring of three amino acids, and some of them are chlorinated. The genes encoding the phomopsins were identified by searching the genome with the canonical core sequence YVIPID. Another 27 related gene clusters were found bioinformatically in other ascomycete fungi. The overall structure of the precursor peptides for the phomopsins is similar in their repeat structure to the ustiloxin precursor peptides (Ding et al. 2016). The term *dikaritin* has been proposed for peptides related to the phomopsins and ustiloxins.

The epichloëcyclins constitute a third family of fungal RiPPs, discovered through transcriptomic analysis of the endophytic fungus *Epichloë festucae* growing on ryegrass, *Lolium perenne* (Johnson et al. 2015). The *gigA* gene, which is highly expressed in planta, encodes a protein with four to eight copies (depending on the species of *Epichloë*) of an imperfect 27-amino acid repeat. The precursor peptide has a signal peptide, like the ustiloxins. The proteins are processed, presumably by a kex-like protease (kexin), to release nonapeptides that are then processed further by cyclization and dimethylation at a conserved Lys (K) residue. The nonapeptides have the approximate consensus (P/I/V/L)NFK*(I/M)(P/I)Y(S/R/K)G, where K* indicates the dimethylated Lys.

Like the ustiloxins, the epichloëcyclins are not head-to-tail (homodetic) cyclic peptides, but rather are cyclized through the first amino acid to the seventh amino acid, which is a conserved Tyr. The exact nature of the coupling is still unclear. Epichloëcyclins are therefore technically cyclic isopeptides. Masses corresponding to the epichloëcyclins were detected in extracts of infected plants, along with variants lacking the terminal Gly. Additional work is necessary to definitively establish the structures of the epichloëcyclins, including stereochemistry of the amino acid α-carbons.

Genes for the epichloëcyclins are found only in the family *Clavicipitaceae* (class Sordariomycetes, phylum Ascomycota). No biological activities for the epichloëcyclins have yet been reported, but their conservation across endophytic *Epichloë* species and their high expression in planta implies an important biological function (Johnson et al. 2015).

Omphalotin A, an N-methylated cyclic dodecapeptide from the agaric *Omphalotus olearius*, is also a RiPP (Ramm et al. 2017; van der Velden et al. 2017). Although N-methylation is considered a hallmark of nonribosomal peptides such as cyclosporin, it had also previously been found in the ustiloxins (Umemura et al. 2014). The amino acids found in mature omphalotin A are fused at the C-terminus to an N-methyltransferase (OphMA) that catalyzes the N-methylations. As discussed above, the gene for OphMA is clustered in the genome with a gene, OphA, encoding a putative prolyl oligopeptidase (POP). In OphMA, the core peptide is preceded immediately upstream by a Pro residue, consistent with a role for OphA in cleaving and cyclizing the core dodecapeptide. Following the core peptide sequence are six additional amino acids (SVMSTE), which must be somehow removed, but there is no Pro residue at the end of the core peptide to suggest the involvement of OphA in catalyzing this processing step.

OphMA has only a single copy of the omphalotin A sequence, similar to the cycloamanides but different from the pattern of tandem core peptide arrays found in the ustiloxin and phomopsin precursor peptides. The genes for OphMA and OphP are clustered with genes predicted to encode an acetyltransferase and two CYP450s in the genome of *O. olearius*. These accessory enzymes probably perform their expected posttranslational modifications to produce other members of the omphalotin family such as omphalotin C and D (Ramm et al. 2017).

The discovery of RiPPs in four fungi, both ascomycetes and basidiomycetes, suggests that RiPPs might be more widespread in the Kingdom Mycota than previously suspected. Currently, our bioinformatics tools are insufficient to find RiPPs de novo, although tools are being developed (Zhou et al. 2013). For example, it might be possible to design search algorithms that can find RiPPs that are composed of short, tandemly repeated sequences of four to ten amino acids like *ustA* and *gigA*. This approach would not work to find RiPPs composed of a single core sequence, like AMA1 and OphMA, but at least for RiPP families like the cycloamanides it might be possible to find such genes by searching for small (<40 amino acids) predicted genes and then sub-searching within this set for families of genes containing a combination of conserved and variable motifs.

References

Aertgeerts K, Ye S, Tennant MG, Kraus ML, Rogers J, Sang BC, Skene RJ, Webb DR, Prasad GS (2004) Crystal structure of human dipeptidyl peptidase IV in complex with a decapeptide reveals details on substrate specificity and tetrahedral intermediate formation. Protein Sci 13:412–421. https://doi.org/10.1110/ps.03460604

Agarwal V, Pierce E, McIntosh J, Schmidt EW, Nair SK (2012) Structures of cyanobactin maturation enzymes define a family of transamidating proteases. Chem Biol 19:1411–1422. https://doi.org/10.1016/j.chembiol.2012.09.012

Arnison PG, Bibb MJ, Bierbaum G, Bowers AA, Bugni TS, Bulaj G, Camarero JA, Campopiano DJ, Challis GL, Clardy J, Cotter PD, Craik DJ, Dawson M, Dittmann E, Donadio S, Dorrestein PC, Entian KD, Fischbach MA, Garavelli JS, Göransson U, Gruber CW, Haft DH, Hemscheidt TK, Hertweck C, Hill C, Horswill AR, Jaspars M, Kelly WL, Klinman JP, Kuipers OP, Link AJ, Liu W, Marahiel MA, Mitchell DA, Moll GN, Moore BS, Müller R, Nair SK, Nes IF, Norris GE, Olivera BM, Onaka H, Patchett ML, Piel J, Reaney MJ, Rebuffat S, Ross RP, Sahl HG, Schmidt EW, Selsted ME, Severinov K, Shen B, Sivonen K, Smith L, Stein T, Süssmuth RD, Tagg JR, Tang GL, Truman AW, Vederas JC, Walsh CT, Walton JD, Wenzel SC, Willey JM, van der Donk WA (2013) Ribosomally synthesized and post-translationally modified peptide natural products: overview and recommendations for a universal nomenclature. Nat Prod Rep 30:108–160. https://doi.org/10.1039/c2np20085f

Bansal PS, Torres AM, Crossett B, Wong KK, Koh JM, Geraghty DP, Vandenberg JI, Kuchel PW (2008) Substrate specificity of platypus venom L-to-D-peptide isomerase. J Biol Chem 283:8969–8975. https://doi.org/10.1074/jbc.M709762200

Barber CJS, Pjuara PT, Reed DW, Chiwocha S, Zhang H, Covello PS (2013) The two-step biosynthesis of cyclic peptides from linear precursors in a member of the plant family Caryophyllaceae involves cyclization by a serine protease-like enzyme. J Biol Chem 288:12500–12510. https://doi.org/10.1074/jbc.M112.437947

Bastos IMD, Motta FN, Grellier P, Santana JM (2013) Parasite prolyl oligopeptidases and the challenge of designing chemotherapeuticals for Chagas disease, leishmaniasis, and African trypanosomiasis. Curr Med Chem 20:3103–3115. https://doi.org/10.2174/0929867311320250006

Bezerra GA, Dobrovetsky E, Viertlmayr R, Dong A, Binter A, Abramic M, Macheroux P, Dhe-Paganon S, Gruber K (2012) Entropy-driven binding of opioid peptides induces a large domain motion in human dipeptidyl peptidase III. Proc Natl Acad Sci U S A 109:6525–6530. https://doi.org/10.1073/pnas.1118005109

Brandt I, Scharpé S, Lambeir AM (2007) Suggested functions for prolyl oligopeptidase: a puzzling paradox. Clin Chim Acta 377:50–61. https://doi.org/10.1016/j.cca.2006.09.001

Buczek O, Bulaj G, Olivera BM (2005) Conotoxins and the posttranslational modification of secreted gene products. Cell Mol Life Sci 62:3067–3079. https://doi.org/10.1007/s00018-005-5283-0

Canning P, Rea D, Morty RE, Fülöp V (2013) Crystal structures of *Trypanosoma brucei* oligopeptidase B broaden the paradigm of catalytic regulation in prolyl oligopeptidase family enzymes. PLoS One 8:e79349. https://doi.org/10.1371/journal.pone.0079349

Ceccoli RD, Bianchi DA, Rial DV (2014) Flavoprotein monooxygenases for oxidative biocatalysis: recombinant expression in microbial hosts and applications. Front Microbiol 5:25. https://doi.org/10.3389/fmicb.2014.00025

Chekan JR, Estrada P, Covello PS, Nair SK (2017) Characterization of the macrocyclase involved in the biosynthesis of RiPP cyclic peptides in plants. Proc Natl Acad Sci U S A 114:6551–6556. https://doi.org/10.1073/pnas.1620499114

Cho HJ, Cho HY, Kim KJ, Kim MH, Kim SW, Kang BS (2011) Structural and functional analysis of bacterial flavin-containing monooxygenase reveals its ping-pong-type reaction mechanism. J Struct Biol 175:39–48. https://doi.org/10.1016/j.jsb.2011.04.007

Choi HS, Kim JK, Cho EH, Kim YC, Kim JI, Kim SW (2003) A novel flavin-containing monooxygenase from *Methylophaga* sp. strain SK1 and its indigo synthesis in *Escherichia coli*. Biochem Biophys Res Commun 306:930–936

Conticello SG, Pilpel Y, Glusman G, Fainzilber M (2000) Position-specific codon conservation in hypervariable gene families. Trends Genet 16:57–59. doi:dx.doi.org/10.1016/S0168-9525(99)01956-3

Craik DJ, Mylne JS, Daly NL (2010) Cyclotides: macrocyclic peptides with applications in drug design and agriculture. Cell Mol Life Sci 67:9–16. https://doi.org/10.1007/s00018-009-0159-3

Czekster CM, Naismith JH (2017) Kinetic landscape of a peptide bond-forming prolyl oligopeptidase. Biochemistry 56:2086–2095. https://doi.org/10.1021/acs.biochem.7b00012

Czekster CM, Ge Y, Naismith JH (2016) Mechanisms of cyanobactin biosynthesis. Curr Opin Chem Biol 35:80–88. https://doi.org/10.1016/j.cbpa.2016.08.029

Czekster CM, Ludewig H, McMahon SA, Naismith JH (2017) Characterization of a dual function macrocyclase enables design and use of efficient macrocyclization substrates. Nat Commun 8:1045. https://doi.org/10.1038/s41467-017-00862-4

Deng WQ, Li TH, Xi PG, Gan LX, Xiao ZD, Jiang ZD (2011) Peptide toxin components of *Amanita exitialis* basidiocarps. Mycologia 103:946–949. https://doi.org/10.3852/10-319

Dijkman WP, de Gonzalo G, Mattevi A, Fraaije MW (2013) Flavoprotein oxidases: classification and applications. Appl Microbiol Biotechnol 97:5177–5188. https://doi.org/10.1007/s00253-013-4925-7

Ding W, Liu WQ, Jia Y, Li Y, van der Donk WA, Zhang Q (2016) Biosynthetic investigation of phomopsins reveals a widespread pathway for ribosomal natural products in Ascomycetes. Proc Natl Acad Sci U S A 113:3521–3526. https://doi.org/10.1073/pnas.1522907113

von Döhren H (2009) A survey of nonribosomal peptide synthetase (NRPS) genes in *Aspergillus nidulans*. Fungal Genet Biol 46:S45–S52. https://doi.org/10.1016/j.fgb.2008.08.008

Donia MS, Hathaway BJ, Sudek S, Haygood MG, Rosovitz MJ, Ravel J, Schmidt EW (2006) Natural combinatorial peptide libraries in cyanobacterial symbionts of marine ascidians. Nat Chem Biol 2:729–735. https://doi.org/10.1038/nchembio829

Donia MS, Ruffner DE, Cao S, Schmidt EW (2011) Accessing the hidden majority of marine natural products through metagenomics. ChemBioChem 12:1230–1236. https://doi.org/10.1002/cbic.201000780

Faulstich H, Buku A, Bodenmüller H, Wieland T (1980) Virotoxins: actin-binding cyclic peptides of *Amanita virosa* mushrooms. Biochemistry 19:3334–3343. https://doi.org/10.1021/bi00555a036

Fülöp V, Bocskei Z, Polgár L (1998) Prolyl oligopeptidase: an unusual beta-propeller domain regulates proteolysis. Cell 94:161–170

Fülöp V, Szeltner Z, Renner V, Polgár L (2001) Structures of prolyl oligopeptidase substrate/inhibitor complexes. Use of inhibitor binding for titration of the catalytic histidine residue. J Biol Chem 276:1262–1266. https://doi.org/10.1074/jbc.M007003200

Gorres KL, Raines RT (2010) Prolyl 4-hydroxylase. Crit Rev Biochem Mol Biol 45:106–124. https://doi.org/10.3109/10409231003627991

Gorson J, Ramrattan G, Verdes A, Wright EM, Kantor Y, Rajaram Srinivasan R, Musunuri R, Packer D, Albano G, Qiu WG, Holford M (2015) Molecular diversity and gene evolution of the venom arsenal of Terebridae predatory marine snails. Genome Biol Evol 7:1761–1778. https://doi.org/10.1093/gbe/evv104

Haefelé C, Bonfils C, Sauvaire Y (1997) Characterization of a dioxygenase from *Trigonella foenum-graecum* involved in 4-hydroxyisoleucine biosynthesis. Phytochemistry 44:563–566

Hallen HE, Luo H, Scott-Craig JS, Walton JD (2007) Gene family encoding the major toxins of lethal *Amanita* mushrooms. Proc Natl Acad Sci U S A 104:19097–19101. https://doi.org/10.1073/pnas.0707340104

Hansen BG, Kliebenstein DJ, Halkier BA (2007) Identification of a flavin-monooxygenase as the S-oxygenating enzyme in aliphatic glucosinolate biosynthesis in *Arabidopsis*. Plant J 50:902–910. https://doi.org/10.1111/j.1365-313X.2007.03101.x

Hara R, Kino K (2009) Characterization of novel 2-oxoglutarate dependent dioxygenases converting L-proline to *cis*-4-hydroxy-L-proline. Biochem Biophys Res Commun 379:882–886. https://doi.org/10.1016/j.bbrc.2008.12.158

Harris KS, Durek T, Kaas Q, Poth AG, Gilding EK, Conlan BF, Saska I, Daly NL, van der Weerden NL, Craik DJ, Anderson MA (2015) Efficient backbone cyclization of linear peptides by a recombinant asparaginyl endopeptidase. Nat Commun 16:10199. https://doi.org/10.1038/ncomms10199

Hibi M, Kawashima T, Sokolov PM, Smirnov SV, Kodera T, Sugiyama M, Shimizu S, Yokozeki K, Ogawa J (2013) L-leucine 5-hydroxylase of *Nostoc punctiforme* is a novel type of Fe(II)/α-ketoglutarate-dependent dioxygenase that is useful as a biocatalyst. Appl Microbiol Biotechnol 97:2467–2472. https://doi.org/10.1007/s00253-012-4136-7

Houssen WE, Wright SH, Kalverda AP, Thompson GS, Kelly SM, Jaspars M (2010) Solution structure of the leader sequence of the patellamide precursor peptide, $PatE_{1-34}$. ChemBioChem 11:1867–1873. https://doi.org/10.1002/cbic.201000305

Houwaart S, Youssar L, Hüttel W (2014) Pneumocandin biosynthesis: involvement of a trans-selective proline hydroxylase. ChemBioChem 15:2365–2369. https://doi.org/10.1002/cbic.201402175

Jagadish K, Camarero JA (2010) Cyclotides, a promising molecular scaffold for peptide-based therapeutics. Biopolymers (Pept Sci) 94:611–616. https://doi.org/10.1002/bip.21433

Jalkanen AJ, Leikas JV, Forsberg MM (2014) Prolyl oligopeptidase inhibition decreases extracellular acetylcholine levels in rat hippocampus and prefrontal cortex. Neurosci Lett 579:110–113. https://doi.org/10.1016/j.neulet.2014.07.026

Jensen CN, Cartwright J, Ward J, Hart S, Turkenburg JP, Ali ST, Allen MJ, Grogan G (2012) A flavoprotein monooxygenase that catalyses a Baeyer-Villiger reaction and thioether oxidation using NADH as the nicotinamide cofactor. ChemBioChem 13:872–878. https://doi.org/10.1002/cbic.201200006

Jetté L, Harvey L, Eugeni K, Levens N (2009) 4-Hydroxyisoleucine: a plant-derived treatment for metabolic syndrome. Curr Opin Investig Drugs 10:353–358

Jilek A, Mollay C, Tippelt C, Grassi J, Mignogna G, Mullegger J, Sander V, Fehrer C, Barra D, Kreil G (2005) Biosynthesis of a D-amino acid in peptide linkage by an enzyme from frog skin secretions. Proc Natl Acad Sci U S A 102:4235–4239. https://doi.org/10.1073/pnas.0500789102

Johnson RD, Lane GA, Koulman A, Gao M, Fraser K, Fleetwood DJ, Voisey CR, Dyer JM, Pratt J, Christensen M, Simpson WR, Bryan GT, Johnson LJ (2015) A novel family of cyclic oligopeptides derived from ribosomal peptide synthesis of an in planta-induced gene, *gigA*, in *Epichloë* endophytes of grasses. Fungal Genet Biol 85:14–24. https://doi.org/10.1016/j.fgb.2015.10.005

Kaas Q, Yu R, Jin AH, Dutertre S, Craik DJ (2012) ConoServer: updated content, knowledge, and discovery tools in the conopeptide database. Nucleic Acids Res 40:D325–D330. https://doi.org/10.1093/nar/gkr886. Website:www.conoserver.org

Kaukinen K, Lindfors K (2015) Novel treatments for celiac disease: glutenases and beyond. Dig Dis 33:277–281. https://doi.org/10.1159/000369536

Koehnke J, Bent A, Houssen WE, Zollman D, Morawitz F, Shirran S, Vendome J, Nneoyiegbe AF, Trembleau L, Botting CH, Smith MC, Jaspars M, Naismith JH (2012) The mechanism of patellamide macrocyclization revealed by the characterization of the PatG macrocyclase domain. Nat Struct Mol Biol 19:767–772. https://doi.org/10.1038/nsmb.2340

Kohler A, Kuo A, Nagy LG, Morin E, Barry KW, Buscot F, Canbäck B, Choi C, Cichocki N, Clum A, Colpaert J, Copeland A, Costa MD, Doré J, Floudas D, Gay G, Girlanda M, Henrissat B, Herrmann S, Hess J, Högberg N, Johansson T, Khouja HR, LaButti K, Lahrmann U, Levasseur A, Lindquist EA, Lipzen A, Marmeisse R, Martino E, Murat C, Ngan CY, Nehls U, Plett JM, Pringle A, Ohm RA, Perotto S, Peter M, Riley R, Rineau F, Ruytinx J, Salamov A, Shah F, Sun H, Tarkka M, Tritt A, Veneault-Fourrey C, Zuccaro A, Mycorrhizal Genomics Initiative Consortium, Tunlid A, Grigoriev IV, Hibbett DS, Martin F (2015) Convergent losses of decay mechanisms and rapid turnover of symbiosis genes in mycorrhizal mutualists. Nat Genet 47:410–415. https://doi.org/10.1038/ng.3223

Koiso Y, Li Y, Iwasaki S, Hanaoka K, Kobayashi T, Sonoda R, Fujita Y, Yaegashi H, Sato Z (1994) Ustiloxins, antimitotic cyclic peptides from false smut balls on rice panicles caused by *Ustilaginoidea virens*. J Antibiot 47:765–773

Li M, Chen C, Davies DR, Chiu TK (2010) Induced-fit mechanism for prolyl endopeptidase. J Biol Chem 285:21487–21495. https://doi.org/10.1074/jbc.M109.092692

Li P, Deng WQ, Li TH, Song B, Shen YH (2013) Illumina-based de novo transcriptome sequencing and analysis of *Amanita exitialis* basidiocarps. Gene 532:63–71. https://doi.org/10.1016/j.gene.2013.09.014

Li P, Deng W, Li T (2014) The molecular diversity of toxin gene families in lethal *Amanita* mushrooms. Toxicon 83:59–68. https://doi.org/10.1016/j.toxicon.2014.02.020

Lopez-Vera E, Walewska A, Skalicky JJ, Olivera BM, Bulaj G (2008) Role of hydroxyprolines in the in vitro oxidative folding and biological activity of conotoxins. Biochemistry 47:1741–1751. https://doi.org/10.1021/bi701934m

Luo H, Hallen-Adams HE, Walton JD (2009) Processing of the phalloidin proprotein by prolyl oligopeptidase from the mushroom *Conocybe albipes*. J Biol Chem 284:18070–18077. https://doi.org/10.1074/jbc.M109.006460

Luo H, Hallen-Adams HE, Scott-Craig JS, Walton JD (2010) Co-localization of amanitin and a candidate toxin-processing prolyl oligopeptidase in *Amanita* basidiocarps. Eukaryot Cell 9:1891–1900. https://doi.org/10.1128/EC.00161-10

Luo H, Hallen-Adams HE, Scott-Craig JS, Walton JD (2012) Ribosomal biosynthesis of α-amanitin in *Galerina marginata*. Fungal Genet Biol 49:123–129. https://doi.org/10.1016/j.fgb.2011.12.005

Luo H, Hong SY, Sgambelluri RM, Angelos E, Li X, Walton JD (2014) Peptide macrocyclization catalyzed by a prolyl oligopeptidase involved in α-amanitin biosynthesis. Chem Biol 21:1610–1617. https://doi.org/10.1016/j.chembiol.2014.10.015

Matsumoto T, Shishido A, Morita H, Itokawa H, Takeya K (2001) Cyclolinopeptides F-I, cyclic peptides from linseed. Phytochemistry 57:251–260

Maurya CK, Singh R, Jaiswal N, Venkateswarlu K, Narender T, Tamrakar AK (2014) 4-Hydroxyisoleucine ameliorates fatty acid-induced insulin resistance and inflammatory response in skeletal muscle cells. Mol Cell Endocrinol 395:51–60. https://doi.org/10.1016/j.mce.2014.07.018

May JP, Perrin DM (2007) Tryptathionine bridges in peptide synthesis. Biopolymers 88:714–724. https://doi.org/10.1002/bip.20807

May JP, Perrin DM (2008) Intraannular Savige–Fontana reaction: one-step conversion of one class of monocyclic peptides into another class of bicyclic peptides. Chemistry 14:3404–3409. https://doi.org/10.1002/chem.200701088

McCluskey K, Paterson NG, Bland ND, Isaacs NW, Mottram JC (2010) Crystal structure of *Leishmania* major oligopeptidase B gives insight into the enzymatic properties of a trypanosomatid virulence factor. J Biol Chem 285:39249–39259. https://doi.org/10.1074/jbc.M110.156679

Meunier B, de Visser SP, Shaik S (2004) Mechanism of oxidation reactions catalyzed by cytochrome p450 enzymes. Chem Rev 104:3947–3980. https://doi.org/10.1021/cr020443g

Morita H, Shishido A, Kayashita T, Shimomura M, Takeya K, Itokawa H (1994) Two novel cyclic peptides, yunnanins A and B from *Stellaria yunnanensis*. Chem Lett 1994:2415–2418

Moroni P, Buronfosse T, Longin-Sauvageon L, Delatour P, Benoit E (1995) Chiral sulfoxidation of albendazole by the flavin adenine dinucleotide-containing and cytochrome P450-dependent monooxygenases from rat liver. Drug Metab Dispos 23:160–170

Mulvenna JP, Wang C, Craik DJ (2006) CyBase: a database of cyclic protein sequence and structure. Nucleic Acids Res 34:D192–D194. Website: www.cybase.org.au. https://doi.org/10.1093/nar/gkj005

Nagano N, Umemura M, Izumikawa M, Kawano J, Ishii T, Kikuchi M, Tomii K, Kumagai T, Yoshimi A, Machida M, Abe K, Shin-Ya K, Asai K (2016) Class of cyclic ribosomal peptide synthetic genes in filamentous fungi. Fungal Genet Biol 86:58–70. https://doi.org/10.1016/j.fgb.2015.12.010

Nagatsu T (2017) Prolyl oligopeptidase and dipeptidyl peptidase II/dipeptidyl peptidase IV ratio in the cerebrospinal fluid in Parkinson's disease: historical overview and future prospects. J Neural Transm (Vienna) 124:739–744. https://doi.org/10.1007/s00702-016-1604-8

Nguyen GK, Wang S, Qiu Y, Hemu X, Lian Y, Tam JP (2014) Butelase 1 is an Asx-specific ligase enabling peptide macrocyclization and synthesis. Nat Chem Biol 10:732–738. https://doi.org/10.1038/nchembio.1586

Nguyen GK, Kam A, Loo S, Jansson AE, Pan LX, Tam JP (2015) Butelase 1: a versatile ligase for peptide and protein macrocyclization. J Am Chem Soc 137:15398–15401. https://doi.org/10.1021/jacs.5b11014

Ollivaux C, Soyez D, Toullec JY (2014) Biogenesis of D-amino acid containing peptides/proteins: where, when and how? J Pept Sci 20:595–612. doi:10.1002/psc.2637

Oman TJ, van der Donk WA (2010) Follow the leader: the use of leader peptides to guide natural product biosynthesis. Nat Chem Biol 6:9–18. https://doi.org/10.1038/nchembio.286

Ortega MA, van der Donk WA (2016) New insights into the biosynthetic logic of ribosomally synthesized and post-translationally modified peptide natural products. Cell Chem Biol 23:31–44. https://doi.org/10.1016/j.chembiol.2015.11.012

Partida-Martinez LP, Hertweck C (2005) Pathogenic fungus harbours endosymbiotic bacteria for toxin production. Nature 437:884–888. https://doi.org/10.1038/nature03997

Peltonen I, Myöhänen TT, Männistö PT (2011) Association of prolyl oligopeptidase with conventional neurotransmitters in the brain. CNS Neurol Disord Drug Targets 10:311–318

Pisarewicz K, Mora D, Pflueger FC, Fields GB, Marí F (2005) Polypeptide chains containing D-gamma-hydroxyvaline. J Am Chem Soc 127:6207–6215. https://doi.org/10.1021/ja050088m

Podust LM, Sherman DH (2012) Diversity of P450 enzymes in the biosynthesis of natural products. Nat Prod Rep 29:1251–1266. https://doi.org/10.1039/c2np20020a

Pulman JA, Childs KL, Sgambelluri RM, Walton JD (2016) Expansion and diversification of the MSDIN family of cyclic peptide genes in the poisonous agarics *Amanita phalloides* and *A. bisporigera*. BMC Genomics 17:1038. https://doi.org/10.1186/s12864-016-3378-7

Ramm S, Krawczyk B, Mühlenweg A, Poch A, Mösker E, Süssmuth RD (2017) A self-sacrificing N-methyltransferase is the precursor of the fungal natural product omphalotin. Angew Chem Int Ed Eng 56:9994–9997. https://doi.org/10.1002/anie.201703488

Rep M, Kistler HC (2010) The genomic organization of plant pathogenicity in *Fusarium* species. Curr Opin Plant Biol 13:420–426. https://doi.org/10.1016/j.pbi.2010.04.004

Riley R, Salamov AA, Brown DW, Nagy LG, Floudas D, Held BW, Levasseur A, Lombard V, Morin E, Otillar R, Lindquist EA, Sun H, LaButti KM, Schmutz J, Jabbour D, Luo H, Baker SE, Pisabarro AG, Walton JD, Blanchette RA, Henrissat B, Martin F, Cullen D, Hibbett DS, Grigoriev IV (2014) Extensive sampling of basidiomycete genomes demonstrates inadequacy of the white-rot/brown-rot paradigm for wood decay fungi. Proc Natl Acad Sci U S A 111:9923–9928. https://doi.org/10.1073/pnas.1400592111

Roy RS, Kim S, Baleja JD, Walsh CT (1998) Role of the microcin B17 propeptide in substrate recognition: solution structure and mutational analysis of $McbA_{1-26}$. Chem Biol 5:217–228

Saska I, Gillon AD, Hatsugai N, Dietzgen RG, Hara-Nishimura I, Anderson MA, Craik DJ (2007) An asparaginyl endopeptidase mediates in vivo protein backbone cyclization. J Biol Chem 282:29721–29728. https://doi.org/10.1074/jbc.M705185200

Seo JA, Proctor RH, Plattner RD (2001) Characterization of four clustered and coregulated genes associated with fumonisin biosynthesis in *Fusarium verticillioides*. Fungal Genet Biol 34:155–165. https://doi.org/10.1006/fgbi.2001.1299

Sgambelluri RM, Epis S, Sassera D, Luo H, Angelos ER, Walton JD (2014) Profiling of amatoxins and phallotoxins in the genus *Lepiota* by liquid chromatography combined with UV absorbance and mass spectrometry. Toxins 6:2336–2347. https://doi.org/10.3390/toxins6082336

Sgambelluri RM, Smith MO, Walton JD (2018) Versatility of prolyl oligopeptidase B in peptide macrocyclization. ACS Synth Biol 7:145–152. https://doi.org/10.1021/acssynbio.7b00264

Shan L, Matthews II, Khosla C (2005) Structural and mechanistic analysis of two prolyl endopeptidases: role of interdomain dynamics in catalysis and specificity. Proc Natl Acad Sci U S A 102:3599–3604. https://doi.org/10.1073/pnas.0408286102

Sivonen K, Leikoski N, Fewer DP, Jokela J (2010) Cyanobactins – ribosomal cyclic peptides produced by cyanobacteria. Appl Microbiol Biotechnol 86:1213–1225. https://doi.org/10.1007/s00253-010-2482-x

Smirnov SV, Sokolov PM, Kotlyarova VA, Samsonova NN, Kodera T, Sugiyama M, Torii T, Hibi M, Shimizu S, Yokozeki K, Ogawa J (2013) A novel L-isoleucine-4′-dioxygenase and L-isoleucine dihydroxylation cascade in *Pantoea ananatis*. MicrobiologyOpen 2:471–481. https://doi.org/10.1002/mbo3.87

Sprules T, Kawulka KE, Gibbs AC, Wishart DS, Vederas JC (2004) NMR solution structure of the precursor for carnobacteriocin B2, an antimicrobial peptide from *Carnobacterium piscicola*. Eur J Biochem 271:1748–1756. https://doi.org/10.1111/j.1432-1033.2004.04085.x

Stone BL, Gray WR (1982) Occurrence of hydroxyproline in a toxin from the marine snail *Conus geographus*. Arch Biochem Biophys 216:765–767

Suda S, Lawton EM, Wistuba D, Cotter PD, Hill C, Ross RP (2012) Homologues and bioengineered derivatives of LtnJ vary in ability to form D-alanine in the lantibiotic lacticin 3147. J Bacteriol 194:708–714. https://doi.org/10.1128/JB.06185-11

Szeltner Z, Polgár L (2008) Structure, function and biological relevance of prolyl oligopeptidase. Curr Protein Pept Sci 9:96–107

Szeltner Z, Juhász T, Szamosi I, Rea D, Fülöp V, Módos K, Juliano L, Polgár L (2013) The loops facing the active site of prolyl oligopeptidase are crucial components in substrate gating and specificity. Biochim Biophys Acta 1834:98–111. https://doi.org/10.1016/j.bbapap.2012.08.012

Tokai T, Koshino H, Takahashi-Ando N, Sato M, Fujimura M, Kimura M (2007) *Fusarium* Tri4 encodes a key multifunctional cytochrome P450 monooxygenase for four consecutive oxygenation steps in trichothecene biosynthesis. Biochem Biophys Res Commun 353:412–417. https://doi.org/10.1016/j.bbrc.2006.12.033

Torres AM, Tsampazi M, Tsampazi C, Kennett EC, Belov K, Geraghty DP, Bansal PS, Alewood PF, Kuchel PW (2006) Mammalian L-to-D-amino-acid-residue isomerase from platypus venom. FEBS Lett 580:1587–1591. https://doi.org/10.1016/j.febslet.2006.01.089

Tsukui T, Nagano N, Umemura M, Kumagai T, Terai G, Machida M, Asai K (2015) Ustiloxins, fungal cyclic peptides, are ribosomally synthesized in *Ustilaginoidea virens*. Bioinformatics 31:981–985. https://doi.org/10.1093/bioinformatics/btu753

Tsunematsu Y, Ishikawa N, Wakana D, Goda Y, Noguchi H, Moriya H, Hotta K, Watanabe K (2013) Distinct mechanisms for spiro-carbon formation reveal biosynthetic pathway crosstalk. Nat Chem Biol 9:818–825. https://doi.org/10.1038/nchembio.1366

Tudzynski B, Rojas MC, Gaskin P, Hedden P (2002) The gibberellin 20-oxidase of *Gibberella fujikuroi* is a multifunctional monooxygenase. J Biol Chem 277:21246–21253. https://doi.org/10.1074/jbc.M201651200

Umemura M, Nagano N, Koike H, Kawano J, Ishii T, Miyamura Y, Kikuchi M, Tamano K, Yu J, Shin-ya K, Machida M (2014) Characterization of the biosynthetic gene cluster for the ribosomally synthesized cyclic peptide ustiloxin B in *Aspergillus flavus*. Fungal Genet Biol 68:23–30. https://doi.org/10.1016/j.fgb.2014.04.011

van der Velden NS, Kälin N, Helf MJ, Piel J, Freeman MF, Künzler M (2017) Autocatalytic backbone N-methylation in a family of ribosomal peptide natural products. Nat Chem Biol 13:833–835. https://doi.org/10.1038/nchembio.2393

Venäläinen JI, Juvonen RO, Männistö PT (2004) Evolutionary relationships of the prolyl oligopeptidase family enzymes. Eur J Biochem 271:2705–2715. https://doi.org/10.1111/j.1432-1033.2004.04199.x

Völler GH, Krawczyk B, Ensle P, Süssmuth RD (2013) Involvement and unusual substrate specificity of a prolyl oligopeptidase in class III lanthipeptide maturation. J Am Chem Soc 135:7426–7429. https://doi.org/10.1021/ja402296m

Walton JD (1987) Two enzymes involved in biosynthesis of the host selective phytotoxin HC toxin. Proc Natl Acad Sci U S A 84:8444–8447

Walton JD (2000) Horizontal gene transfer and the evolution of secondary metabolite gene clusters in fungi: an hypothesis. Fungal Genet Biol 30:167–171. https://doi.org/10.1006/fgbi.2000.1224

Walton JD, Panaccione DG, Hallen H (2004) Peptide synthesis without ribosomes. In: Tkacz J, Lange L (eds) Advances in fungal biotechnology for industry, agriculture, and medicine. Kluwer Academic, New York, pp 127–162

Wieland T (1986) Peptides of poisonous *Amanita* mushrooms. Springer, New York

Wight WD, Kim KH, Lawrence CB, Walton JD (2009) Biosynthesis and role in virulence of the histone deacetylase inhibitor depudecin from *Alternaria brassicicola*. Mol Plant-Microbe Interact 22:1258–1267. https://doi.org/10.1094/MPMI-22-10-1258

Wolfe BE, Kuo M, Pringle A (2012) *Amanita thiersii* is a saprotrophic fungus expanding its range in the United States. Mycologia 104:22–33. https://doi.org/10.3852/11-056

Woodward SR, Cruz LJ, Olivera BM, Hillyard DR (1990) Constant and hypervariable regions in conotoxin propeptides. EMBO J 9:1015–1020

Yoshimi A, Umemura M, Nagano N, Koike H, Machida M, Abe K (2016) Expression of ustR and the Golgi protease KexB are required for ustiloxin B biosynthesis in *Aspergillus oryzae*. AMB Express 6:9. https://doi.org/10.1186/s13568-016-0181-4

Zhang P, Chen Z, Hu J, Wei B, Zhang Z, Hu W (2005) Production and characterization of Amanitin toxins from a pure culture of *Amanita exitialis*. FEMS Microbiol Lett 252:223–228. https://doi.org/10.1016/j.femsle.2005.08.049

Zhou P, Silverstein KA, Gao L, Walton JD, Nallu S, Guhlin J, Young ND (2013) Detecting small plant peptides using SPADA (Small Peptide Alignment Discovery Application). BMC Bioinformatics 14:335. https://doi.org/10.1186/1471-2105-14-335

Chapter 5
Biological Activities of the *Amanita* Peptide Toxins

Nature alone is antique, and the oldest art a mushroom.

—Thomas Carlyle

The amatoxins, phallotoxins, cycloamanides, and antamanide belong to the special class of secondary metabolites known as *cyclic peptides* or *cyclopeptides* – strings of amino acids in which the N and C termini are joined head-to-tail by amide (peptide) bonds (Chap. 2). Cyclic peptides possess many chemical attributes that contribute to high biological activity, including chemical stability, membrane permeability, and rigidity. The importance of these attributes is discussed in more detail in Chap. 7. The focus of this chapter is the specific biological activities of the characterized cyclic peptides of *Amanita* and other mushrooms. The majority of the studies have concerned the amatoxins, followed by the phallotoxins and, distantly, the other peptides. The organization of this chapter proceeds from the whole organism level to the intra-organismal level, concluding with toxin action at the molecular level.

5.1 Survival and Uptake from the Digestive Tract

In all documented cases of mushroom poisoning, the route of toxin ingestion is oral – through the mouth, esophagus, stomach, and intestines. Therefore, from the human toxicological point of view, one of the most critical attributes of the toxins is their ability to survive the mammalian digestive tract and be efficiently absorbed into the bloodstream. Their chemical stability is also manifested by their resistance to heat and extremes of pH. From the documented history of human poisonings, they are known to survive all types of cooking, including baking, frying, and boiling. They also survive all of the sub-compartments of the digestive tracts of at least some animals such as guinea pigs and dogs (Chap. 6).

An effect of amatoxins directly on the mucosal cells of the duodenum and ileum is consistent with the observed early (6–8 h) symptoms of *Amanita* cyclic peptide

J. Walton, *The Cyclic Peptide Toxins of Amanita and Other Poisonous Mushrooms*, https://doi.org/10.1007/978-3-319-76822-9_5

toxin poisoning, which include vomiting (emesis), abdominal cramps, and diarrhea (Faulstich and Zilker 1994). The etiology of these early symptoms is not well understood, but amatoxins are probably the cause, as discussed in Chap. 7.

The ileal cells must be able to transport at least the amatoxins across both of their membranes and into the bloodstream. The transport protein(s) responsible for the uptake and excretion into the bloodstream of amatoxins by ileal cells are not yet known. Transport might be mediated by the ileal Na^+-dependent bile salt transporter (also known as ISBT or SLC10A2) (Trauner and Boyer 2003). Epithelial intestinal cells also have transporters for small peptides, such as SLC15A1 (Kramer 2011). α-Amanitin is not absorbed through the skin of mice (Kaya et al. 2014).

In regard to phallotoxins, according to an oral communication from T. Wieland, cited by Litten (1975), "No animal has been demonstrably poisoned by the oral administration of phalloidin." Phallotoxins were not transported into isolated small intestinal cells from guinea pig nor into brush border membrane vesicles isolated from rat ileal cells, whereas the same cells could transport endogenous substrates such as cholate and taurocholate (Petzinger et al. 1982). This lack of uptake by ileal cells is a plausible explanation for why phallotoxins are not toxic when ingested orally, because once in the bloodstream they are known to be highly hepatotoxic. Petzinger et al. (1982) did not attempt to study uptake of amatoxins by ileal cells, which would have made an interesting comparison because amatoxins, unlike phallotoxins, certainly do somehow cross the ileum to enter the bloodstream.

5.2 Uptake by the Liver

The amatoxins and phallotoxins are actively and rapidly taken up by the liver and by isolated hepatocytes despite their overall poor membrane permeability (Anderl et al. 2012; Wieland 1986). The mechanism of uptake by specific transporters has been the subject of multiple studies. Earlier work concluded that the transport of the *Amanita* cyclic peptides utilized the same transport system as the bile salts such as taurocholate and glycocholate, but this was before the high diversity and overlapping activities of mammalian membrane transporters were fully appreciated. Later, when individual transport genes and proteins had been identified and could be expressed individually in systems such as *Xenopus* oocytes or transfected human cells (e.g., HEK-293, originally from embryonic kidneys), it was shown that phallotoxins (e.g., [^{3}H]demethylphalloin) are specifically transported into human hepatocytes by OATP1B1 and OATP1B3, which are members of the organic anion transporting superfamily (Fehrenbach et al. 2003; Meier-Abt et al. 2004) (Fig. 5.1). OATP1B1 and OATP1B3 have also been known as SLC21A6 and SLC21A8, but since renaming of the SLC21 family are now known as SLCO1B1 and SLCO1B3, respectively (Hagenbuch and Meier 2004; He et al. 2009). OATP1B1 is the major phallotoxin transporter in humans; OATP1B3 is less efficient (Fig. 5.1c). The K_m of human OATP1B1 for demethylphalloin is ~6 μM (Meier-Abt et al. 2004). The rat protein Oatp1b2, which also transports phallotoxins, is equivalent to human

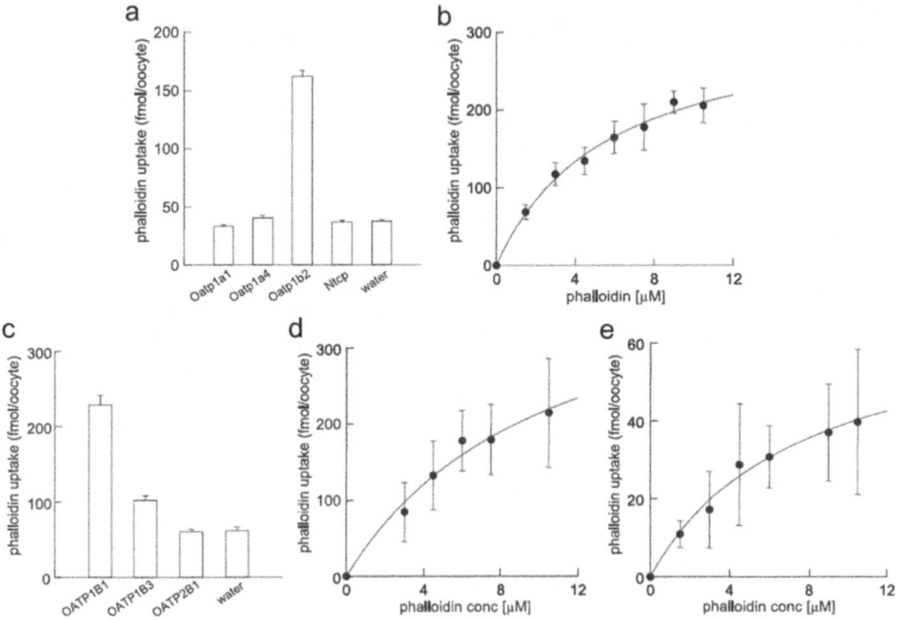

Fig. 5.1 (**a**) Transport of demethylphalloin by *Xenopus* oocytes expressing different members of the rat Oatp1 family, (**b**) kinetics of uptake by oocytes expressing rat Oatp1b2, (**c**) transport in oocytes expressing different members of the human OATP1 family, (**d**) kinetics of transport by human OATP1B1, (**e**) kinetics of transport by human OATP1B3. (Reprinted from Meier-Abt et al. 2004 with permission from Elsevier)

OATP1B1 (Fig. 5.1a). Other tested transporters, including the sodium-dependent Na^+/taurocholate cotransporting polypeptide (Ntcp, also known as SLC10A1), do not transport demethylphalloin (Fehrenbach et al. 2003; Meier-Abt et al. 2004).

Mice genetically engineered to specifically lack Oatp1b2 are healthy but completely resistant to phalloidin-induced hepatotoxicity (Lu et al. 2008). Oatp1b2-null mice are still sensitive to α-amanitin, supporting the earlier studies indicating that the two major classes of *Amanita* cyclic peptides enter hepatocytes through different transporters.

Bile salts move from the liver back into the small intestine via the bile duct as part of the enterohepatic circulatory system. The bile canaliculi (bile capillaries) collect bile secreted by liver cells (hepatocytes). The canaliculi merge to form the common hepatic duct, which eventually drains into the duodenum of the small intestine. Part of the reason that the *Amanita* cyclic peptides are so toxic is that they, too, recirculate, resulting in their re-introduction into the liver without any opportunity to be cleared by the kidneys. Using genetic knockout mice and rats, Gavrilova et al. (2007) showed that multidrug resistance-associated protein 2 (Mrp2), but not the ABC transporters Mdr1 and Bcrp, transports demethylphalloin out of hepatocytes into the bile duct. To account for the lack of accumulation of demethylphalloin

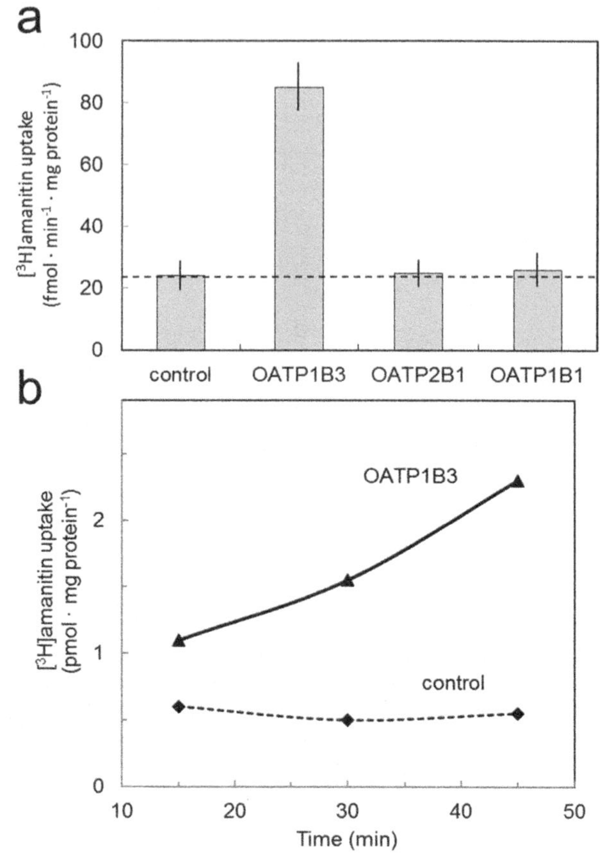

Fig. 5.2 (**a**) Uptake of [^{3}H]-amanitin by MDCKII control cells and MDCKII cells expressing OATP1B3, OATP2B1, or OATP1B1, (**b**) time course of [^{3}H]-amanitin uptake by cells expressing OATP1B3. (Redrawn from Letschert et al. 2006)

in the knockout rat livers, the authors proposed the existence of some degree of retrograde transport from hepatocytes back into the blood (Gavrilova et al. 2007).

To identify the mechanism of uptake of amatoxins by hepatocytes, Gundala et al. (2003) transfected hepatoblastoma cell line HepG2, which has lost native Ntcp (SLC10A1) expression, with the rat gene for Ntcp. mRNA synthesis was more sensitive to α-amanitin in the Ntcp-transfected cells than in the control cells, consistent with Ntcp being a transporter for this toxin. A more direct test using uptake of radiolabeled α-amanitin (specifically, [^{3}H]-O-methyl-dehydroxymethyl-α-amanitin) into Madin-Darby canine kidney (MDCK II) cells transfected with individual transporter genes showed that the organic acid transporter OATP1B3 (SLCO1B3) efficiently transports α-amanitin with high affinity (K_m ~3.7 μM), whereas OATP1B1 (SLCO1B1) and OATP2B1 (SLCO2B1) do not (Letschert et al. 2006) (Fig. 5.2). Cells expressing OATP1B3 are also more sensitive to killing by α-amanitin. In these experiments, Ntcp was not tested as a possible α-amanitin transporter. As discussed above, phallotoxin transport into hepatocytes relies primarily on OATP1B1

(SLCO1B1) with OATP1B3 having a secondary role. α-Amanitin is only a weak inhibitor of phallotoxin uptake mediated by SLCO1B1 (Fehrenbach et al. 2003). These results further support the conclusion that phallotoxins and amatoxins are taken up by liver cells through different but related transporters of the OATP (SLCO) family.

OATP1B1 (the major phallotoxin transporter in humans) and OATP1B3 (the major amatoxin transporter) are predominantly expressed in the liver (Hagenbuch and Meier 2004; Petryszak et al. 2016). Native substrates for OATP1B1 and OATP1B3 are bile salts and organic anions, but many exogenous compounds are also substrates or competitive inhibitors (Hagenbuch and Meier 2004). The broad specificity of these transport proteins provides a mechanistic rationale for the ability of many compounds, such as antamanide and silymarin, to protect against mushroom poisoning by competitive inhibition of amatoxin and phallotoxin uptake by hepatocytes (Chap. 7).

The cyclic heptapeptide microcystins are also taken up by mammalian cells through the OATP1B1 and OATP1B3 transporters (Niedermeyer et al. 2014). Microcystins (not to be confused with microcins; see below) are potent serine/threonine protein phosphatase inhibitors that are biosynthesized by nonribosomal peptide synthetases. The strong expression of these two transporters in liver cells contributes to the high hepatotoxicity of microcystins. OATP1B3 (but not OATP1B1) is strongly expressed in some cancer cells, raising the possibility of developing anticancer therapies based on microcystin-like molecules that have been modified to be preferentially transported by OATP1B3 (Niedermeyer et al. 2014).

Although studies with individual transporters show dramatic specificity for uptake of the *Amanita* peptide toxins, it seems that other transporters, including probably Ntcp, can also transport amatoxins and/or phallotoxins to a lesser yet still toxicologically significant degree. OATP1B1 and OATP1B3 are expressed almost exclusively in liver cells (Hagenbuch and Meier 2004), yet it is well-documented that other mammalian cells are equally or almost equally sensitive to the amatoxins and phallotoxins (Fiume and Barbanti-Brodano 1974; Wieland 1986). Alternative routes to amatoxin uptake are implied from the results of Letschert et al. (2006), who found that MDCK II control cells (i.e., not expressing OATP1B3) eventually succumbed to 10 μg/ml α-amanitin. The involvement of an alternate transporter, or no transporter at all, is also indicated by the fact that plant cells are sensitive to amatoxins (Holtorf et al. 1999; Theologis et al. 1985).

For reasons not yet explicated, newborn animals (e.g., mice, rats, rabbits, and hamsters) are tolerant to high doses of phallotoxins delivered parenterally (Wieland 1986). Tolerance is correlated with lack of uptake by liver cells. Substances that damage liver cells, such as carbon tetrachloride, galactosamine, or nitrosamine, protect animals against phalloidin. This was postulated to be due to interference of these compounds with normal carrier-mediated uptake mechanisms (Wieland 1986). Several studies have shown that protein conjugates of phallotoxins and amatoxins are not only toxic to liver cells, but even 50–100 times more toxic than the native compounds (Barbanti-Brodano et al. 1974; Faulstich and Fiume 1985).

The mechanism of uptake of toxin-protein conjugates is not fully understood, but is probably through endocytosis (Fiume and Barbanti-Brodano 1974). This attribute has been exploited to make antibody-amanitin conjugates with high specificity and toxicity to cancer cells (Chap. 7).

5.3 RNA Polymerase II as the Site of Action of Amatoxins

There is definitive evidence that the amatoxins are potent and specific inhibitors of eukaryotic DNA-dependent RNA polymerase II, also known as pol II or RNAP II. Due to the strong evidence for this site of action, and in the absence of any evidence for alternative sites, the amatoxins can be considered to be defining inhibitors of pol II. That is, inhibition of a biological process by α-amanitin serves as presumptive evidence for the involvement of pol II in that process. For example, inhibition of the hormonal induction of an enzyme by α-amanitin is a strong evidence that the induction requires de novo gene transcription.

Pol II makes a single-stranded RNA copy from a double-stranded DNA template. It is responsible for transcription of all messenger RNA (mRNA) in eukaryotes and also microRNA (miRNA). After processing, mRNA transcripts are exported to the cytoplasm where they are translated on ribosomes into proteins. All cells have multiple DNA-dependent RNA polymerases. Pol I transcribes ribosomal RNA (rRNA) and pol III transcribes transfer RNA (tRNA) and some other small RNA. Pol IV and pol V are found only in plants and transcribe small interfering RNA (siRNA). Pol IV and Pol V from *Arabidopsis* are not inhibited by 5 μg/ml α-amanitin and can therefore be classified as amanitin-resistant (Haag et al. 2012). Although in eukaryotes only pol II is strongly inhibited by amatoxins, pol II enzymes from different organisms differ considerably in their sensitivity (see below).

α-Amanitin inhibition of DNA-dependent RNA polymerase activity was first reported by Stirpe and Fiume (1967) and confirmed by Jacob et al. (1970). The specificity of amanitin for pol II over pol I and pol III was demonstrated in sea urchin and rat liver extracts by Lindell et al. (1970) and in calf thymus by Kedinger et al. (1970) (Fig. 5.3). A brief history of this discovery was recounted by Lindell (1984).

In humans and yeast, pol II is a 550 kDa complex of 12 subunits. The gene for the largest subunit, Rpb1, is called *POLR2A* in humans and *RPO21* in yeast (Sainsbury et al. 2015). This is the main site of binding of α-amanitin, although the toxin also makes contact with the second largest subunit, Rpb2 (see below). Inhibition of pol II leads to cessation of new mRNA synthesis. As the pool of mRNA turns over, protein synthesis gradually ceases, and as essential proteins themselves turn over and are not replenished, the cells die.

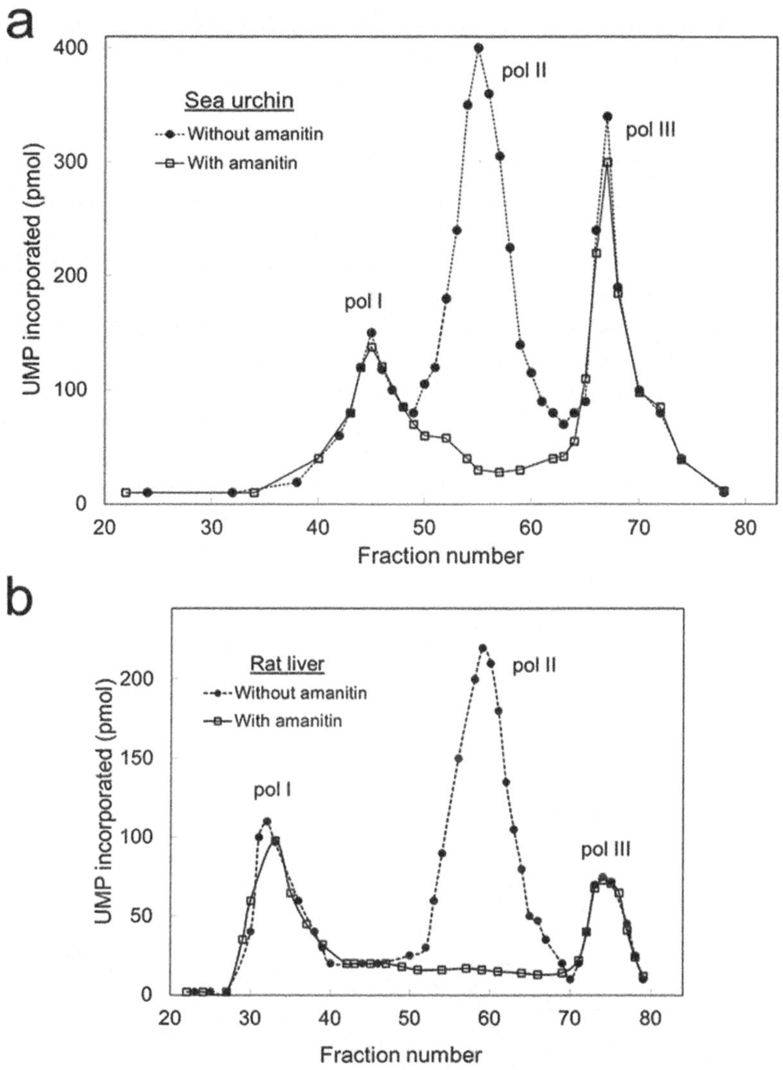

Fig. 5.3 Specific inhibition of (**a**) sea urchin and (**b**) rat liver RNA polymerase II by α-amanitin. Nuclear extracts were separated by anion exchange chromatography (DEAE-Sephadex) and assayed for incorporation of [^{3}H]UMP into RNA from a DNA template. α-Amanitin concentration was 3.4 μM. Open and closed circles indicate activity in the presence or absence of amanitin, respectively. (Data taken from Lindell et al. 1970)

5.3.1 Molecular and Structural Basis of Inhibition of Pol II by Amatoxins

Amanitin is a valuable reagent for studying all aspects of mRNA biosynthesis, including understanding the molecular mechanisms of RNA polymerase II (pol II). X-ray crystallography of pol II complexed with α-amanitin has yielded important insights into pol II action (Brueckner et al. 2009; Brueckner and Cramer 2008; Bushnell et al. 2002; Kaplan et al. 2008; Lu et al. 2015). Bushnell et al. (2002) solved the structure of co-crystals of amanitin and the 10-subunit form of yeast pol II at 2.8 Å resolution. The previously solved structure of a derivative of amanitin, 6′-O-methyl-α-amanitin (S)-sulfoxide, by itself (Wieland et al. 1983) helped to solve the structure of native α-amanitin bound to pol II (after removing the O-methyl group on the Trp residue and changing the *S* configuration of the sulfoxide to the *R* configuration). Amanitin was shown to bind to a site beneath the "bridge helix" (Fig. 5.4). Numerous direct and indirect hydrogen bonds were observed between five of the

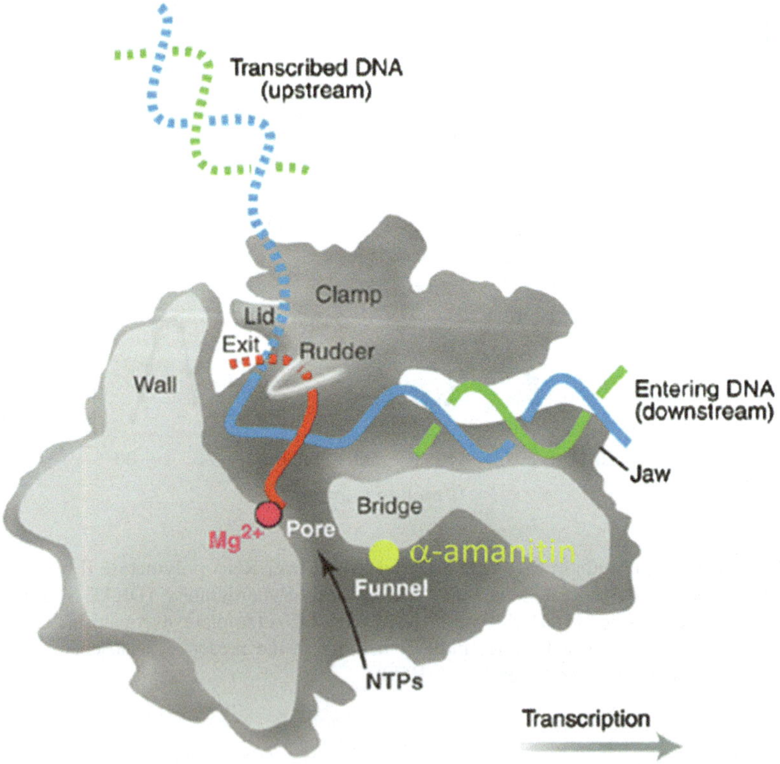

Fig. 5.4 Model of RNA polymerase II (pol II). Double-stranded DNA enters from the right and exits from the upper left. The transcribed RNA copy is shown in red. The binding site of α-amanitin is shown in yellow. (Adapted from Klug 2001 and Bushnell et al. 2002. Reprinted with permission from AAAS Copyright (2002) National Academy of Sciences, USA)

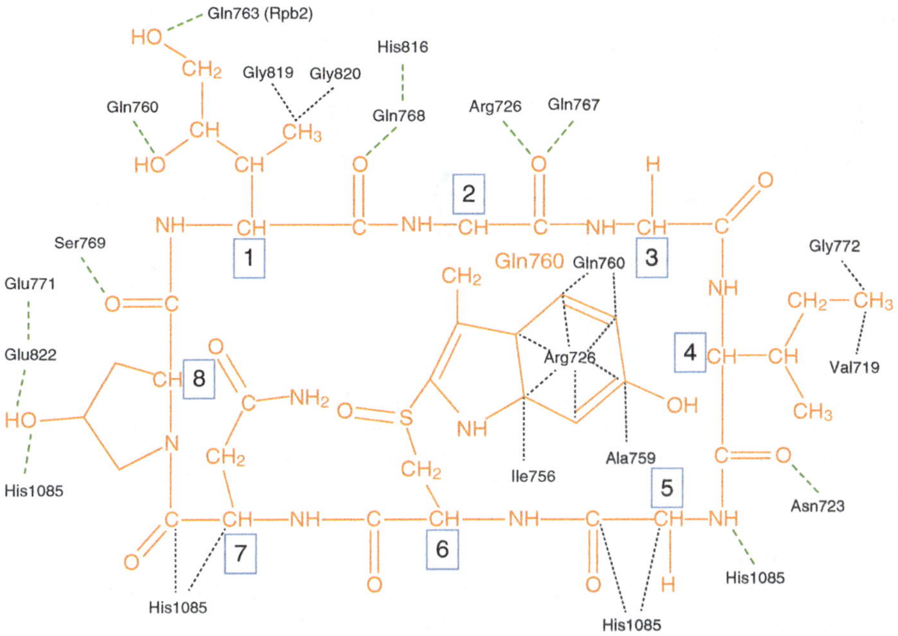

Fig. 5.5 Hydrogen bonds (shown in dashed green lines) and hydrophobic contacts (shown as dotted black lines) between α-amanitin and yeast pol II. All pol II residues are in subunit Rpb1 except Gln763, which is in Rpb2. The numbering of the amanitin residues has been changed from the original to correspond to the conventions used in this book; in the original numbering scheme, Ile #1 was amino acid #3. (Reprinted from Brueckner and Cramer 2008, by permission from Macmillan Publishers Ltd)

eight amino acids of α-amanitin and specific backbone and side chain atoms of pol II (Fig. 5.5) (Brueckner and Cramer 2008; Bushnell et al. 2002). All of the interactions of α-amanitin are with amino acid residues in the Rpb1 subunit of pol II, except for one in Rpb2 (Gln763) (Fig. 5.5). The observed binding site and specific interactions with pol II are consistent with earlier amanitin structure/activity studies of both the toxin and the enzyme. For example, the 4-hydroxyl group of Pro #8, which is essential for toxin activity, forms hydrogen bonds with Glu822 and His1085 (Brueckner and Cramer 2008; Kaplan et al. 2008). Furthermore, most mutations in pol II that confer α-amanitin resistance map to the amanitin binding site in pol II (see below).

The initial structure of the amanitin/pol II co-crystal suggested that there was no hydrogen bond involving the 4-hydroxyl group (also known as the γ hydroxyl) of Ile #1, which is known to be important for potent α-amanitin activity (Wieland 1986). Bushnell et al. (2002) proposed that the failure of this bond to form in the complex might account for why yeast pol II is less sensitive to α-amanitin compared to pol II enzymes from some other organisms, notably mammals. However, in the later refined crystal structure, a hydrogen bond between the γ-hydroxyl and Glu760 was established (Brueckner and Cramer 2008) (Fig. 5.5). There is also a hydrogen bond between the 5-hydroxyl (also known as the δ-hydroxyl) group of Ile #1 with Gln763 of Rpb2,

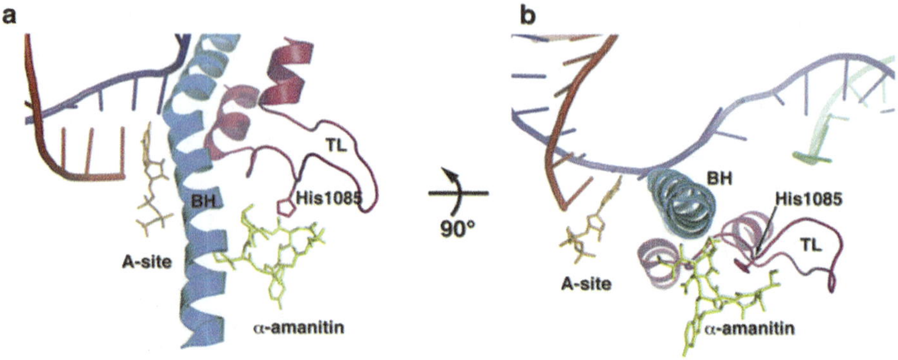

Fig. 5.6 Details of the interaction between α-amanitin and His1085 of the trigger loop (TL) in the Rpb1 subunit of yeast pol II. (**a**) Overall view of α-amanitin and the TL in relation to the bridge helix (BH). (**b**) A 90° rotation showing the α-amanitin position in relation to the BH and His1085. (Reprinted from Kaplan et al. 2008 with permission from Elsevier)

although this hydroxyl group is not required for activity (α-amanitin lacking it is known as γ-amanitin and is fully active; Fig. 2.6). The crystal structures also indicate that the side chain of amino acid #7 does not interact with pol II, which explains why β-amanitin (which has Asp as amino acid #7 instead of Asn as in α-amanitin) is fully active. Finally, the 6-hydroxyl group of Trp does not interact with pol II, which is consistent with the high potency of amanin and amaninamide (Fig. 2.6).

Brueckner and Cramer (2008) produced pol II crystals trapped in pre- and post-translocation states. α-Amanitin locked the elongation complex into a new state. Both they and Kaplan et al. (2008) identified by mutational and structural analysis an additional important residue, His1085, involved in α-amanitin binding and also in both nascent RNA elongation and nucleotide (NTP) substrate selection. His1085 also forms a hydrophobic contact with the backbone carbons of amanitin Gly #5 and a hydrogen bond with the 4-hydroxyl group of Pro #8 (Fig. 5.5) (Brueckner and Cramer 2008). This interaction is important because His1085 is part of the highly conserved and mobile "trigger loop" proposed to be critical for substrate selection and catalysis (Fig. 5.6).

The location of the binding site of α-amanitin in pol II is consistent with studies showing that amanitin blocks translocation of pol II along the transcribed DNA and does not interfere directly with NTP entry or binding (Chafin et al. 1995; Gong et al. 2004; Rudd and Luse 1996). However, Brueckner and Cramer (2008) concluded that amanitin inhibition involves not only stabilization of the conformation of the elongation complex but also interference with NTP incorporation. Collectively, the structural studies of pol II complexed with α-amanitin are consistent with a model in which the toxin inhibits transcription by trapping the bridge helix and the trigger loop such that both NTP incorporation and translocation are compromised (Brueckner et al. 2009; Wang et al. 2006). The importance of trigger loop mobility in proper pol II function is supported by mutational studies with RNA polymerase subunit Rpb9, because α-amanitin suppresses the effect of Rpb9 on NTP misincorporation (Kaster et al. 2016) (Fig. 5.7). A movie of the pol II transcription cycle is

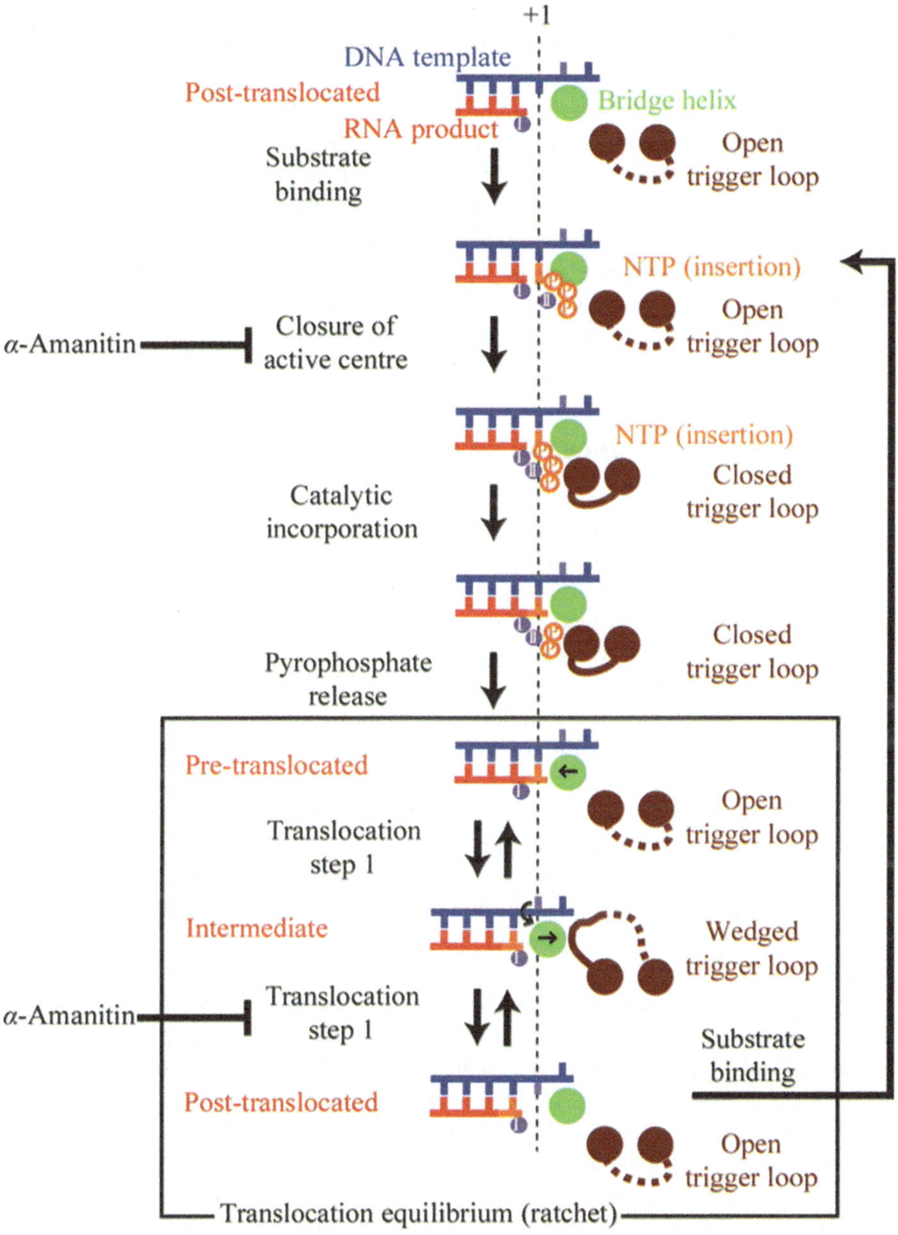

Fig. 5.7 Model of the nucleotide-addition cycle of RNA pol II. The two sites where amanitin interferes are indicated. (From Brueckner et al. 2009, Creative Commons Attribution License)

available from the laboratory of Patrick Cramer at the Max Planck Institute for Biophysical Chemistry (Cheung and Cramer 2012); α-amanitin makes its appearance at minute 4:15.

It is interesting to note that microcin J25, a bicyclic 21-amino acid peptide made by several enterobacterial species, including some strains of *Escherichia coli*, is an inhibitor of bacterial RNA polymerases, which are insensitive to amanitin (Delgado et al. 2001). Microcin J25 is a *lasso peptide* in which the C-terminal tail is threaded through a loop formed by an internal isopeptide bond (Maksimov et al. 2012). Microcin J-25 shares some significant overall structural similarity to α-amanitin, and molecular modeling indicates that it binds to the site on bacterial pol II corresponding to the amanitin binding site in yeast pol II (Gong et al. 2004).

5.4 Genetic Resistance to α-Amanitin

Some organisms are naturally resistant to α-amanitin, notably prokaryotes, unicellular eukaryotes, and perhaps some fungi (see below). In some cases the basis of this resistance can be attributed to intrinsic resistance of pol II, but in most studies the possible existence of extrinsic resistance factors, such as detoxifying enzymes, amanitin-sequestering proteins, or proteins that block amanitin from binding to pol II, cannot be excluded.

A number of organisms and cell lines that are insensitive to α-amanitin have been developed by artificial selection, including strains of mammalian cells, *Drosophila*, yeast, and nematodes. Resistance to α-amanitin due to natural selection in the genus *Drosophila* is considered in Chap. 6. The first successful efforts to induce resistance to amanitin used mammalian cell lines. Despite the fact that most mammalian cell lines lack the specific amanitin transporter, OATP1B3, all mammalian cells are eventually killed by α-amanitin. Addition of amphotericin makes cells more permeable, and hence more sensitive, to amanitin (e.g., Amati et al. 1975; Ben-Zeev and Becker 1977).

Examples of mammalian cells induced to become resistant to amanitin include Chinese hamster ovary (CHO), rat myoblast, and mouse lines (Amati et al. 1975; Bartolomei and Corden 1987, 1995; Chan et al. 1972; Ingles 1978; Somers et al. 1975; Wulf and Bautz 1976). The enhanced resistance of pol II ranged from 5- to 600-fold in different selected cell lines. These lines have been exploited for studying several aspects of pol II biology and biochemistry including regulation of pol II expression, catalytic mechanism, and role in viral replication (Ben-Zeev and Becker 1977; Guialis et al. 1977; Somers et al. 1975). Steeg et al. (1990) demonstrated gene targeting in mammalian cells by introduction of the N792D point mutation conferring amanitin resistance that had been discovered by Bartolomei and Corden (1987).

In every case that has been investigated, the in vitro pol II activity in resistant cell lines was found to be less sensitive to inhibition by α-amanitin, i.e., the basis of resistance of the cells was an alteration of the intrinsic properties of the target enzyme (Lobban et al. 1976). Only small differences in the catalytic properties of the resistant pol II enzymes were detected, e.g., an increased preference for poly[d(AT)] as substrate and somewhat greater thermal sensitivity (Bryant et al. 1977; Lobban et al. 1976).

Artificial selection for amanitin resistance has also been conducted at the whole organism level. After feeding amanitin (8 μg/ml) to 600,000 EMS-mutagenized *Drosophila melanogaster* (fruit fly), nine mutant flies were recovered (Greenleaf et al. 1979). The pol II activity from fly mutant strain C4 was 270-fold less sensitive than the wild-type pol II activity, and the mutation mapped to the gene for the largest subunit of pol II, RpII215 (Greenleaf 1983). Although thermal stability of pol II from the C4 line was unaltered, its activity in the presence of Mg^{2+} instead of Mn^{2+} was lower, its K_m's for the substrates UTP and GTP were higher by twofold, and its elongation rate was reduced (Coulter and Greenleaf 1982, 1985). The catalytic deficiencies of the pol II from the C4 strain of *Drosophila* are manifested in reduced fitness of the C4 strain compared to the wild type in the absence of α-amanitin (Jaenike et al. 1983; Chap. 6).

Like *Drosophila*, the nematode *Caenorhabditis elegans* has proven to be highly suitable for studying mutations that confer α-amanitin resistance in whole organisms. Nematodes were fed α-amanitin at 20 or 40 μg/ml, and multiple alleles at two loci, *ama-1* and *ama-2*, were ultimately identified as well as EMS-induced revertants to sensitivity (Rogalski et al. 1988; Sanford et al. 1983). The *ama-1* locus encodes the largest subunit of pol II, but *ama-2* has apparently not yet been characterized at the molecular level. One of the resistant strains (DR432 with the m118 allele) was subjected to a further round of α-amanitin selection using 100–200 μg/ml amanitin, which resulted in a doubly mutated strain (DR1099 m118 m526) that was 20,000-fold more resistant than the original wild type (Rogalski et al. 1990). The double mutant developed slowly and had low fecundity; the authors suggested that complete abolishment of amanitin binding is irreconcilable with normal pol II function (*A. nidulans* and *N. crassa* may be exceptions to this rule; see below). Both mutations mapped to the gene encoding the largest subunit of pol II, known as *ama-1* or *RPB1*. The phenotype of the single m526 mutant is apparently still uncharacterized (Rogalski et al. 1990).

The specific mutations that give amanitin resistance have been determined in mammalian cells, *Drosophila*, and *C. elegans* (Tables 5.1 and 5.2) (Bartolomei and Corden 1987, 1995; Bowman et al. 2011; Chen et al. 1993). Most of the known mutations confer amanitin resistance map to a region corresponding to yeast amino acids 718–778 in the largest subunit of pol II, which comprises the "funnel" between the bridge helix (BH) and the trigger loop (TL) (Bushnell et al. 2002; Seshadri et al. 2003; Figs. 5.4 and 5.9). Some of the resistance-conferring amino acids in this region are among those that form hydrogen bonds and/or hydrophobic interactions with α-amanitin, i.e., Arg726, Ile756, Gly772, and Ser769 (yeast numbering) (Brueckner and Cramer 2008; Seshadri et al. 2003) (Fig. 5.5). Bartolomei and Corden (1995) suggested that substitutions with the bulkier Phe at Leu745 and Ile779 in mouse (equivalent to Leu722 and Ile756 in yeast) interfere with the hydrophobic binding pocket for amanitin. The *C. elegans* C777Y mutation also causes replacement of a small side chain with a bulky hydrophobic residue (Bowman et al. 2011).

Glu822, which forms a hydrogen bond with the hydroxyl group of amanitin Pro #8, is outside this region in the primary sequence of pol II but also forms part of the amanitin binding site. Another important amino acid outside the 718–780 region is

Table 5.1 Mutations in pol II that confer resistance to amanitin

Organism	Mutation (native numbering)	Known interaction with amanitin	Notes	References
Mouse 1	N792D	Yes		Bartolomei and Corden (1987)
Mouse 2	L745F	No		Bartolomei and Corden (1995)
Mouse 3	R749P	Yes		Bartolomei and Corden (1995)
Mouse 4	I779F	Yes		Bartolomei and Corden (1995)
Nematode 1	C777Y	No	Strain DR432; allele m118	Bowman et al. (2011)
Nematode 2	R739H	Yes	Strain DR786; allele m322	Bowman et al. (2011)
Nematode 3	G785E	Yes	Allele m526; phenotype of single m526 mutant unknown	Rogalski et al. (1990) and Bowman et al. (2011)
Fruit fly	R741H	Yes	C4 strain	Chen et al. (1993)
Yeast	H1085Y	Yes	Enzyme catalytically defective	Kaplan et al. (2008)

Table 5.2 Amino acids in pol II that interact with α-amanitin and/or confer insensitivity when mutated

pol II amino acid (yeast numbering)	Resistance mutation (organism)	Hydrogen bond with amanitin	Hydrophobic contact with amanitin
Val719			x
Leu722	x (mouse)		
Asn723		x	
Arg726	x (mouse, fly, nematode)	x	x
Ile756	x (mouse)		x
Ala759			x
Gln760		x	x
Cys764	x (nematode)		
Gln767		x	
Gln768		x	
Ser769	x (N792 in mouse)	x	
Gly772	x (nematode)		x
Gly819			x
Gly820			x
Glu822		x	
His1085	x (yeast)	x	x
Gln763 (Rpb2)		x	

All pol II amino acids are in subunit Rpb1 except Gln763, which is in Rpb2

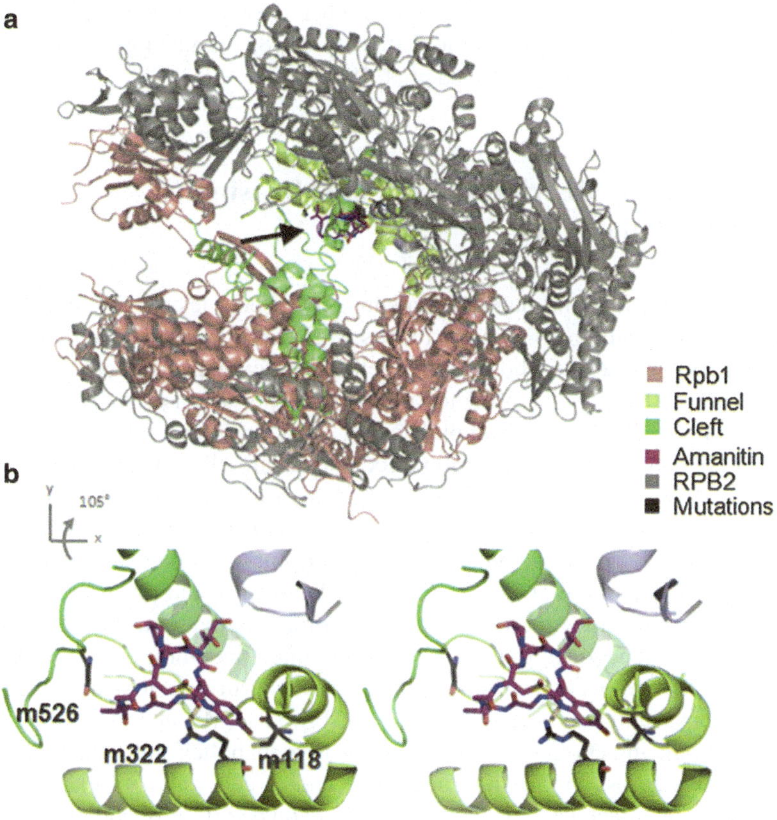

Fig. 5.8 Positions of mutations in *C. elegans* that confer resistance to α-amanitin. (**a**) Location of α-amanitin binding site in yeast Pol II (PDB ID: 3cqz), based on Kaplan et al. (2008). The black arrow points to the α-amanitin binding site between the "funnel" and "cleft" domains shown in light olive and bright green, respectively. (**b**) Stereo image showing the location of resistance-inducing mutations in *C. elegans* (m526, m322, and m118), all of which are within the universally conserved α-amanitin binding site (From Bowman et al. 2011, Creative Commons Attribution, used by the author's permission)

His1052. His1052 is part of the trigger loop and forms a hydrogen bond with α-amanitin. When changed to Tyr in yeast by targeted mutation, the enzyme is crippled, and its residual activity is resistant to amanitin (Brueckner and Cramer 2008; Kaplan et al. 2008).

Bowman et al. (2011) characterized the DNA sequences of 12 of the previously known mutations affecting pol II activity in *C. elegans*. Three of the mutations confer amanitin resistance (Tables 5.1 and 5.2). The m118 allele results in a substitution of Tyr (Y) for Cys (C) at position 777 (equivalent to Cys764 in yeast), the m322 allele has His (H) instead of Arg (R) at position 739, and the m526 allele results in substitution of Glu (E) for Gly (G) at position 785 (Table 5.1, Fig. 5.8). The m322 mutation has also been found in mouse and fly. The m526 allele is especially intriguing because mutations at this site have not been found in other screens for amanitin resistance. It confers super-resistance when combined with m118; the phenotype of the single m526 mutant is not known (Rogalski et al. 1990).

5.5 Taxonomic Distribution of Pol II Sensitivity to the Amatoxins

RNA polymerases from many organisms have been tested for sensitivity to α-amanitin. For a summary of early results, see Wieland (1986). Neither prokaryotic RNA polymerases nor those of organelles (mitochondria and chloroplasts) are inhibited at <100 μM. Some pol I and pol III RNA polymerases show some degree of inhibition at >20 μM. For example, the pol III of *Schizosaccharomyces pombe* is inhibited 50% by 0.4 mg/ml amanitin (Rödicker et al. 1999). (Since the molecular weight of amanitin is 919, 0.4 mg/ml is equivalent to 0.43 mM.) All tested mammalian pol II enzymes are inhibited by 1–30 nM α-amanitin (Wieland and Faulstich 1991). The pol II from nematodes (e.g., *C. elegans*), amphibians, insects (e.g., *Drosophila*), slime molds (*Dictyostelium* and *Physarum*), and plants are also sensitive at nanomolar to low micromolar concentrations. In the gastropod *Aplysia californica*, pol II-dependent transcription is inhibited in whole cells by 2 μg/ml amanitin (Montarolo et al. 1986), which is potentially ecologically significant because many snails and slugs feed on mushrooms (Maunder and Voitk 2010) (Chap. 6). Some intact organisms are less sensitive than their pol II enzymes due to lack of uptake across the plasma membrane and/or cell wall. For example, RNA transcription in intact yeast (*Saccharomyces cerevisiae*) cells and protoplasts is insensitive to amanitin (13% inhibition at 50 μM) although in cell free extracts its pol II is inhibited at ~1 μM (Giami and Simchen 1977).

The pol II enzymes of some organisms are strikingly less sensitive than mammalian pol II to amatoxins. The pol II's of *Tetrahymena pyriformis*, *Trichomonas vaginalis*, *Entamoeba histolytica*, and *Giardia lamblia* are inhibited only partially at >1 μM, >250 μM, >1 mM, and >1 mM, respectively (Quon et al. 1996; Seshadri et al. 2003; Vaňáčová et al. 2001). Protein-encoding gene transcription in nuclear extracts of the causative agent of malaria, *Plasmodium falciparum*, was inhibited by 100 μg/ml α-amanitin, but lower doses were not tested (Militello et al. 2005). All of these organisms are early-branching, unicellular eukaryotes, commonly known as protozoans.

5.6 The Basis of Natural Resistance Among DNA-Dependent RNA Polymerases

Pol II of *Giardia lamblia*, which is naturally highly resistant to α-amanitin, has multiple differences in many of the amino acids that have been shown by mutational studies to be important for sensitivity to amanitin (Seshadri et al. 2003). For example, mutation of Leu722 (yeast numbering) to Phe in mouse cells confers amanitin resistance (Table 5.1); this amino acid is conserved in all pol II's but is altered to Val or Thr in *Giardia* or *Trichomonas vaginalis* (whose pol II is also resistant), respectively (Fig. 5.9). Mutation of Arg726 to Pro or His in mouse or *Drosophila* confers resistance; Arg726 is conserved in all pol II's but is replaced by Leu or Ser in *Trichomonas* or *Giardia* (Fig. 5.9). The same trend generally holds for Ile756 and Gly772 (in the latter amino acid, *Giardia* has Ser, but *Trichomonas* is still Gly). The pattern is more complex for Cys764 and Ser769, where there is less conservation among all pol II's (Fig. 5.9). The correspondence of critical amino acid differences between naturally and artificially selected pol II's suggests that differences in the pol II's of *Giardia* (and perhaps *Trichomonas*) might provide a molecular basis for the resistance of these pol II's to α-amanitin. Interestingly, the pol III of *Giardia* is more sensitive to α-amanitin than pol II (85% inhibition at 50 μg/ml vs. no inhibition at 1 mg/ml) (Seshadri et al. 2003). The possible significance of the correlation between amino acid substitutions and natural and artificial resistance is tempered by the fact that some of the resistance-conferring amino acid substitutions do not occur in the pol III's of *S. cerevisiae* and *H. sapiens*, which are naturally resistant to amanitin (Fig. 5.9). Specifically, these two pol III's have Leu722, Arg726, Ile/Leu756, and Gly772 (yeast numbering), which are characteristic of sensitive pol II's (Fig. 5.9). Furthermore, the overall conservation of the *Giardia* pol II sequence is low compared to other pol II's, including those from other basal eukaryotic lineages, so it cannot be excluded that other amino acid substitutions in *Giardia* might account for, or at least contribute to, its natural resistance to amanitin (Seshadri et al. 2003).

A similar comparison was performed on the five DNA-dependent RNA polymerases of *Arabidopsis* (Haag et al. 2012). In addition to pol I, II, and III found in other eukaryotes, plants also have pol IV and pol V. These two DNA-dependent RNA polymerases generate small interfering RNA (siRNA) involved in gene silencing. Even though they evolved from pol II, pol IV and pol V are resistant to 5 μg/ml α-amanitin, at which concentration *Arabidopsis* pol II is strongly inhibited. Resistance was correlated with differences in several otherwise highly conserved

```
yeast II          718 VVRFLNEARDKAGRLAEVNLKDLNNVKQMVMAGSKGSFINIAQMSACVGQQSVEGKRIAFG 778
mouse II          741 VNRILNDARDKTGSSAQKSLSEYNNFKSMVVSGAKGSKINISQVIAVVGQQNVEGKRIPFG 801
nematode II       731 VNQILNDARDRTGSSAQKSLSEFNNFKSMVVSGAKGSKINISQVIACVGQQNVEGKRIPFG 791
Drosophila II     733 VNRILNDARDKTGGSAKKSLTEYNNLKAMVVSGSKGSNINISQVIACVGQQNVEGKRIPFG 793
Arabidopsis II    737 VNQVLNKARDDAGSSAQKSLAETNNLKAMVTAGSKGSFINISQMTACVGQQNVEGKRIPFG 797
Trichomonas II    736 VNNLTNEILSKTYKVINAKIRGDNSLSEMLSAGSKGADTNMSQIIGVVGQQNMEGKRVKFG 796
Giardia II        843 VISKVSGTSLALEKVITDAAPHRNALLVMINAGSKGKKFNMMQISSSLGQQFLQSKRMPHR 903
Giardia III      1007 LNQILSNVRESCAQIALKELHFTNKPLIMSLCGSKGSPINIAQMIIILGQQSFGGSRAPDD 1067
yeast III         766 IGGLLSKVREEVGDVCINELDNWNAPLIMATCGSKGSTLNVSQMVAVVGQQIISGNRVPDG 826
human III         752 ILKELSVIRDHAGSACLRELDKSNSPLTMALCGSKGSFINISQMIACVGQQAISGSRVPDG 812
                        |         |         |         |         |         |
                       720       730       740       750       760       770
```

Fig. 5.9 Alignment of amanitin binding region of pol II and corresponding regions of pol III from human, yeast, mouse, nematode (*C. elegans*), *Drosophila*, *Arabidopsis*, *Trichomonas*, and *Giardia*. The amanitin-sensitive enzymes are underlined. Amino acid residues in red and underlined are of particular interest

amino acids in the amanitin binding pocket (Haag et al. 2012). Especially intriguing is the finding that pol IV and pol V have Ser898 or Pro849, respectively, in place of His1085 (yeast numbering), because yeast is intolerant of substitutions at this amino acid, which interacts with the trigger loop (Kaplan et al. 2008). However, the low overall conservation (<20%) between the *Arabidopsis* RNA polymerases in the region around His1085 makes it difficult to identify homologous amino acids with any certainty. In addition, like for the study of *Giardia* pol II resistance to amanitin (Seshadri et al. 2003), one or more amino acid differences that are not immediately in the amanitin binding pocket could also influence protein tertiary structure and hence resistance to amanitin. This possibility can only be addressed by the construction and testing of additional specific targeted mutations.

In the pol I complex of yeast, two zinc ribbons of subunit A12.2 are located inside the NTP entry pore of the 190 kDa subunit (RPA190), which is the homolog of Rpb1 in the pol II complex. The C-terminal zinc ribbon of A12.2 blocks the binding site of amanitin and was therefore proposed to account for the relative insensitivity of pol I to α-amanitin (Engel et al. 2013; Fernández-Tornero et al. 2013). However, considering that RPA190 and Rpb1 share only 27% overall amino acid identity, other intrinsic differences between these two proteins might account for, or contribute to, their differential amanitin sensitivity. Of the 16 residues of Rpb1 that are involved in amanitin sensitivity and/or binding (Table 5.2), 9 are conserved between yeast pol I (RPA190) and yeast pol II (Rpb1) (Fig. 5.10). Of the seven differences, one is conservative (I756L), and another (C764V) is also found in several sensitive pol II's including mammalian Rpb1 (Fig. 5.9) (all amino acid numbers are based on yeast Rpb1). It cannot be excluded that the other amino acid differences (e.g., V719G, N723S, A759S, S769I, G819S) contribute to the relative insensitivity of pol I to amanitin, either instead of or in addition to occlusion of the amanitin binding site by subunit A12.2 of pol I (Engel et al. 2013; Fernández-Tornero et al. 2013).

```
pol II   713   SFEDNVVRFLNEARDKAGRLAEVNLKDLNNVKQMVMAGSKGSFINIAQMSACVGQQSVEG   772
               + E  +   L++ R++ G +    L + N    M   GSKGS +N++QM A VGQQ + G
pol I    761   TLEAKIGGLLSKVREEVGDVCINELDNWNAPLIMATCGSKGSTLNVSQMVAVVGQQIISG   820

pol II   773   KRIAFGFVDRTLPHFSKDDYSPESKGFVENSYLRGLTPQEFFFHAMGGREGLIDTAVKTA   832
                R+  GF DR+LPHF K+  +P+SKGFV NS+  GL+P EF FHA+ GREGL+DTAVKTA
pol I    821   NRVPDGFQDRSLPHFPKNSKTPQSKGFVRNSFFSGLSPPEFLFHAISGREGLVDTAVKTA   880

-----------------------------------------

pol II 1072   IGEPATQMTLNTFHFAGVASKKVTSGVPRLKEILNVAKNMKTPSLTVYLEPGHAADQEQA   1131
               IGEP TQMTL TFHFAGVAS  VT GVPR+KEI+N +K + TP +   L   +  D+  A
pol I  1097   IGEPGTQMTLKTFHFAGVASMNVTLGVPRIKEIINASKVISTPIINAVLVNDN--DERAA   1154
```

Fig. 5.10 Alignment of the two major amanitin binding regions of the largest subunit of yeast pol II (Rpb1) with the corresponding regions in yeast pol I. Residues of particular interest are shown in red and underlined. Residues that confer resistance of pol II to amanitin when mutated in at least one organism are shown in green

5.7 Sensitivity of Pol II from Fungi

The question of whether pol II enzymes in fungi are sensitive to amatoxins is relevant to the question of self-protection, i.e., why don't amatoxin-producing mushrooms kill themselves? This question can be subdivided into three. First, are pol II's of fungi in general resistant or sensitive to amanitin? Second, are the pol II's of mushrooms (i.e., fungi in the Agaricales), but not other fungi, resistant? Third, are the pol II's of the cyclic peptide toxin-producing mushrooms uniquely resistant? A positive answer to the first two questions would suggest that the toxin-producing fungi were "preadapted" in this trait prior to evolving the biosynthetic capacity to make the toxins. A positive answer to the third question would suggest that toxin biosynthesis and toxin self-protection evolved concurrently. In natural product biosynthetic pathways in bacteria and filamentous fungi, biosynthetic pathway genes and genes endowing self-protection against the product of that pathway are frequently clustered and therefore co-inherited (Desjardins and Proctor 2007; Walton 2000).

The pol II's of the early-branching fungus *Blastocladiella emersonii* (phylum Blastocladiomycota, order Blastocladiales) (Fig. 1.3) and its close relative *Allomyces arbuscula* are highly sensitive to α-amanitin. Partially purified enzymes (i.e., separated from pol I and pol III by anion exchange chromatography) are strongly inhibited by 0.27 μM α-amanitin (Cain and Nester 1973; Horgen and Griffin 1971). Pol II of *Mucor rouxii* (a member of the early-branching fungal order *Mucorales*) is inhibited 90% by 5 μM α-amanitin and therefore qualifies as sensitive (Young and Whiteley 1975). The sensitivity of partially or completely purified pol II of the single-cell ascomycete *Saccharomyces cerevisiae* (budding yeast; phylum Ascomycota, order Saccharomycetales) has been measured in several studies. Schultz and Hall (1976) found an IC_{50} of 1.1 μM, whereas Hager et al. (1977) found an IC_{50} of 10 μM. That is, yeast pol II qualifies as sensitive but 10–100 times less so than mammalian pol II. The purified pol II of the fission yeast *Candida utilis* (also in the order Saccharomycetales of the phylum Ascomycota) is inhibited at 2 μM (Patturajan 1995).

Within subphylum Pezizomycotina of the Ascomycota (Fig. 1.3) (often referred to inaccurately as the "filamentous fungi", inasmuch as the growth habit of many fungi in the Basidiomycota is also predominantly filamentous), pol II sensitivity to amanitin has apparently been studied only in *Neurospora crassa* and *Aspergillus nidulans*. Tellez de Iñon et al. (1974) separated the RNA polymerase activities of *N. crassa* by anion exchange chromatography into four peaks. Two peaks were sensitive to α-amanitin; one was inhibited by 56% and the other (presumably pol II) by 97% at 14 μM α-amanitin. In another study, RNA polymerase activities from *N. crassa* were separated into two peaks, identified as pol I and pol II (Timberlake and Turian 1974). The pol II peak was inhibited 94% by 44 μM α-amanitin, whereas the other peak, probably pol I, was inhibited only 6% at the same concentration. Tyler and Giles (1985) found that pol II-mediated transcription of two protein-encoding genes in soluble nuclear extracts from *N. crassa* was inhibited only by 50–80% at the very high α-amanitin concentration of 1.1 mM. Loros and Dunlap (1991) found inhibition of mRNA transcription of two clock-controlled genes in crude nuclear extracts of *N. crassa* at 1.1 mM but not by 54 μM α-amanitin. Collectively, the studies with *N. crassa* suggest that its pol II is much less sensitive compared to the pol II's of lower fungi, yeast, insects, and mammals.

The most convincing evidence that the pol II of at least some filamentous ascomycetes is strongly resistant to amatoxins comes from the work of Stunnenberg et al. (1981). Pol II from *A. nidulans* that had been carefully purified to homogeneity was not inhibited by 440 μM α-amanitin nor did it bind O-[^{14}C]-methyl-γ-amanitin (i.e., γ-amanitin with a [^{14}C]methyl group on the 6-hydroxyl group of Trp, a compound shown to be biologically active; Chap. 2). In parallel experiments, rat liver pol II showed the expected half-maximal inhibition by α-amanitin at ~20 ng/ml.

In conclusion, several studies convincingly indicate that the pol II's of early-branching fungi and unicellular yeasts are sensitive to amanitin, but that the pol II's of filamentous ascomycetes are somewhat or very resistant to amatoxins. The study with highly purified pol II from *A. nidulans* constitutes the most convincing evidence that pol II in at least this one fungus is intrinsically highly resistant to amanitin (Stunnenberg et al. 1981). However, in most of the other relevant studies, intrinsic vs. indirect resistance, such as restricted ability of amanitin to enter nuclei or the presence of proteins that bind or otherwise inactivate the toxins, cannot be differentiated (see Tyler and Giles 1985). It is not known if *Neurospora* and *Aspergillus* are representative of other filamentous ascomycetes. Additional studies with highly purified pol II from additional species in the Pezizomycotina will be necessary to answer this question.

5.7.1 Sensitivity of Pol II from Agarics (Mushrooms)

There have been several studies on the effects of amatoxins on the RNA polymerases of the Agaricales, which taxon includes all of the known amatoxin-producing fungi. The answer to this question is important for understanding the basis of self-protection and the coevolution of amatoxin biosynthesis and its biochemical target. Horgen et al. (1978) tested total RNA polymerase activity in isolated nuclei from *Amanita muscaria* and *Agaricus bisporus* (neither of which produces amatoxins) and *A. phalloides* (which does). *Achlya ambisexualis* and rabbit nuclei were used as amanitin-sensitive controls (IC_{50}'s <0.25 μg/ml). All three agarics were inhibited only at α-amanitin concentrations above 25 μg/ml (i.e., 27 μM). At 125 μg/ml, the RNA polymerase activity of *A. bisporus* was inhibited by 70%, and even at 300 μg/ml, the activity of *A. phalloides* was inhibited by only 49%. In light of later studies described below, it is significant that Horgen et al. (1978) found no major difference in sensitivity between the toxin-producing and nonproducing agarics. Because RNA polymerase activity was measured in whole nuclei, interpretation of these results is confounded by the presence of pol I and pol III, which contributed an unknown proportion of intrinsically amanitin-resistant activity to the total polymerase activity. Also, as the authors discuss, the experiments could not exclude alternate explanations for the apparent resistance, e.g., the nuclei of agarics may be poorly permeable to amanitin.

Pol II was purified to homogeneity from the common button mushroom *Agaricus bisporus* and shown to be half maximally inhibited by 7 μM α-amanitin (Vaisius and Horgen 1979). Although this IC_{50} qualifies as resistant compared to mammalian pol II, it is in the range for other fungi including budding yeast (IC_{50} 1–10 μM). This excellent work convincingly demonstrates that at least some pol II enzymes from agarics are rather insensitive to α-amanitin compared to mammals and lower fungi, although more sensitive than the pol II from *A. nidulans*.

Of particular interest to the evolution of toxin production and the mechanism of self-protection is whether the pol II's of amatoxin-producing and nonproducing agarics differ in sensitivity. Johnson and Preston (1979) compared total RNA polymerase activity in extracts of isolated nuclei from basidiocarps (fruiting bodies) or cultures of several species of *Amanita*. The RNA polymerase activities of two toxin-producing fungi (*A. suballiacea* and *A. hygroscopica*, both in sect. *Phalloideae*) were inhibited half maximally at 1.4–1.5 mM α-amanitin. In contrast, the activities of two toxin-nonproducing fungi (*A. brunnescens* and *A. solitaria*, in sect. *Validae* and *Lepidella*, respectively) were inhibited at 10–32 μM. In other words, the pol II activities of the toxin-producing mushrooms were ~45 to 150-fold more resistant than those from the nonproducers. This study did not attempt to address the resistance mechanism, i.e., whether it is intrinsic to pol II or whether it is due to another factor such as a nuclear amanitin binding protein.

In the study of Johnson and Preston (1979), RNA polymerase activity was assayed in isolated nuclei and therefore could not distinguish between pol I, II, and III. In a later study (Johnson and Preston 1980), the authors fractionated the RNA polymerase activities of four species of *Amanita* by the anion exchange method developed by Lindell et al. (1970). The peaks of activity corresponding to pol II from the two toxin-producing species (*A. suballiacea* and *A. hygroscopica*) were very resistant (IC_{50}'s 2.0–3.3 mM), whereas the pol II's from the toxin nonproducers (*A. solitaria* and *A. alliacea*) were about as sensitive as the pol II's of yeast and *Agaricus bisporus* (9.8 μM and 10 μM, respectively). That is, the pol II's from the two toxin producers were ~240- to 300-fold less sensitive than the two toxin nonproducers. The sensitivity of the nonproducers was comparable to *Agaricus bisporus* and yeast, and therefore it is the pol II's of the toxin-producing fungi that show anomalous behavior. The high degree of purity obtained for the pol II of *A. suballiacea* argues that resistance is probably an intrinsic property of the pol II.

Johnson and Preston (1979) reported that their culture of *A. hygroscopica* grew quickly on defined medium, which is atypical behavior for a species of *Amanita*. As discussed in Chap. 3, this culture was actually a toxin-nonproducing polypore, not a species of *Amanita*. This misidentification undermines the validity of the conclusion that the traits of pol II resistance and cyclic peptide toxin production are strictly correlated among the agarics.

In conclusion, in regard to amanitin sensitivity of pol II among the fungi, lower fungi are almost as sensitive as mammals (IC_{50}'s ~10 to 50 nM). Fungi in the phylum Ascomycota, including yeast, are more resistant than mammals, but the published values range widely (between 10 and >400 μM). Agarics (which belong to the phylum Basidiomycota) show a similar degree of sensitivity as yeast (~10 μM), with the possible exception of agarics that produce α-amanitin. However, the evidence that the pol II's of α-amanitin-producing species of *Amanita* are highly resistant (IC_{50} >1 mM) is far from conclusive. Furthermore, in most studies, intrinsic vs. extrinsic mechanisms of resistance cannot be distinguished.

5.7.2 Resistance of Agaric Pol II Revisited

As discussed above, the pol II activity of the mushroom *Agaricus bisporus* is somewhat insensitive to amanitin (IC_{50} 7 μM), and the pol II of *A. nidulans* is highly insensitive (IC_{50} >400 μM) (Stunnenberg et al. 1981; Vaisius and Horgen 1979). In addition, the pol II activities of two amanitin-producing species of *Amanita* were claimed to be resistant to >2 mM α-amanitin (Johnson and Preston 1980). Since these studies were published, there has been much progress on elucidating the details of the interaction between amanitin and pol II and on the identification of pol II amino acids that are important for binding to α-amanitin and/or that confer resistance to amanitin. It is therefore now possible to ask if the pol II's of agarics, especially amanitin-producing species, differ in key amino acids that are known to confer insensitivity when mutated, or differ in key amino acids that form contacts with α-amanitin.

Hydrogen bonding and hydrophobic contacts between pol II and α-amanitin cluster in four regions of pol II (Brueckner and Cramer 2008; Figs. 5.5 and 5.11). In the largest subunit of pol II, these clusters are between Val718 (yeast numbering) and Gly778, in the vicinity of Glu822, and at His1085. In addition, α-amanitin contacts Rpb2 at Gln763. All known mutations that confer α-amanitin resistance map to the Val718-Gly778 region and His1085. Alignment of these critical amino acids and surrounding regions indicates a high degree of conservation among all fungal pol II's (Fig. 5.11). Of particular significance to the question of whether α-amanitin-producing fungi are insensitive to amanitin, there are zero amino acid differences between agarics that make amanitin (i.e., *Amanita bisporigera*, *A. phalloides*, *Galerina marginata*, and *Lepiota subincarnata*) and those that do not (*Agaricus bisporus*, *A. muscaria*, and *A. thiersii*). All seven agarics have identical amino acids at every position known to be involved in binding or sensitivity to amanitin (Fig. 5.11).

The pol II activities of the ascomycetes *Neurospora crassa* and *Aspergillus nidulans* are insensitive to amanitin (Stunnenberg et al. 1981; Tyler and Giles 1985). Comparisons of the known amanitin-responsive sites in these two fungi provide no evidence for intrinsic resistance (Fig. 5.11).

In conclusion, if the pol II's of any fungi are, in fact, insensitive to amanitin, it cannot be due to amino acid substitutions at sites known from structural and mutational studies to be important for binding or sensitivity to amanitin. Of course, insensitivity could be due to any of a number of other factors, such as amino acid

A. Core amanitin binding region (Rpb1)

```
yeast        718  VVRFLNEARDKAG...(17)...MVMAGSKGSFINIAQMSACVGQQSVEGKRIAFG 778
mouse             VNRILNDARDKTG           MVVSGAKGSKINISQVIAVVGQQNVEGKRIPFG
worm              VNQILNDARDRTG           MVVSGAKGSKINISQVIACVGQQNVEGKRIPFG
Giardia           VISKVSGTSLALE           MINAGSKGKKFNMMQISSSLGQQFLQSKRMPHR
Entamoeba         NECTSNVIKECIP           MTTTGAKGSKVNHSQITCLLGQQELEGRRVPMA
A. nidulans       VSVALNTARDEAG           MSRSGSKGSIINISQMTAVVGQQMVEGKRIAFG
N. crassa         VSKALNTARDKAG           MALSGSKGSSINISQMTALVGQQIVEGKRIPF
Agaricus          VERQLNLARDTSG           MVVAGSKGSFINISQMSVCVGQQSVEGRRIPFG
A. muscaria       VERELNLARDRSG           MVVAGSKGSFINISQMSVCVGQQSVEGRRIPFG
A. thiersii       VERELNLARDRSG           MVVAGSKGSFINISQMSVCVGQQSVEGRRIPFG
A. bisporigera    AERELNLARDRSG           MVVAGSKGSFINISQMSVCVGQQSVEGRRIPFG
A. phalloides     AERELNLARDRSG           MVVAGSKGSFINISQMSVCVGQQSVEGRRIPFG
Galerina          VERQLNLARDKSG           MVVAGSKGSFINISQMSVCVGQQSVEGRRIPFG
Lepiota           VERELNLARDQPG           MVVAGSKGSFINISQMSVCVGQQSVEGRRIPFG
```

C. Region around His1085 (Rpb1)

```
yeast       1071 SIGEPATQMTLNTFHFAGVASKKVTSGVPRLKEIL 1105
mouse            SLGEPATQMTLNTFHYAGVSAKNVTLGVPRLKELI
worm             SLGEPATQMTLNTFHYAGVSAKNVTLGVPRLKEII
Giardia          STGEPSTQLTLNTFHLAGHGGVNATMGVPRLREII
Entamoeba        SIGAPATQMTLNTFHHAGISSASVTEGVPRLKELL
A. nidulans      SIGEPATQMTLNTFHFAGVSSKNVTLGVPRLKEIL
N. crassa        SIGEPATQMTLNTFHFAGVSSKNVTLGVPRLKEIL
Agaricus         SIGEPATQMTLNTFHYAGVSSKNVTLGVPRLKEII
A. muscaria      SIGEPATQMTLNTFHYAGVSSKNVTLGVPRLKEII
A. thiersii      SIGEPATQMTLNTFHYAGVSSKNVTLGVPRLKEII
A. bisporigera   SIGEPATQMTLNTFHYAGVSSKNVTLGVPRLKEII
A. phalloides    SIGEPATQMTLNTFHYAGVSSKNVTLGVPRLKEII
Galerina         SIGEPATQMTLNTFHYAGVSSKNVTLGVPRLKEII
Lepiota          SIGEPATQMTLNTFHYAGVSSKNVTLGVPRLKEII
```

B. Region around Glu822 (Rpb1)

```
yeast        802 NSYLRGLTPQEFFFHAMGGREGLIDTAVKTAETGYIQRRLV 842
mouse            NSYLAGLTPTEFFFHAMGGREGLIDTAVKTAETGYIQRRLI
worm             NSYLAGLTPSEFFFHAMGGREGLIDTAVKTAETGYIQRRLI
Giardia          NSFIRGLNPAEFYFHAVGGREGVCDTAVKTADSGYVERKLL
Entamoeba        NSYVMGLSPQEFFFHAMAGRQGIIDTAVKTSDTGYIQRKMV
A. nidulans      NSYLRGLTPTEFFFHAMAGREGLIDTAVKTAETGYIQRKLV
N. crassa        NSYLRGLTPSEFFFHAMAGREGLIDTAVKTAETGYIQRRLV
Agaricus         NSYLRGLTPQEFFFHAMAGREGLIDTAVKTAETGYIQRRLV
A. muscaria      NSYLRGLTPQEFFFHAMAGREGLIDTAVKTAETGYIQRRLV
A. thiersii      NSYLRGLTPQEFFFHAMAGREGLIDTAVKTAETGYIQRRLV
A. bisporigera   NSYLRGLTPQEFFFHAMAGREGLIDTAVKTAETGYIQRRLV
A. phalloides    NSYLRGLTPQEFFFHAMAGREGLIDTAVKTAETGYIQRRLV
Galerina         NSYLRGLTPQEFFFHAMAGREGLIDTAVKTAETGYIQRRLV
Lepiota          NSYLRGLTPQEFFFHAMAGREGLIDTAVKTAETGYIQRRLV
```

D. Region around Gln763 (Rpb2)

```
yeast        745 PSMILGVAASIIPFPDHNQSPRNTYQSAMGKQAMG 779
mouse            PSMILGVCASIIPFPDHNQSPRNTYQSAMGKQAMG
worm             PAMILGVCASIIPFPDHNQSPRNTYQSAMGKQAMG
Giardia          PSMVLSAITSLIPFPDHNQSPRNLYQASMGKQAVG
Entamoeba        PALVMGVLGSIIPFPDHNQSPRNTYQCAMGKQAIG
A. nidulans      PSMILGVCASIIPFPDHNQSPRNTYQSAMGKQAMG
N. crassa        PSMLLGICASIIPFPDHNQSPRNTYQSAMGKQAMG
Agaricus         PSMILGICASIIPFPDHNQSPRNTYQSAMAKQAMG
A. muscaria      PSMILGICASIIPFPDHNQSPRNTYQSAMGKQAMG
A. thiersii      PSMILGICASIIPFPDHNQSPRNTYQSAMGKQAMG
A. bisporigera   PSMILGICASIIPFPDHNQSPRNTYQSAMGKQAMG
A. phalloides    PSMILGICASIIPFPDHNQSPRNTYQSAMGKQAMG
Galerina         PSMILGICASIIPFPDHNQSPRNTYQSAMGKQAMG
Lepiota          PSMILGICASIIPFPDHNQKSPRNTYQSAMGQAMG
```

Fig. 5.11 Alignment of native pol II sequences from regions of the enzyme that contain the amino acids that interact with α-amanitin and/or are known, when mutated, to confer insensitivity to α-amanitin. Important residues are highlighted in yellow. All sequences were obtained from GenBank except *A. thiersii* (JGI), *A. bisporigera* and *A. phalloides* (Pulman et al. 2016), and *Lepiota subincarnata* (unpublished results from the author's lab). Yeast is *Saccharomyces cerevisiae*; mouse is *Mus musculus*; worm is *Caenorhabditis elegans*; *Giardia* is *G. lamblia*; *Entamoeba* is *E. histolytica*; *A. nidulans* is *Aspergillus*; *N. crassa* is *Neurospora*; *Agaricus* is *A. bisporus*; *A. muscaria*, *A. thiersii*, *A. bisporigera*, and *A. phalloides* are all *Amanita*; *Galerina* is *G. marginata*; and *Lepiota* is *L. subincarnata*

changes outside the immediate amanitin binding region that influence overall tertiary structure of pol II, extrinsic amanitin binding proteins, or detoxifying enzymes.

5.8 Actin as the Site of Action of the Phallotoxins

As discussed elsewhere, the phallotoxins such as phallacidin and phalloidin do not contribute to the toxicity of mushrooms ingested orally in any organism. However, the phallotoxins are very toxic when injected into whole animals and to isolated cells. In fact, when injected they kill animals faster (~2 to 5 h) than the amatoxins (>30 h) (Wieland 1986). The ED_{50} of phalloidin for rats and mice is 1.2–2.1 mg/kg, which is intermediate between the concentration of α-amanitin needed to kill mice (0.6 mg/kg) and the concentration of amanitin needed to kill rats (3.1 mg/kg (Wieland 1986). The reason that phallotoxins are not toxic when ingested orally is probably due to lack of uptake by ileal cells of the small intestine (Petzinger et al. 1982) (see above).

The toxicity of the phallotoxins has been clearly established to be due to its ability to bind to actin. Actin is a protein that forms microfilaments of the cytoskeleton and the contractile apparatus of muscle and other cells and is involved in many types of cellular movement, including cell motility, cytokinesis, vesicle and organelle movement, and overall cell shape. Actin exists in a dynamic equilibrium between the monomeric form (G-actin) and the filamentous form (F-actin). In the presence of phalloidin, the equilibrium between G-actin and F-actin is shifted toward F-actin (Cooper 1987). That is, phalloidin acts as an enhancer of actin polymerization.

Due to its specific and dramatic effects on actin, phalloidin rivals α-amanitin for its value as a reagent for basic research. It has been used in thousands of studies to explore the manifold roles of actin in cell biology. Fluorescent derivatives of phalloidin have been especially useful to visualize the actin cytoskeleton in permeabilized fixed cells (Faulstich et al. 1988; Small et al. 1999) (Figs. 5.12, 5.13, and 5.14). Gold-labeled derivatives of phalloidin are used in electron microcopy, and biotinylated phalloidin can be visualized with streptavidin coupled to colloidal gold or quantum dots (Faulstich et al. 1989; Izdebska et al. 2013; Lachapelle and Aldrich 1988).

The amino acid sequence of actin is highly conserved throughout the eukaryotes, and therefore phalloidin can be used as a reagent with many cell types, including plants, fungi, algae, oomycetes, apicomplexan parasites, mammals, and slime molds (e.g., Hoch and Staples 1983; Holzinger and Blaas 2016; Lovy-Wheeler et al. 2005; Nothnagel et al. 1981; Opalski et al. 2005; Skillman et al. 2011; Van Gestel et al. 2001; Verderame et al. 1980; Von Olenhusen and Wohlfarth-Bottermann 1979; Wulf et al. 1979; Yanagida et al. 1984; Yu et al. 2004; see www.cytoskeleton.com). Phalloidin binds to all forms of actin in all cell types and locations, including muscular, cytoplasmic, and nuclear (Hendzel 2014). It also promotes actin polymerization in vitro (Lengsfeld et al. 1974). However, it has been argued that the precise

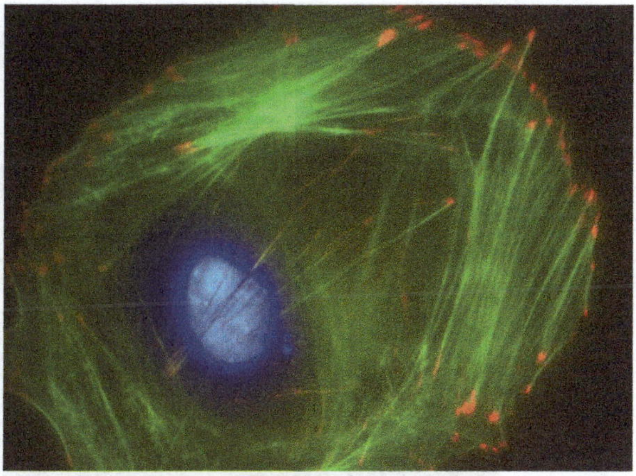

Fig. 5.12 Visualization of actin stress fibers in a Swiss 3T3 (mouse) cell with Acti-stain™ 488 phalloidin (green). Vinculin visualized with anti-vinculin antibody appears red and the nucleus stained with DAPI appears blue. (Photo credit: Cytoskeleton, Inc. (www.cytoskeleton.com), used with permission)

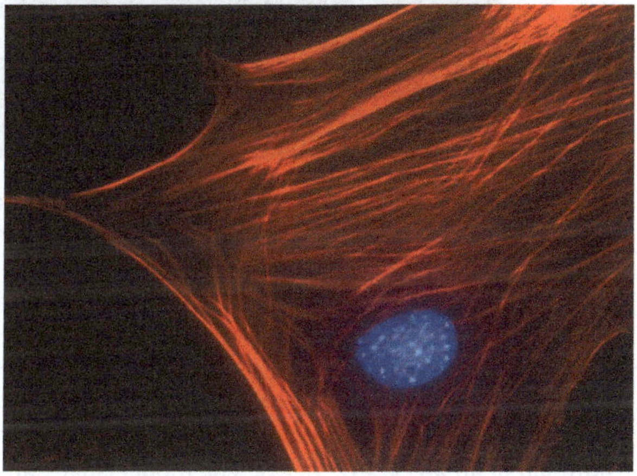

Fig. 5.13 Visualization of actin stress fibers in a Swiss 3T3 (mouse) cell with Acti-stain™ 555 phalloidin (red). The nucleus is visualized with DAPI (blue). (Photo credit: Cytoskeleton, Inc. (www.cytoskeleton.com), used with permission)

reason why binding of phalloidin to actin is so cytotoxic remains to be fully understood (Cooper 1987; Frimmer 1987).

Phalloidin binds to actin with an affinity (K_D) of ~20 nM. It reacts with amino acids Glu117, Met119, and Met355, which are in a highly conserved region in the cleft between the two domains of the actin monomer (Vandekerckhove et al. 1985). X-ray fiber diffraction, electron microscopy of gold-tagged phalloidin, and molecular modeling and computation docking have been applied to the study of the interaction of phalloidin with actin (Falcigno et al. 2001; Oda et al. 2005; Skillman et al. 2011; Steinmetz et al. 1998). The structural studies are in good agreement with each other and with earlier work. Amino acids Ala #1, Trp #2, Cys #6, and Pro #7, which form one face of the phalloidin molecule, were earlier concluded to be more important than the amino acids on the other face (Leu #3, Val/Ala #4, and Asp/Thr #5) on the other side (Wieland 1986). The more recent studies also predict maximum interaction between the Ala-Trp-Cys-Pro face and actin. The essential hydroxyl group of

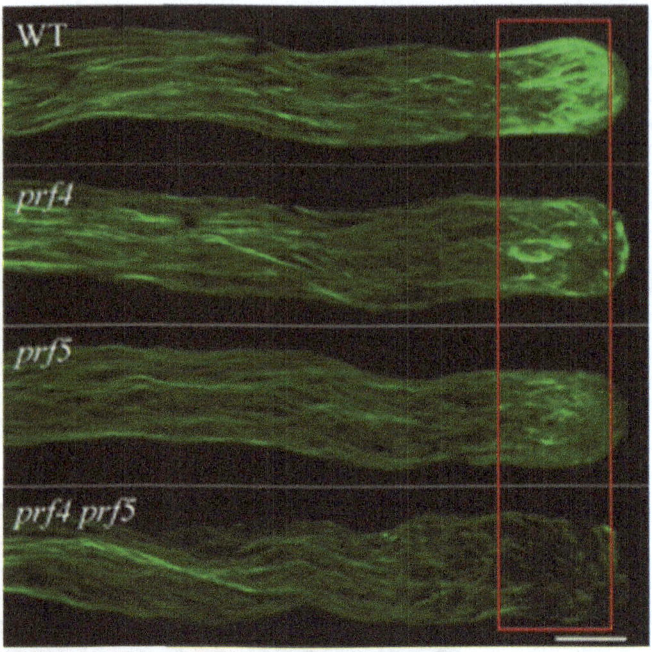

Fig. 5.14 Staining of actin filaments with Alexa Fluor™ 488 phalloidin in pollen tubes of *Arabidopsis* wild-type and profilin mutants. Scale bar is 5 μm. (Reprinted from Liu et al. 2015 with permission from Elsevier)

Pro #7 is predicted to form hydrogen bonds with amino acids in the actin from mammalian muscle, *Toxoplasma,* and *Plasmodium*, whereas the inessential γ-hydroxyl group of Leu #3 faces away from the binding pocket and into the solvent (Skillman et al. 2011; Fig. 5.15). The models also show hydrogen bonding between actin and the hydroxyl group of D-Thr, which is consistent with earlier work showing that synthetic phalloidins in which D-Thr is replaced with D-amino-isobutyric acid (Abu) have reduced actin binding (Falcigno et al. 2001).

The virotoxins, which are monocyclic variants of the phallotoxins, also bind and stabilize F-actin and in all other ways that have been studied behave biologically like the phallotoxins (Wieland 1986).

5.9 Biological Activities of the Cycloamanides

In addition to the bicyclic amatoxins and phallotoxins, *A. phalloides* also makes cycloamanides, which are homodetic cyclic hexa- to decapeptides (Table 2.2). Originally, the cycloamanide family included CylA, CylB, CylC, and CylD and the cyclic decapeptide antamanide (Wieland 1986). Additional cycloamanides from

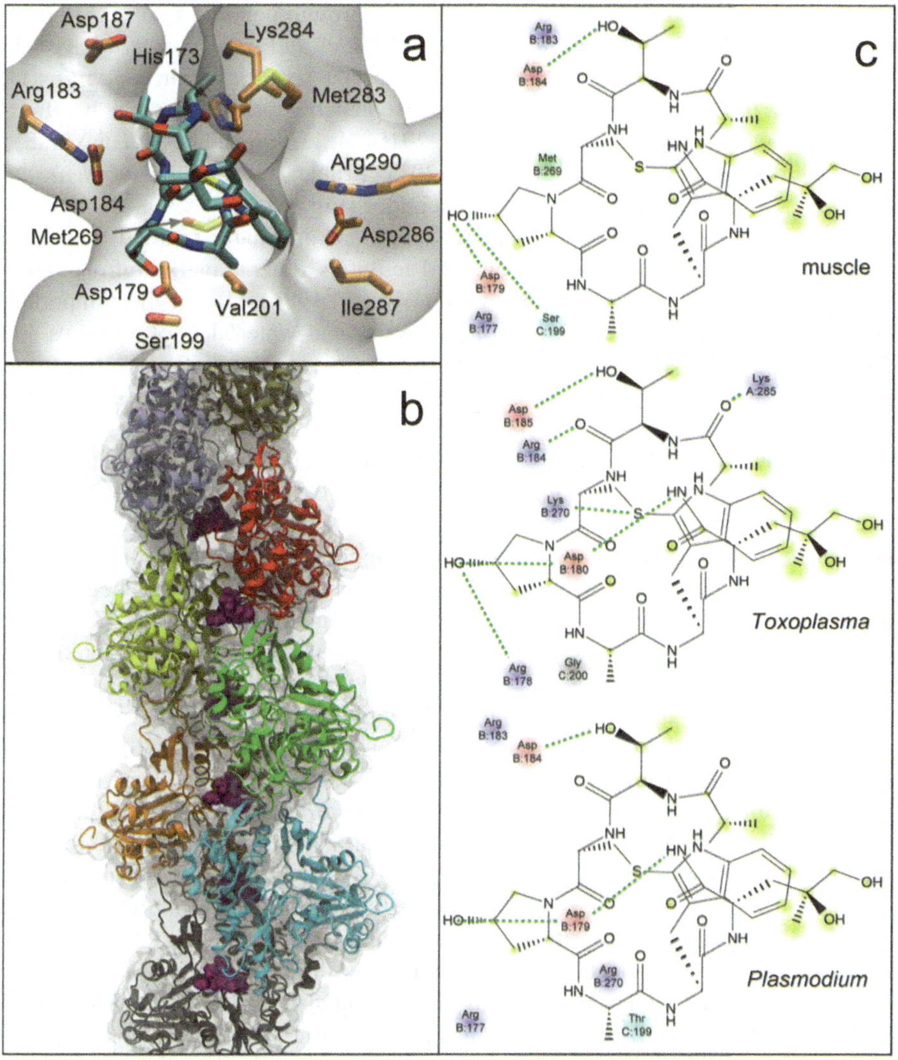

Fig. 5.15 Predicted binding site of phalloidin in actin filaments based on molecular modeling and computational docking. (**a**) Details of the binding pocket for phalloidin in actin, showing side chains of nearby amino acids. (**b**) Binding position of phalloidin (small purple structures) in the actin filament. (**c**) 2D interaction diagrams showing hydrogen bonds between phalloidin and actin (dashed green lines) and solvent-accessible residues (yellow). The three actins are from mammalian muscle, *Toxoplasma gondii* (TgACTI), or *Plasmodium falciparum* (PfACTII), respectively. (From Skillman et al. 2011, with permission of the author. Reprinted under Creative Commons Attribution license)

A. phalloides (CylE and CylF) were discovered through genome analysis (Pulman et al. 2016) (Table 2.2). Between them, *A. phalloides* and *A. bisporigera* have ~55 unique cycloamanide genes, suggesting that the 70+ species within *Amanita* sect. *Phalloideae* of *Amanita* may have the collective genetic potential to make hundreds of additional cycloamanides (Pulman et al. 2016). *A. exitialis* makes amanexitide, a homodetic cyclic nonapeptide (Xue et al. 2011) (Table 2.2). *Lepiota subincarnata* also makes a family of homodetic cyclic peptides (Fig. 4.13).

The cycloamanides other than antamanide were originally purified with no a priori knowledge of their possible biological activities, other than the ability of antamanide to protect mice against phalloidin (Wieland 1986). One of their important chemical attributes is reduced solubility in water and water/alcohol mixtures and enhanced solubility in nonpolar solvents such as acetone and ethyl acetate compared to the amatoxins and phallotoxins. This increased lipophilicity can be expected to have a major impact on their behaviors in living cells and organisms.

There have been relatively few studies on the biological activities of antamanide and the other homodetic cycloamanides. The protective effect of antamanide is due to competitive inhibition of phalloidin uptake through the OATP1B1 transporter of hepatocytes (Fehrenbach et al. 2003). Antamanide does not protect mice against injection with α-amanitin (Wieland 1986), even though it does protects hepatocytes as well as kidney cells expressing the major amanitin transporter, OATP1B3 (Letschert et al. 2006). A possible explanation for this paradox is that excretion of amanitin from the liver into the bile, as part of the enterohepatic circulatory system, is also inhibited by antamanide, and thus although uptake of α-amanitin is blocked by antamanide, so is the ability of the liver to remove it (Wieland 1986).

A number of derivatives of antamanide have been synthesized and studied by physicochemical methods. Structural requirements for anti-phalloidin activity have been elucidated by the chemical synthesis of compounds with amino acid replacements. Antamanide forms complexes with alkali metals such as sodium and calcium (Wieland 1986). Azzolin et al. (2011) showed that antamanide inhibits the mitochondrial permeability transition pore, which blocks cell death mediated by cyclophilin D. The permeability transition pore is implicated in various neurodegenerative diseases such as muscular dystrophy and also some cancers. Since phalloidin does not cause the same pore transition, the effect on pores is distinct from antamanide's inhibition of phalloidin uptake by hepatocytes (Azzolin et al. 2011).

Nielsen (1986) reported that when given at a dose of 3.5 mg per day for 4 days, antamanide prolonged survival of a mouse strain (AKR) that is prone to spontaneous leukemia, but there have apparently been no follow-up studies on this observation. Ruzza et al. (1999) synthesized a number of antamanide derivatives and tested them for cytotoxicity against F10 cells from B16 murine melanoma. In their assay, antamanide had an IC_{50} of ~10 μM. Antamanide and phalloidin reduce edema in rats induced by interleukin 2 (IL-2) (Welbourn et al. 1985).

Cycloamanides including antamanide and some synthetic derivatives show immunosuppressive activity in both in vivo and in vitro assays (Siemion et al. 1992; Thell et al. 2014; Wieczorek et al. 1993). Assays included graft-versus-host reaction, delayed-type hypersensitivity, in vitro humoral immune response, and an autologous rosette formation cell test with thymocytes. In some assays the potencies of some of the cycloamanides, especially CylA, were comparable to cyclosporin A, which is an immunosuppressant in wide clinical use (Wieczorek et al. 1993).

5.10 Conclusion

It is just one of the many remarkable traits of the *Amanita* cyclic peptides that the two major groups, the amatoxins and the phallotoxins, made by the same organisms through the same biosynthetic pathway and having a strong degree of chemical similarity, should attack completely different cellular targets with high potency and specificity.

The importance for cell viability of the targets of the amatoxins and phallotoxins makes them very toxic to most eukaryotic cells and organisms. Neither group of compounds has been shown to have any effects on prokaryotes. The very specific and unique modes of action of these compounds have made them valuable research tools for studying gene transcription and cell motility. Little is known about the biological activities, if any, of the majority of the members of the extended cycloamanide family of ribosomally encoded peptides, whose numbers might reach into hundreds within *Amanita* sect. *Phalloideae* (Pulman et al. 2016). It is plausible that some of them will eventually be shown to have similar levels of potency and specificity comparable to the better known bicyclic peptides.

References

Amati P, Blasi F, Di Porzio U, Ricchio A, Trabone C (1975) Hamster α-amanitin-resistant RNA polymerase II able to transcribe polyoma virus genome in somatic cell hybrids. Proc Natl Acad Sci USA 72:753–757

Anderl J, Echner H, Faulstich H (2012) Chemical modification allows phallotoxins and amatoxins to be used as tools in cell biology. Beilstein J Org Chem 8:2072–2084. https://doi.org/10.3762/bjoc.8.233

Azzolin L, Antolini N, Calderan A, Ruzza P, Sciacovelli M, Marin O, Mammi S, Bernardi P, Rasola A (2011) Antamanide, a derivative of *Amanita phalloides*, is a novel inhibitor of the mitochondrial permeability transition pore. PLoS One 6:e16280. https://doi.org/10.1371/journal.pone.0016280

Barbanti-Brodano G, Derenzini M, Fiume L (1974) Toxic action of a phalloidin-albumin conjugate on cells with a high protein uptake. Nature 248:63–65

Bartolomei MS, Corden JL (1987) Localization of an α-amanitin resistance mutation in the gene encoding the largest subunit of mouse RNA polymerase II. Mol Cell Biol 7:586–594

Bartolomei MS, Corden JL (1995) Clustered α-amanitin resistance mutations in mouse. Mol Gen Genet 246:778–782

Ben-Zeev A, Becker Y (1977) Requirement of host cell RNA polymerase II in the replication of herpes simplex virus in α-amanitin-sensitive and –resistant cell lines. Virology 76:246–253

Bowman EA, Riddle DL, Kelly W (2011) Amino acid substitutions in the *Caenorhabditis elegans* RNA polymerase II large subunit AMA-1/RPB-1 that result in α-amanitin resistance and/or reduced function. G3 1:411–416. https://doi.org/10.1534/g3.111.000968

Brueckner F, Cramer P (2008) Structural basis of transcription inhibition by α-amanitin and implications for RNA polymerase II translocation. Nat Struct Mol Biol 15:811–818. https://doi.org/10.1038/nsmb.1458

Brueckner F, Armache KJ, Cheung A, Damsma GE, Kettenberger H, Lehmann E, Sydow J, Cramer P (2009) Structure-function studies of the RNA polymerase II elongation complex. Acta Crystallogr D Biol Crystallogr 65:112–120. https://doi.org/10.1107/S0907444908039875

Bryant RE, Adelberg EA, Magee PT (1977) Properties of an altered RNA polymerase II activity from an α-amanitin-resistant mouse cell line. Biochemistry 16:4237–4244

Bushnell DA, Cramer P, Kornberg RD (2002) Structural basis of transcription: α-amanitin-RNA polymerase II cocrystal at 2.8 Å resolution. Proc Natl Acad Sci USA 99:1218–1222. https://doi.org/10.1073/pnas.251664698

Cain AK, Nester EW (1973) Ribonucleic acid polymerase in *Allomyces arbuscula*. J Bacteriol 115:769–776

Chafin DR, Guo H, Price DH (1995) Action of α-amanitin during pyrophosphorolysis and elongation by RNA polymerase II. J Biol Chem 270:19114–19119. https://doi.org/10.1074/jbc.270.32.19114

Chan VL, Whitmore GF, Siminovitch L (1972) Mammalian cells with altered forms of RNA polymerase II. Proc Natl Acad Sci USA 69:3119–3123. https://doi.org/10.1073/pnas.69.11.3119

Chen Y, Weeks J, Mortin MA, Greenleaf AL (1993) Mapping mutations in genes encoding the two large subunits of *Drosophila* RNA polymerase II defines domains essential for basic transcription functions and for proper expression of developmental genes. Mol Cell Biol 13:4214–4222

Cheung ACM, Cramer P (2012) A movie of RNA polymerase II transcription. Cell 149:1431–1437. https://doi.org/10.1016/j.cell.2012.06.006. http://www.mpibpc.mpg.de/12602574/download and https://www.youtube.com/watch?v=WlMV_l88Lus

Cooper JA (1987) Effects of cytochalasin and phalloidin on actin. J Cell Biol 105:1473–1478

Coulter DE, Greenleaf AL (1982) Properties of mutationally altered RNA polymerases II of *Drosophila*. J Biol Chem 257:1945–1952

Coulter DE, Greenleaf AL (1985) A mutation in the largest subunit of RNA polymerase II alters RNA chain elongation in vitro. J Biol Chem 260:13190–13198

Delgado MA, Rintoul MR, Farías RN, Salomón RA (2001) *Escherichia coli* RNA polymerase is the target of the cyclopeptide antibiotic microcin J25. J Bacteriol 183:4543–4550. https://doi.org/10.1128/JB.183.15.4543-4550.2001

Desjardins AE, Proctor RH (2007) Molecular biology of *Fusarium* mycotoxins. Int J Food Microbiol 119:47–50. https://doi.org/10.1016/j.ijfoodmicro.2007.07.024

Engel C, Sainsbury S, Cheung AC, Kostrewa D, Cramer P (2013) RNA polymerase I structure and transcription regulation. Nature 502:650–655. https://doi.org/10.1038/nature12712

Falcigno L, Costantini S, D'Auria G, Bruno BM, Zobeley S, Zanotti G, Paolillo L (2001) Phalloidin synthetic analogues: structural requirements in the interaction with F-actin. Chemistry 7:4665–4673

Faulstich H, Fiume L (1985) Protein conjugates of fungal toxins. Methods Enzymol 112:225–237

Faulstich H, Zilker TR (1994) Amatoxins. In: Spoerke DG, Rumack BH (eds) Handbook of mushroom poisoning: diagnosis and treatment. CRC Press, Boca Raton, pp 233–248

Faulstich H, Zobeley S, Rinnerthaler G, Small JV (1988) Fluorescent phallotoxins as probes for filamentous actin. J Muscle Res Cell Motil 9:370–383

Faulstich H, Zobeley S, Bentrup U, Jockusch BM (1989) Biotinylphallotoxins: preparation and use as actin probes. J Histochem Cytochem 37:1035–1045. https://doi.org/10.1177/37.7.2499619

Fehrenbach T, Cui Y, Faulstich H, Keppler D (2003) Characterization of the transport of the bicyclic peptide phalloidin by human hepatic transport proteins. Naunyn Schmiedeberg's Arch Pharmacol 368:415–420. https://doi.org/10.1007/s00210-003-0814-4

Fernández-Tornero C, Moreno-Morcillo M, Rashid UJ, Taylor NM, Ruiz FM, Gruene T, Legrand P, Steuerwald U, Müller CW (2013) Crystal structure of the 14-subunit RNA polymerase I. Nature 502:644–649. https://doi.org/10.1038/nature12636

Fiume L, Barbanti-Brodano G (1974) Selective toxicity of amanitin-albumin conjugates for macrophages. Experientia 30:76–77. https://doi.org/10.1007/BF01921609

Frimmer M (1987) What we have learned from phalloidin. Toxicol Lett 35:169–182

Gavrilova O, Geyer J, Petzinger E (2007) In vivo relevance of Mrp2-mediated biliary excretion of the *Amanita* mushroom toxin demethylphalloin. Biochim Biophys Acta 1768:2070–2077. https://doi.org/10.1016/j.bbamem.2007.07.006

Giami S, Simchen G (1977) Incorporation of [^{3}H]UMP into yeast spheroplasts in a hypotonic solution. Exp Cell Res 106:450–454

Gong XQ, Nedialkov YA, Burton ZF (2004) α-Amanitin blocks translocation by human RNA polymerase II. J Biol Chem 279:27422–27427. https://doi.org/10.1074/jbc.M402163200

Greenleaf AL (1983) Amanitin-resistant RNA polymerase II mutations are in the enzyme's largest subunit. J Biol Chem 258:13403–13406

Greenleaf AL, Borsett LM, Jiamachello PF, Coulter DE (1979) α-Amanitin-resistant *D. melanogaster* with an altered RNA polymerase II. Cell 18:613–622. https://doi.org/10.1016/0092-8674(79)90116-8

Guialis A, Beatty BG, Ingles CJ, Crerar MM (1977) Regulation of RNA polymerase II activity in α-amanitin-resistant CHO hybrid cells. Cell 10:53–60. https://doi.org/10.1016/0092-8674(77)90139-8

Gundala S, Wells LD, Milliano MT, Talkad V, Luxon BA, Neuschwander-Tetri BA (2003) The hepatocellular bile acid transporter Ntcp facilitates uptake of the lethal mushroom toxin α-amanitin. Arch Toxicol 78:68–73. https://doi.org/10.1007/s00204-003-0527-y

Haag JR, Ream TS, Marasco M, Nicora CD, Norbeck AD, Paša-Tolić, Pikaard CS (2012) In vitro transcription activities of Pol IV, Pol V, and RDR2 reveal coupling of Pol IV and RDR2 for dsRNA synthesis in plant RNA silencing. Mol Cell 48:811–818. https://doi.org/10.1016/j.molcel.2012.09.027

Hager GL, Holland MJ, Rutter WJ (1977) Isolation of ribonucleic acid polymerases I, II, and III from *Saccharomyces cerevisiae*. Biochemistry 16:1–8

Hagenbuch B, Meier PJ (2004) Organic anion transporting polypeptides of the OATP/SLC21 family: phylogenetic classification as OATP/SLCO superfamily, new nomenclature and molecular/functional properties. Pflugers Arch 447:653–665. https://doi.org/10.1007/s00424-003-1168-y

He L, Vasiliou K, Nebert DW (2009) Analysis and update of the human solute carrier (SLC) gene superfamily. Hum Genomics 3:195–206

Hendzel MJ (2014) The F-act's of nuclear actin. Curr Opin Cell Biol 28:84–89. https://doi.org/10.1016/j.ceb.2014.04.003

Hoch HC, Staples RC (1983) Visualization of actin in situ by rhodamine-conjugated phalloin in the fungus *Uromyces phaseoli*. Eur J Cell Biol 32:52–58

Holtorf H, Schöb H, Kunz C, Waldvogel R, Meins F Jr (1999) Stochastic and nonstochastic post-transcriptional silencing of chitinase and β-1,3-glucanase genes involves increased RNA turnover – possible role for ribosome-independent RNA degradation. Plant Cell 11:471–483

Holzinger A, Blaas K (2016) Actin-dynamics in plant cells: the function of actin-perturbing substances: jasplakinolide, chondramides, phalloidin, cytochalasins, and latrunculins. Meth Mol Biol 1365:243–261. https://doi.org/10.1007/978-1-4939-3124-8_13

Horgen PA, Griffin DH (1971) Specific inhibitors of the three RNA polymerases from the aquatic fungus *Blastocladiella emersonii*. Proc Natl Acad Sci USA 68:338–341

Horgen PA, Vaisius AC, Ammirati JF (1978) The insensitivity of mushroom nuclear RNA polymerase activity to inhibition by amatoxins. Arch Microbiol 118:317–319

Ingles CJ (1978) Temperature-sensitive RNA polymerase II mutations in Chinese hamster ovary cells. Proc Natl Acad Sci USA 75:405–409

Izdebska M, Gagat M, Grzanka D, Grzanka A (2013) Ultrastructural localization of F-actin using phalloidin and quantum dots in HL-60 promyelocytic leukemia cell line after cell death induction by arsenic trioxide. Acta Histochem 115:487–495. https://doi.org/10.1016/j.acthis.2012.11.005

Jacob ST, Sajdel EM, Munro HN (1970) Specific action of α-amanitin on mammalian RNA polymerase protein. Nature 225:60–62

Jaenike J, Grimaldi DA, Sluder AE, Greenleaf AL (1983) α-Amanitin tolerance in mycophagous *Drosophila*. Science 221:165–167. https://doi.org/10.1126/science.221.4606.165

Johnson BC, Preston JF (1979) Unique amanitin resistance of RNA synthesis in isolated nuclei from *Amanita* species accumulating amanitins. Arch Microbiol 122:161–167

Johnson BC, Preston JF (1980) α-Amanitin-resistant RNA polymerase II from carpophores of *Amanita* species accumulating amatoxins. Biochim Biophys Acta 607:102–114

Kaplan CD, Larsson KM, Kornberg RD (2008) The RNA polymerase II trigger loop functions in substrate selection and is directly targeted by α-amanitin. Mol Cell 30:547–556. https://doi.org/10.1016/j.molcel.2008.04.023

Kaster BC, Knippa KC, Kaplan CD, Peterson DO (2016) RNA polymerase II trigger loop mobility: indirect effects of Rpb9. J Biol Chem 291:14883–14895. https://doi.org/10.1074/jbc.M116.714394

Kaya E, Surmen MG, Yaykasli KO, Karahan S, Oktay M, Turan H, Colakoglu S, Erdem H (2014) Dermal absorption and toxicity of alpha amanitin in mice. Cutan Ocul Toxicol 33:154–160. https://doi.org/10.3109/15569527.2013.802697

Kedinger C, Gniazdowski M, Mandel JL, Gissinger F, Chambon P (1970) α-Amanitin: a specific inhibitor of one of the two DNA-dependent RNA polymerase activities from calf thymus. Biochem Biophys Res Commun 38:165–171

Klug A (2001) Structural biology: a marvellous machine for making messages. Science 292:1844–1846. https://doi.org/10.1126/science.1062384

Kramer W (2011) Transporters, Trojan horses and therapeutics: suitability of bile acid and peptide transporters for drug delivery. Biol Chem 392:77–94. https://doi.org/10.1515/BC.2011.017

Lachapelle M, Aldrich HC (1988) Phalloidin-gold complexes: a new tool for ultrastructural localization of F-actin. J Histochem Cytochem 36:1197–1202. https://doi.org/10.1177/36.9.3403970

Lengsfeld AM, Löw I, Wieland T, Dancker P, Hasselbach W (1974) Interaction of phalloidin with actin. Proc Natl Acad Sci USA 71:2803–2807

Letschert K, Faulstich H, Keller D, Keppler D (2006) Molecular characterization and inhibition of amanitin uptake into human hepatocytes. Toxicol Sci 91:140–149. https://doi.org/10.1093/toxsci/kfj141

Lindell TJ (1984) This week's citation classic. Current contents no. 29, July 16, 1984. Accessed 5 Jul 2017 at: garfield.library.upenn.edu/classics1984/A1984SY56700002.pdf

Lindell TJ, Weinberg F, Morris PW, Roeder RG, Rutter WJ (1970) Specific inhibition of nuclear RNA polymerase II by α-amanitin. Science 170:447–449

Litten W (1975) The most poisonous mushrooms. Sci Am 232:90–101

Liu X, Qu X, Jiang Y, Chang M, Zhang R, Wu Y, Fu Y, Huang S (2015) Profilin regulates apical actin polymerization to control polarized pollen tube growth. Mol Plant 8:1694–1709. https://doi.org/10.1016/j.molp.2015.09.013

Lobban PE, Siminovitch L, Ingles CJ (1976) The RNA polymerase II of an α-amanitin-resistant Chinese hamster ovary cell line. Cell 8:65–70

Loros JJ, Dunlap JC (1991) *Neurospora crassa* clock-controlled genes are regulated at the level of transcription. Mol Cell Biol 11:558–563

Lovy-Wheeler A, Wilsen KL, Baskin TI, Hepler PK (2005) Enhanced fixation reveals the apical cortical fringe of actin filaments as a consistent feature of the pollen tube. Planta 221:95–104. https://doi.org/10.1007/s00425-004-1423-2

Lu H, Choudhuri S, Ogura K, Csanaky IL, Lei X, Cheng X, Song PZ, Klaassen CD (2008) Characterization of organic anion transporting polypeptide 1b2-null mice: essential role in hepatic uptake/toxicity of phalloidin and microcystin-LR. Toxicol Sci 103:35–45. https://doi.org/10.1093/toxsci/kfn038

Lu J, Trnka MJ, Roh SH, Robinson PJ, Shiau C, Fujimori DG, Chiu W, Burlingame AL, Guan S (2015) Improved peak detection and deconvolution of native electrospray mass spectra from large protein complexes. J Am Soc Mass Spectrom 26:2141–2151. https://doi.org/10.1007/s13361-015-1235-6

Maksimov MO, Pan SJ, James Link A (2012) Lasso peptides: structure, function, biosynthesis, and engineering. Nat Prod Rep 29:996–1006. https://doi.org/10.1039/c2np20070h

Maunder JE, Voitk AJ (2010) What we don't know about slugs and mushrooms. Fungi 3:36–44

Meier-Abt F, Faulstich H, Hagenbuch B (2004) Identification of phalloidin uptake systems of rat and human liver. Biochim Biophys Acta 1664:64–69. https://doi.org/10.1016/j.bbamem.2004.04.004

Militello KT, Patel V, Chessler AD, Fisher JK, Kasper JM, Gunasekera A, Wirth DF (2005) RNA polymerase II synthesizes antisense RNA in *Plasmodium falciparum*. RNA 11:365–370. https://doi.org/10.1261/rna.7940705

Montarolo PG, Goelet P, Castellucci VF, Morgan J, Kandel ER, Schacher S (1986) A critical period for macromolecular synthesis in long-term heterosynaptic facilitation in *Aplysia*. Science 234:1249–1254

Niedermeyer TH, Daily A, Swiatecka-Hagenbruch M, Moscow JA (2014) Selectivity and potency of microcystin congeners against OATP1B1 and OATP1B3 expressing cancer cells. PLoS One 10:e91476. https://doi.org/10.1371/journal.pone.0091476

Nielsen O (1986) Antamanide antagonizes phalloidin-induced human lymphocyte aggregation and prevents leukaemic mice from death: a pilot study. Acta Pharmacol Toxicol 59:249–251

Nothnagel EA, Barak LS, Sanger JW, Webb WW (1981) Fluorescence studies on modes of cytochalasin B and phallotoxin action on cytoplasmic streaming in *Chara*. J Cell Biol 88:364–372

Oda T, Namba K, Maéda Y (2005) Position and orientation of phalloidin in F-actin determined by X-ray fiber diffraction analysis. Biophys J 88:2727–2736. https://doi.org/10.1529/biophysj.104.047753

Opalski KS, Schultheiss H, Kogel KH, Huckelhoven R (2005) The receptor-like MLO protein and the RAC/ROP family G-protein RACB modulate actin reorganization in barley attacked by the biotrophic powdery mildew fungus *Blumeria graminis* f.sp. *hordei*. Plant J 41:291–303. https://doi.org/10.1111/j.1365-313X.2004.02292.x

Patturajan M (1995) Purification and characterization of DNA-dependent RNA polymerase II from *Candida utilis*. Biochem Mol Biol Int 37:295–304

Petryszak R, Keays M, Tang YA, Fonseca NA, Barrera E, Burdett T, Füllgrabe A, Fuentes AM, Jupp S, Koskinen S, Mannion O, Huerta L, Megy K, Snow C, Williams E, Barzine M, Hastings E, Weisser H, Wright J, Jaiswal P, Huber W, Choudhary J, Parkinson HE, Brazma A (2016) Expression atlas update – an integrated database of gene and protein expression in humans, animals, and plants. Nucl Acids Res 44:D746–D752. https://doi.org/10.1093/nar/gkv1045

Petzinger E, Burckhardt G, Schwenk M, Faulstich H (1982) Lack of intestinal transport of [^{3}H]-demethylphalloin: comparative studies with phallotoxins and bile acids on isolated small intestinal cells and ileal brush border membrane vesicles. Naunyn Schmiedeberg's Arch Pharmacol 320:196–200

Pulman JA, Childs KL, Sgambelluri RM, Walton JD (2016) Expansion and diversification of the MSDIN family of cyclic peptide genes in the poisonous agarics *Amanita phalloides* and *A. bisporigera*. BMC Genomics 17:1038. https://doi.org/10.1186/s12864-016-3378-7

Quon DV, Delgadillo MG, Johnson PJ (1996) Transcription in the early diverging eukaryote *Trichomonas vaginalis*: an unusual RNA polymerase II and α-amanitin-resistant transcription of protein-coding genes. J Mol Evol 43:253–262

Rödicker F, Ossenbühl F, Michels D, Benecke BJ (1999) Faithful in vitro transcription by fission yeas tRNA polymerase III reveals unique alpha-amanitin sensitivity. Gene Expr 8:165–174

Rogalski TM, Bullerjahn AME, Riddle DL (1988) Lethal and amanitin-resistance mutations in the *Caenorhabditis elegans ama-1* and *ama-2* genes. Genetics 120:409–422

Rogalski TM, Golomb M, Riddle DL (1990) Mutant *Caenorhabditis elegans* RNA polymerase II with a 20,000-fold reduced sensitivity to α-amanitin. Genetics 126:889–898

Rudd MD, Luse DS (1996) Amanitin greatly reduces the rate of transcription by RNA polymerase II ternary complexes but fails to inhibit some transcript cleavage modes. J Biol Chem 271:21549–21558

Ruzza P, Calderan A, Biondi B, Carrara M, Tancredi T, Borin G (1999) Ion-binding and pharmacological properties of Tyr^6 and Tyr^9 antamanide analogs. J Pep Res 53:442–452

Sainsbury S, Bernecky C, Cramer P (2015) Structural basis of transcription initiation by RNA polymerase II. Nat Rev Mol Cell Biol 16:129–143. https://doi.org/10.1038/nrm3952

Sanford T, Golomb M, Riddle DL (1983) RNA polymerase II from wild-type and amanitin-resistant strains of *Caenorhabditis elegans*. J Biol Chem 258:12804–12809

Schultz LD, Hall B (1976) Transcription in yeast: α-amanitin sensitivity and other properties which distinguish between RNA polymerases I and III. Proc Natl Acad Sci U S A 73:1029–1033

Seshadri V, McArthur AG, Sogin ML, Adam RD (2003) *Giardia lamblia* RNA polymerase II: amanitin-resistant transcription. J Biol Chem 278:27804–27810. https://doi.org/10.1074/jbc.M303316200

Siemion IZ, Pedyczak A, Trojnar J, Zimecki M, Wieczorek Z (1992) Immunosuppressive activity of antamanide and some of its analogues. Peptides 13:1233–1237

Skillman KM, Diraviyam K, Khan A, Tang K, Sept D, Sibley LD (2011) Evolutionarily divergent, unstable filamentous actin is essential for gliding motility in apicomplexan parasites. PLoS Pathog 7:e1002280. https://doi.org/10.1371/journal.ppat.1002280

Small J, Rottner K, Hahne P, Anderson KI (1999) Visualising the actin cytoskeleton. Microsc Res Tech 47:3–17. https://doi.org/10.1002/(SICI)1097-0029(19991001)47:1<3::AID-JEMT2>3.0.CO;2-2

Somers DG, Pearson ML, Ingles CJ (1975) Regulation of RNA polymerase II activity in a mutant rat myoblast cell line resistant to α-amanitin. Nature 253:372–374

Steeg CM, Ellis J, Bernstein A (1990) Introduction of specific point mutations into RNA polymerase II by gene targeting in mouse embryonic stem cells: evidence for a DNA mismatch repair mechanism. Proc Natl Acad Sci U S A 87:4680–4684

Steinmetz MO, Stoffler D, Müller SA, Jahn W, Wolpensinger B, Goldie KN, Engel A, Faulstich H, Aebi U (1998) Evaluating atomic models of F-actin with an undecagold-tagged phalloidin derivative. J Mol Biol 276:1–6. https://doi.org/10.1006/jmbi.1997.1529

Stirpe F, Fiume L (1967) Effect of α-amanitin on ribonucleic acid synthesis and on ribonucleic acid polymerase in mouse liver. Biochem J 103:67P–68P

Stunnenberg HG, Wennekes LM, Spierings T, van den Broek HW (1981) An α-amanitin-resistant DNA-dependent RNA polymerase II from the fungus *Aspergillus nidulans*. Eur J Biochem 117:121–129

Tellez de Iñon MT, Leoni PD, Torres HN (1974) RNA polymerase activities in *Neurospora crassa*. FEBS Lett 39:91–95

Thell K, Hellinger R, Schabbauer G, Gruber CW (2014) Immunosuppressive peptides and their therapeutic applications. Drug Discov Today 19:645–653. https://doi.org/10.1016/j.drudis.2013.12.002

Theologis A, Huynh TV, Davis RW (1985) Rapid induction of specific mRNAs by auxin in pea epicotyl tissue. J Mol Biol 183:53–68

Timberlake WE, Turian G (1974) Multiple DNA-dependent RNA polymerases of *Neurospora*. Experientia 30:1236–1238

Trauner M, Boyer JL (2003) Bile salt transporters: molecular characterization, function, and regulation. Physiol Rev 83:633–671. https://doi.org/10.1152/physrev.00027.2002

Tyler BM, Giles NH (1985) Accurate transcription of cloned *Neurospora* RNA polymerase II-dependent genes in vitro by homologous soluble extracts. Proc Natl Acad Sci U S A 82:5450–5454

Vaisius AC, Horgen PA (1979) Purification and characterization of RNA polymerase II resistant to α-amanitin from the mushroom *Agaricus bisporus*. Biochemistry 18:795–803

Van Gestel K, Le J, Verbelen JP (2001) A comparison of F-actin labeling methods for light microscopy in different plant specimens: multiple techniques supplement each other. Micron 32:571–578

Vaňáčová S, Tachezy J, Ullu E, Tschudi C (2001) Unusual diversity in α-amanitin sensitivity of RNA polymerases in trichomonads. Mol Biochem Parasitol 115:239–247

Vandekerckhove J, Deboben A, Nassal M, Wieland T (1985) The phalloidin binding site of F-actin. EMBO J 4:22815–22818

Verderame M, Alcorta D, Egnor M, Smith K, Pollack R (1980) Cytoskeletal F-actin patterns quantitated with fluorescein isothiocyanate-phalloidin in normal and transformed cells. Proc Natl Acad Sci USA 77:6624–6628

Von Olenhusen KG, Wohlfarth-Bottermann KE (1979) Evidence for actin transformation during the contraction-relaxation cycle of cytoplasmic actomyosin: cycle blockade by phalloidin injection. Cell Tissue Res 196:455–470

Walton JD (2000) Horizontal gene transfer and the evolution of secondary metabolite gene clusters in fungi: an hypothesis. Fung Genet Biol 30:167–171. https://doi.org/10.1006/fgbi.2000.1224

Wang D, Bushnell DA, Westover KD, Kaplan CD, Kornberg RD (2006) Structural basis of transcription: role of the trigger loop in substrate specificity and catalysis. Cell 127:941–954. https://doi.org/10.1016/j.cell.2006.11.023

Welbourn R, Goldman G, Kobzik L, Valeri CR, Hechtman HB, Shepro D (1985) Attenuation of IL-2-induced multisystem organ edema by phalloidin and antamanide. J Appl Physiol 70:1364–1368. https://doi.org/10.1152/jappl.1991.70.3.1364

Wieczorek Z, Siemion IZ, Zimecki M, Bolewska-Pedyczak E, Wieland T (1993) Immunosuppressive activity in the series of cycloamanide peptides from mushrooms. Peptides 14:1–5

Wieland T (1986) Peptides of poisonous *Amanita* mushrooms. Springer, New York

Wieland T, Faulstich H (1991) Fifty years of amanitin. Experientia 47:1186–1193

Wieland T, Götzendörfer C, Dabrowski J, Lipscomb WN, Shoham G (1983) Unexpected similarity of the structures of the weakly toxic amanitin (*S*)-sulfoxide and the highly toxic (*R*)-sulfoxide and sulfone as revealed by proton nuclear magnetic resonance and X-ray analysis. Biochemistry 22:1264–1271

Wulf E, Bautz L (1976) RNA polymerase B from an α-amanitin resistant mouse myeloma cell line. FEBS Lett 69:6–10

Wulf E, Deboben A, Bautz FA, Faulstich H, Wieland T (1979) Fluorescent phallotoxin, a tool for the visualization of cellular actin. Proc Natl Acad Sci USA 76:4498–4502

Xue JH, Wu P, Chi YL, Xu LX, Wei XY (2011) Cyclopeptides from *Amanita exitialis*. Nat Prod Bioprospect 1:52–66

Yanagida T, Nakase M, Nishiyama K, Oosawa F (1984) Direct observation of motion of single F-actin filaments in the presence of myosin. Nature 307:58–60

Young HA, Whiteley HR (1975) Deoxyribonucleic acid-dependent ribonucleic acid polymerases in the dimorphic fungus *Mucor rouxii*. J Biol Chem 250:479–487

Yu YP, Jackson SL, Garrill A (2004) Two distinct distributions of F-actin are present in the hyphal apex of the oomycete *Achlya bisexualis*. Plant Cell Physiol 45:275–280

Chapter 6
Ecology and Evolution of the *Amanita* Cyclic Peptide Toxins

Miss G. Lister in a letter in September, 1916, wrote: "I have watched a squirrel eating a Tricholoma, I think it may have been terreum...a very pretty sight. I have also watched a rabbit eating Amanita rubescens in our forest, while we were waiting for badgers in the dusk."

(Hastings and Mottram 1916)

6.1 Ecology of the *Amanita* Cyclic Peptide Toxins

Organisms adapt to their environments by the process of natural selection. Whereas abiotic factors such as temperature and water are relatively stable or periodic, the environment created by other organisms is unstable because they themselves are evolving. The coevolution of organisms locked in antagonistic relationships with each other, often referred to as an "evolutionary arms race," can lead to rapid and dramatic changes in the genomes and phenotypes of organisms on both sides of the interaction. (*Antagonist* is used here to mean "other living things that work to make life difficult" [Dawkins 1986]). Although traits such as mimicry, camouflage, and running speed in animals usually receive the most attention as exemplars of the power of the arms race to generate the "organized complexity of the living world" (Dawkins 1986), arguably the most dramatic manifestation of the arms race in some taxa, especially fungi, bacteria, and plants, is their phenomenal skill at chemical synthesis. Collectively, these groups of organisms can biosynthesize tens of thousands of small molecules with highly ornate structures including precise regio- and stereochemistry. To a biochemist, the myriad chemical reactions catalyzed by enzymes are as awe-inspiring as an eagle's eye or a cheetah's speed. (And, of course, the phenomenal properties of eyes and muscles are based on enzyme-catalyzed chemical reactions). Like those macroscopic animal traits, the chemical richness of the natural world arises in large part from the perpetual struggle for survival between

J. Walton, *The Cyclic Peptide Toxins of Amanita and Other Poisonous Mushrooms*, https://doi.org/10.1007/978-3-319-76822-9_6

bacteria, fungi, and plants with each other and with their animal antagonists. It is from this perspective that we will consider the ecological function of the *Amanita* cyclic peptide toxins.

6.1.1 Principles of Secondary Metabolism in Fungi

The *Amanita* cyclic peptide toxins are classic examples of *natural products*, also known as *secondary metabolites* or *specialized metabolites*. (The terms are used interchangeably here.) A precise definition of natural products/secondary metabolites is difficult, but in general they can be considered to be naturally occurring, small (<5 kDa) molecules not involved in the major essential life processes such as DNA, RNA and protein biosynthesis, cellular homeostasis, cell division, or membrane ion transport.

Natural products share several key attributes, which are important evolutionarily and ecologically. First, they show discontinuous taxonomic distribution. Most are restricted to a single taxon, e.g., morphine is found in the opium poppy and nicotine in tobacco. Paradoxically, although restricted taxonomic distribution is a hallmark of secondary metabolites, there are also many examples of the same secondary metabolite being found in widely unrelated species. The *Amanita* cyclic peptide toxins show both kinds of discontinuous distribution. On the one hand, within the genus *Amanita*, they are found only in some species, yet on the other hand they are also found in two unrelated genera of agarics, *Galerina* and *Lepiota* (Chap. 3). There has been a great deal of speculation to explain the discontinuous distribution of secondary metabolites. Possible explanations for the particular case of the cyclic peptide toxins are discussed below.

A second common attribute of secondary metabolites is that they are often inducible by some external signal, such as changing environmental conditions, insect damage, or pathogen attack. This makes biological sense because, insofar as a secondary metabolite has a role in defense, it will therefore be made only when it is needed. It is not clear if the *Amanita* cyclic peptide toxins are inducible (i.e., made in response to an external or developmental signal) or constitutive (i.e., produced all the time), mainly because of the difficulty of culturing the known toxin-producing fungi. They appear not to be under tissue-specific developmental control because they are present in all parts of the fruiting bodies (sporocarps) (Chap. 3). In *G. marginata* the gene for amanitin (*GmAMA1*) is expressed more strongly during growth on low glucose (Luo et al. 2012). Glucose repression is common for fungal genes such as cellulases that are needed when more accessible food is scarce and the substrate of the particular degradative enzyme is present. The induction of *GmAMA1* by low glucose suggests that the toxins are under some level of genetic regulation, but nothing else is known about this aspect of their molecular genetics.

A third common attribute of secondary metabolites is that they often have complex structures whose biosynthesis requires multiple dedicated genes and enzymes.

Some natural products involve more than 20 biosynthetic reactions, and every gene and every enzyme in the pathway are involved in making that one compound and no other. Often, at least in the case of bacterial or fungal metabolites of restricted taxonomic distribution, all of the genes for the entire biosynthetic pathway are absent from the genomes of related species that do not make the metabolite. In the case of the amatoxins and phallotoxins, the genes with a characterized role in their biosynthesis, namely, *AMA1*, *PHA1*, and *POPB*, follow this pattern in being restricted in distribution to species of *Galerina* and *Amanita* that make amanitin and/or phallacidin (Chap. 4).

A fourth general theme of natural products that has emerged in bacteria and fungi (and to a limited extent in plants) is clustering, i.e., the biosynthetic genes of many pathways are often genetically tightly linked, adjacent to each other on the chromosome. Secondary metabolite gene clusters typically contain not only genes for the biosynthetic enzymes that actually build the metabolite but also genes for self-protection, transport, and transcriptional regulation of the cluster genes (Brakhage 2013; Hoffmeister and Keller 2007; Nützmann and Osbourn 2014). The apparent self-contained nature of secondary metabolite clusters has led to speculation that their evolutionary trajectories have distinct features from genes for primary metabolism, perhaps not just exploiting but actually depending on horizontal gene transfer for their long-term survival (Walton 2000). The available evidence on clustering of the genes for the cyclic peptide toxins indicates that there is some clustering, specifically, *AMA1* and *POPB* in *G. marginata*, but to date there is limited information about clustering in *Amanita* or *Lepiota* (Luo et al. 2012, 2014; Pulman et al. 2016).

The consensus among biologists is that the evolutionary raison d'être for most natural products is the *ecological* advantages that they confer on the organisms that produce them. Some endow protection against abiotic environmental factors such as UV light, but most natural products are believed to mediate interactions with other organisms, as either protectants or attractants. The mechanisms by which natural products mediate interactions between organisms seem self-evident in some cases, e.g., bitter-tasting alkaloids deter herbivores and flower pigments attract pollinators, but for most secondary metabolites, we can only speculate about how they increase fitness. Our ability to go beyond speculation and experimentally manipulate both secondary metabolites and ecological interactions in order to obtain definitive answers to this question is more often than not frustratingly limited. One powerful method, but which is often technically unavailable, is to create near-isogenic strains of the producing organism that differ only in their ability to make the metabolite of interest. In this way the ecological functions of some secondary metabolites have been rigorously tested, e.g., nicotine protects tobacco plants against insect herbivory, and HC-toxin is required for plant pathogenic virulence (Panaccione et al. 1992; Steppuhn et al. 2004). However, the advantages that natural products confer might be highly contingent and thus not discernible except in very subtle or long-term experiments. For example, a particular natural product might increase the fitness of an organism only when a rare environmental stress occurs or when a facultative herbivore's preferred food source is absent. Such conditions might occur very infrequently but exert powerful selection when they do. The biologist might be forced to wait a very long time (even centuries?) for the function of a particular secondary metabolite to manifest itself.

For most secondary metabolites, including the *Amanita* cyclic peptides, we are not currently able to experimentally address their possible ecological roles because of the difficulty of growing the producing fungi, much less genetically manipulating them. However, involvement in defense seems the most plausible function for these compounds, because the amatoxins and phallotoxins are poisonous or potentially poisonous to all eukaryotic cells. A possible role of α-amanitin in regulating RNA polymerase has been proposed but seems unlikely in light of our current knowledge of gene regulation in eukaryotes and the molecular genetics of amanitin biosynthesis (Preston et al. 1982).

6.1.2 Why Do Mushrooms Make the Cyclic Peptide Toxins?

Postulating a defensive role for amatoxins and phallotoxins begs the question: defense against what? This gets us finally to the question most frequently asked of *Amanita* toxicologists: Why do the mushrooms make them? In the language of the evolutionary biologist, the question can be phrased: What selective advantage do these compounds confer on the mushrooms? A strict adaptationist point of view would argue that, because biosynthesis of the toxins has a cost (in carbon, nitrogen, and metabolic energy), if they *didn't* confer an advantage natural selection would have long since condemned their biosynthetic genes to oblivion through the accretion of random mutations. A less stringent point of view would argue that they might be relic traits or metabolic anomalies with such little cost to the organism that they persist despite lacking a function, or they are the accidental by-product of some other cellular process under strong selection (Gould and Lewontin 1979). These evolutionary alternatives are discussed in more detail below. However, conceding that the toxins do confer a selective advantage to the producing fungi (the viewpoint favored by the author) does not help us understand *how* they contribute to mushroom survival.

Amanita and some other mushrooms such as *Lepiota* make not only the well-known amatoxins and phallotoxins but also a number of other cyclic peptides known collectively as the cycloamanides (Chap. 2). The evolutionary rationale for these different compounds could well be different from the amatoxins and phallotoxins. However, even in the absence of any definitive experimental evidence, it seems reasonable to conclude that at least α-amanitin must confer a strong selective advantage on the mushrooms that produce it. This is because its concentration in mushroom tissues is high (Chap. 2), it is the most widespread of the *Amanita* cyclic peptides (Chap. 3), and it has potent biological activity against many organisms (Chap. 5). Most of the following discussion will focus on the amatoxins.

6.1.3 Can Any Animals Safely Eat Poisonous Mushrooms?

One approach to the question of why mushrooms make the cyclic peptide toxins is to ask which potential mushroom antagonists (i.e., fungivores, pathogens, parasites, etc.) are sensitive and which are resistant to the toxins. (*Fungivores* are organisms that eat fungi, and *mycophagy* is the niche of obtaining nutrition from mushrooms.) Sensitive organisms could be *potential* antagonists of mushrooms if not for the toxins, whereas insensitivity in an organism might indicate that the toxins have been a significant source of selective pressure at some time in that organism's evolution.

The answer to the question of which organisms can safely eat the cyclic peptide-containing mushrooms is complicated by the innumerable published statements on this point. Unfortunately, the vast majority of these accounts are based on dubious sources, without citation or documentation, and rarely on experimental evidence.

Almost all of the early work on purification and characterization of the *Amanita* cyclic peptide toxins relied on bioassays based on injection, mainly into mice but also other rodents such as guinea pigs, hamsters, and rats. By any parenteral route (i.e., by any route other than oral), the so-called "fast-acting" phallotoxins are as toxic or more toxic than the "slow-acting" amatoxins, yet we now know that only the amatoxins are toxicologically relevant, at least for mushroom-eating mammals (Wieland 1986). Since the early studies relied on an unnatural mode of toxin delivery, they are of little use in addressing the ecological functions of the toxins.

6.1.3.1 Sensitivity of Microbes and Invertebrates Other Than Insects

In regard to the possible role of the amatoxins in defense, it is reasonable to start by asking which organisms *Amanita* and other cyclic peptide toxin-producing mushrooms interact with in nature. Organisms known to inflict damage on mushrooms include pathogenic bacteria, fungi (many fungi are parasitized by other fungi, and some parasitize only species of *Amanita*; Arnold 1976; Roderson and Samuels 1994), nematodes, insects, gastropods (snails and slugs), and a few reptiles such as turtles. Some mammals such as squirrels seek out mushrooms as a major part of their diet, and other mammals, such as cows, probably eat them occasionally by accident.

Neither α-amanitin nor phallotoxins are toxic to bacteria, and their toxicity to other fungi is equivocal (Chap. 5). Although there are reports of nematodes parasitizing fungi, especially agarics, it is more common for fungi to parasitize nematodes. Many fungi physically trap and consume nematodes, sometimes with the aid of nematicidal secondary metabolites (Degenkolb and Vilsinskas 2016; Hsueh et al. 2013). Nematicidal fungi include the mushroom-forming fungi (agarics) *Pleurotus ostreatus* and *Conocybe albipes* (Hutchison et al. 1996; Thorn and Barron 1984). The latter mushroom might produce phallotoxins (Chap. 3). Sometimes fungi chemically kill nematodes but don't consume them. That is, killing nematodes could be a *defensive* strategy as well as a *nutritional* strategy. Examples of fungal

Fig. 6.1 Slug (*Arion subfuscus*) grazing on *Amanita bisporigera*. It is not known if these particular mushrooms contained amatoxins or phallotoxins nor whether the slug survived. (From Maunder and Voitk (2010). Photo credit: Andrus Voitk, Foray Newfoundland and Labrador (www.nlmushrooms.ca), used with permission)

secondary metabolites that are toxic to nematodes include the cyclic octadepsipeptide emodepside (Harder et al. 2003) and omphalotin (Mayer et al. 1999). Omphalotin is interesting because biosynthetically it is a RiPP, like the *Amanita* cyclic peptide toxins (Ramm et al. 2017; Van der Velden et al. 2017) (Chap. 4).

Gastropods are common consumers of mushrooms including "destroying angels," i.e., white species of *Amanita* in sect. *Phalloideae* (Maunder and Voitk 2010) (Fig. 6.1). There is some evidence that some mushrooms make slug antifeedant compounds, e.g., *Clitopilus prunulus* (and many other fungi) makes 1-octen-3-ol, which discourages feeding of banana slugs on otherwise palatable food, and *Clitocybe flaccida* makes the antislug compound clitolactone (Wood et al. 2001, 2004). The author has seen slugs feeding on white *Amanita* mushrooms in Michigan, but it is unknown whether those particular mushrooms contained amatoxins or not. This is a point that cannot be taken for granted, because toxin production can be variable among individual mushrooms, especially in the *Amanita bisporigera* species complex (Chap. 3). Without testing the amatoxin content of the exact same individual mushroom being consumed by a slug, one cannot be sure that the slug actually came into contact with amatoxins. Little is known about the sensitivity of gastropods to amatoxins, but Wieland (1986) reported without a supporting citation that gastropods (snails) are resistant to >20 mg amanitin/kg body weight when injected intraperitoneally, which implies that they would be more tolerant than mammals by a factor of ~200 to amatoxins consumed orally. On the other hand, pol II-dependent transcription in the marine gastropod *Aplysia californica* is inhibited by 2 μg/ml α-amanitin (Montarolo et al. 1986). Thus, protection against mycophagous gastropods is a plausible ecological function for amatoxins.

6.1.3.2 Resistance to Amatoxins in Mycophagous Flies

Mushrooms are sensitive to mycophagy (fungivory) by many insects (Wheeler and Blackwell 1984). In response, mushrooms have evolved insecticidal secondary metabolites (Nakamori and Suzuki 2007; Chap. 3), and in response to that evolutionary development, some insects have evolved resistance to mushroom toxins, including α-amanitin, ibotenic acid, and muscimol (Jaenike et al. 1983; Tuno et al. 2007; for chemical structures, see Chap. 2).

The ecological rationale of the *Amanita* cyclic peptides as a protection against mycophagy has been studied experimentally with flies of the genus *Drosophila*, many of which are mycophagous. Phillips et al. (1982) screened 32 wild and laboratory strains of *Drosophila melanogaster* and found three with significant natural resistance to dietary α-amanitin. Whereas their reference strain was half-maximally killed by 1.2 μg/vial, the three resistant lines (Ama-KLM from Malaysia, Ama-MI from India, and Ama-KTT from Taiwan; all of these countries are in a region of the world where amanitin-containing mushrooms are rather common; Chap. 3) were killed only by 10, 30, and 35 μg/vial, respectively. That is, the degree of resistance was ~8- to ~30-fold. In crosses, resistance was genetically dominant and controlled by two loosely linked genes on chromosome III. Although the responsible genes were mapped, the mechanism of resistance in these strains is apparently still unknown. The pol II activities from the three lines were as sensitive as the control (K_i ~20 ng/ml), and therefore intrinsic pol II insensitivity could be excluded as the basis of resistance. The authors speculated that resistance might be due to overexpression of pol II and not to gut inactivation or reduced toxin uptake from the insect gut (Phillips et al. 1982).

Jaenike et al. (1983) compared larval survival and development of six species of *Drosophila* when reared in the laboratory on a diet containing α-amanitin. Three species (*D. putrida*, *D. recens*, and *D. tripunctata*) were mycophagous (natural mushroom feeders) and lay their eggs in multiple species of mushroom (not just poisonous ones), and three (*D. melanogaster*, *D. immigrans*, and *D. pseudoobscura*) were frugivorous (natural fruit feeders). All three non-mycophagous species showed 100% mortality in response to α-amanitin concentrations below 50 μg/ml, whereas the three species that breed in mushrooms were completely unaffected or only slightly sensitive to 50 μg/ml. The mechanism by which the wild *Drosophila* tolerated high concentrations of α-amanitin was not due to intrinsic pol II resistance, because the pol II activities from all six species were as sensitive to α-amanitin as wild-type *D. melanogaster* and ~100 times less sensitive than the artificially selected strain C4 (Fig. 6.2). It is noteworthy that the C4 flies obtained by artificial selection were much less fit than the wild-type flies in the absence of amanitin, which might explain why spontaneous mutation of pol II to amanitin resistance has not been found in nature.

An intriguing question raised by Jaenike et al. (1983) is the basis of the selective advantage of amanitin resistance to *Drosophila*, since only a small percentage of the mushrooms on which the mycophagous species feed contain amanitin. (As discussed in Chap. 3, reports that other wild and even edible, mushrooms make low

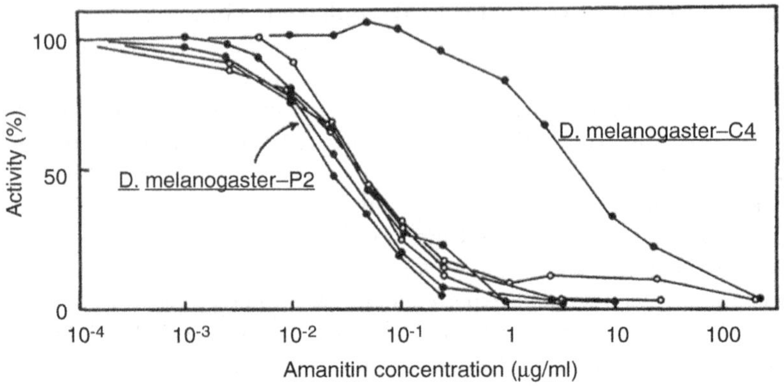

Fig. 6.2 Sensitivity of in vitro RNA polymerase II (pol II) activity from six wild species of *Drosophila* and two lab strains of *D. melanogaster*. The C4 strain of *D. melanogaster* was artificially selected for amanitin resistance from parental strain P2 (Greenleaf et al. 1979). Three wild species were nonmycophagous and three were mycophagous. The data indicate that the pol II activities of all six wild species and the parental lab strain P2 are sensitive to α-amanitin, whereas the pol II activity from C4 is resistant. (From Jaenike et al. 1983. Reprinted with permission from AAAS)

levels of amatoxins have not been confirmed.) Jaenike (1985) addressed this question by studying resistance within the *D. quinaria* species group. Three non-mycophagous species (*D. quinaria, D. palustris*, and *D. subpalustris*) were at least 20-fold more sensitive to α-amanitin compared to three mycophagous species (*D. falleni, D. recens*, and *D. phalerata*). The authors also raised the flies on mushrooms (*A. virosa, A. muscaria*, and *Agaricus bisporus*). The three mycophagous species did comparably well on all three mushrooms, but the non-mycophagous species did consistently worse on *A. virosa*, the only one of these three mushrooms that makes α-amanitin. Jaenike (1985) speculated that toxic mushrooms might not be just an *available* resource for mycophagous flies but actually a *beneficial* one, for two possible reasons. One, the amatoxins might block larval growth of competing insects such as gnats and crane flies. Two, the amatoxins might kill nematodes, such as *Howardula aoronymphium*, that parasitize the flies. Observations supported this second theory. In mushrooms collected in the wild, 500 flies were bred from toxic mushrooms (identified by the author as *A. bisporigera* and *A. virosa*), but only one of them was being parasitized by nematodes. In contrast, the incidence of nematode parasitism was 60-fold higher in the flies collected from cyclic peptide toxin-nonproducing mushrooms (*A. muscaria, A. flavorubescens, A. rubescens, A. brunnescens*, and *Russula emetica*). A reasonable conclusion from this observation was that "the superiority of these [toxic] mushrooms as breeding sites may have accelerated the evolution of amanitin tolerance..." (Jaenike 1985).

An important unanswered question arising from these studies is the mechanism by which amanitin controls the nematode parasites. The nematodes might be exposed to the amanitin while free-living in the mushroom, or, alternatively, they may encounter the toxin inside the flies, i.e., the flies might accumulate amanitin in their bodies. Sequestration into a metabolically inert cellular compartment could therefore serve the flies as both a mechanism to avoid the toxic effects of amanitin as well as a mechanism to protect themselves against parasites and predators. The latter ecological scenario is reminiscent of monarch butterfly larvae, which by accumulating cardiac glycosides from milkweed thereby make the mature butterflies unpalatable to birds (Brower and Fink 1985). One could even look at such a scenario as an example of self-medication. Some animals, including flies, intentionally consume compounds that confer protection against parasites (de Roode et al. 2013). Perhaps mycophagous flies preferentially consume toxin-containing *Amanita* mushrooms when they are infected with parasitic nematodes, or when threatened by other parasites such as wasps (Kacsoh et al. 2013). This theory could be tested by extracting mycophagous flies that have fed on amanitin-containing mushrooms to see if they accumulate nematicidal or insecticidal levels of amanitin, and by preferential feeding studies with toxic and nontoxic mushrooms in the presence of parasitic wasps (Kacsoh et al. 2013).

A subsequent phylogenetic analysis of the *D. quinaria* species group concluded that mycophagy and amanitin tolerance are perfectly positively correlated and both are evolutionarily ancestral to non-mycophagy and amanitin sensitivity (Spicer and Jaenike 1996). Apparently, when flies evolved away from feeding on mushrooms, they lost their resistance to amanitin.

In another study of natural variation in amanitin sensitivity, Begun and Whitley (2000) studied populations of *D. melanogaster* from California fruit orchards. Some isolates were somewhat resistant to α-amanitin (IC_{50} >1 μg/ml) whereas others were killed by tenfold lower concentration. Like Phillips et al. (1982), the authors found evidence for a major resistance gene and a weaker gene, both on chromosome III. The major locus mapped near a gene for a member of the multidrug resistance (MDR) family, Mdr65A. MDR proteins (also known as P-glycoproteins or ABC transporters) are involved in membrane transport (excretion) of cytotoxic and other native and xenobiotic compounds. Since α-amanitin is a cytotoxic xenobiotic, it is reasonable that amanitin resistance might be mediated by an MDR transporter. Begun and Whitley (2000) did not find any amino acid differences in the coding regions of Mdr65A from sensitive and resistant lines of *D. melanogaster*, but they could not exclude differences in noncoding regions that might affect expression levels, i.e., amanitin resistance could be due to upregulation of Mdr65A. In regard to why some flies should be at least moderately resistant to a toxin to which they had never been exposed, Begun and Whitley (2000) speculate that amanitin resistance may be a pleotropic phenomenon, i.e., a side effect of evolved resistance to other natural or synthetic toxins. In regard to this hypothesis, it may be relevant that mycophagous species of *Drosophila* (*D. bizonata*, *D. angularis*, and *D. brachynephros*), but not frugivorous species (*D. immigrans*, and *D. melanogaster*), are also insensitive to the nonpeptidic mushroom toxins ibotenic acid and muscimol (Fig. 2.11) (Tuno et al. 2007). These two related isoxazoles are made by *A. pantherina*, *A. ibotengutake*, and *A. muscaria*, none of which are in sect. *Phalloideae*.

In a comparison of 16 species of *Drosophila*, Stump et al. (2011) also found that mycophagous species (including six newly identified as such) were generally more resistant to amanitin than leaf- and fruit-eating species. Some, but not all, normally tolerant species became more sensitive when the cytochrome P450 (CYP450) inhibitor piperonyl butoxide was added to their diets. One species became more sensitive when the glutathione S-transferase (GST) inhibitor ethacrynic acid was added to its diet. This suggests that resistance in these species is due to detoxification either by a CYP450 or by conjugation to glutathione. Sequencing of the pol II genes of eight tolerant species and one sensitive species did not provide any evidence for the direct involvement of pol II in altered sensitivity to amanitin. The authors hypothesized that natural amanitin tolerance in *Drosophila* is likely due to a general detoxification mechanism that evolved to allow mycophagous flies to cope with a wide range of mushroom toxins (Stump et al. 2011).

Mitchell et al. (2014) used whole genome transcriptional profiling to explore the mechanistic basis of resistance to amanitin in natural populations of *D. melanogaster*. The underlying assumption of this study was that changes in gene expression would be informative of the mechanism(s) of resistance. They compared constitutive differences in gene expression in wild type (sensitive strain Canton-S) vs. the Ama-KTT strain used by Phillips et al. (1982) and also inducible changes in gene expression in Ama-KTT grown on regular diet vs. diet supplemented with α-amanitin. Genes encoding proteins for xenobiotic metabolism were enriched in the upregulated set in both comparisons, especially several members of the CYP450, GST, and UDP-glycosyltransferase (UGT) families. In total, 234 genes were upregulated more than twofold in the resistant strain, 8.5% of which encoded CYP450, GST, or UGT proteins. Certain CYP450 isoforms (Cyp6a2, Cyp12d1-d, and Cyp12d1-p) were 197- to 300-fold upregulated in the resistant stock (Mitchell et al. 2014).

MDR genes, which were implicated by the study of Phillips et al. (1982) and Begun and Whitley (2000) in amanitin resistance, were neither constitutively upregulated in the resistant flies nor induced by amanitin (Mitchell et al. 2014). However, genetic analysis did point to a region of chromosome III near the Mdr65A locus as important for resistance, as found in the earlier studies. Two of the inducible CYP450 genes map near this location.

Based on the observed patterns of differential gene expression, Mitchell et al. (2014) hypothesized that multiple mechanisms are involved in amanitin resistance in *D. melanogaster*, including (1) blocked uptake through the cuticle, (2) detoxification, (3) sequestration into cytoplasmic lipid bodies, and (4) proteolytic cleavage. Sequestration into lipid bodies seems like an unlikely mechanism of resistance because amanitin is not lipid-soluble. Proteolytic cleavage seems unlikely because amanitin, like other cyclic peptides, is highly refractory to peptidases. Metabolism by CYP450, UGT, and GST systems serves mainly to increase the water solubility of xenobiotics and hence enhance their excretion (Jakoby and Ziegler 1990; Lewis and Dickins 2004). Because amanitin is already heavily hydroxylated and quite water-soluble, further hydroxylation by CYP450 seems unlikely to enhance excretion, although it might lower amanitin's affinity for its target, pol II. UGT-mediated glycosylation of the native hydroxyl groups in amanitin might likewise protect flies by reducing binding to pol II rather than by enhancing excretion (Bock 2016).

6.1.3.3 Sensitivity of Vertebrates Including Mammals

The only vertebrates indisputably sensitive to orally ingested amatoxins are humans and dogs. Cats and cows are probably also sensitive orally, but the record is sparse (Beug et al. 2006; Mullenax and Mullenax 1962; Seymour 1932; Tokarz et al. 2012; Yee et al. 2012). Wieland (1986) states unequivocally that mice and rats "cannot be poisoned orally, even with high doses of amatoxins," whereas guinea pigs succumb to oral doses of 0.1 mg/kg. However, this author was not able to find any evidence for any rodents other than squirrels eating mushrooms in nature, at least on a regular basis. If rats, for example, never or rarely eat *any* mushrooms, then their resistance to oral amatoxins is probably not evolutionarily significant, i.e., there has been no arms race between mushrooms and rats driving evolution of toxin biosynthesis on the one hand, and resistance on the other. In regard to phallotoxins and cycloamanides, almost nothing is known about the sensitivity of most animals to these when ingested orally (see below).

There are a plethora of unsupported statements in the scientific and popular literature that many animals, including pigs, skunks, chipmunks, raccoons, horses, rabbits, squirrels, and deer, can eat amatoxin-containing mushrooms with impunity, but for most of these there are not even full anecdotal details much less concrete evidence. Grazing animals probably regularly consume poisonous or other mushrooms accidentally. YouTube (www.utube.com) has videos of squirrels and even turtles eating mushrooms, but it is impossible to determine if the mushrooms being consumed contain amatoxins, and, if so, whether the animals suffered any ill effects (e.g., Hess 2014). Although some sources state that red squirrels of North America can eat poisonous mushrooms, Klugh (1927) more cautiously states "Whether the red squirrel eats the poisonous species, such as *Amanita muscaria* and *A. phalloides*, and whether, having done so, disastrous consequences ensue, is a point upon which there are no exact data." To remind the reader, *A. muscaria* does not make the cyclic peptide toxins (Chap. 2), and *A. phalloides* was not present in eastern North America in 1927 (see below). By *A. phalloides*, early writers such as Klugh (1927) were probably referring to mushrooms in the *A. bisporigera* species complex, which do make amatoxins and phallotoxins as well as many cycloamanides (Chap. 3).

Hastings and Mottram (1916) do not record any observations of rodents eating amatoxin-containing mushrooms in Great Britain. Fogel and Trappe (1978) list hundreds of examples of mycophagy by small animals, but the only amatoxin-containing species mentioned is *A. phalloides* (again, probably the *A. bisporigera* complex) in a report by Hatt (1929). Hatt (1929) himself was of the opinion that red squirrels in eastern North America do not distinguish between poisonous and non-poisonous mushrooms but avoid harm by being immune to mushroom poisons, at least those in *A. muscaria* such as muscimol. Hatt (1929) also mentions a third-party observation that squirrels avoid *A. phalloides* and another observation that they eat "white *Amanita*," but with no information on the outcome. Ballou (1927) records tortoise skeletons found in the wood, "examination having shown that the animals had devoured the deadly *Amanita phalloides*." It is hard to comprehend how any evidence of mushroom consumption could have survived the time it would take (many months or years) for a tortoise to be reduced to a skeleton. Ballou (1927) also witnessed a cow and a flock of hens dying from *A. phalloides*, with no further details. Beug et al. (2006) record an incident of probable poisoning of a horse by amanitin-containing mushrooms.

In the 1930s, feeding hashed stomachs and brains of rabbits to human patients was proposed as a cure for amatoxin poisoning. This protocol was based on the belief (of unknown origin) that rabbits can tolerate a much larger quantity of *A. phalloides* than cats (Limousin and Petit 1932; English translation, Seymour 1932; discussed by Pilát and Ušák 1955). Feeding cats "cooked *phalloides* by the digestive tract" (presumably meaning that the mushrooms entered the cat's stomach through the mouth) resulted in death, whereas rabbits treated with the same amounts in the same way survived. This suggests that rabbits are less sensitive to oral amanitin than cats, but the relative degree of tolerance is unknown.

Piercy et al. (1944) force-fed a rabbit of unspecified weight with "one small mushroom" of the amanitin-containing species *A. virosa* collected in Louisiana, USA. After an initial 4-h period of "depression" and "nausea," the rabbit appeared normal until death in convulsions 2 days later. This progression of symptoms and the postmortem examination of the internal organs were consistent with death caused by α-amanitin. They also fed a 114-kg calf a total of 216 gm of *A. virosa* mushrooms. Assuming 0.4 mg α-amanitin/gm fresh weight, the calf ingested ~85 mg α-amanitin, which would be ~8 times the LD_{50} for a human of the same weight. The calf showed no "abnormalities," presumably meaning that it survived. On the other hand, Yee et al. (2012) described the death of two calves in California with symptoms, including hepatic necrosis, consistent with amanitin poisoning. α-Amanitin was detected by LC/MS/MS in the liver and rumen contents of both calves, so this report can be considered highly reliable evidence that cows are sensitive to oral amanitin poisoning despite their ruminant digestive systems. Cyclic peptide toxin-containing species of *Amanita* are obligately ectomycorrhizal (see below) but nonetheless are often found growing in meadows some distance from their host trees. Cows probably accidentally graze them on a regular basis. The paucity of cases of cows being poisoned (a total of one rigorous report) is probably because a cow would rarely eat more than one or two mushrooms at a time and therefore would normally not receive an injurious much less fatal dose of amanitin.

If some vertebrates can, in fact, safely eat amanitin-containing mushrooms, it is almost certainly not because of intrinsic resistance of DNA-dependent RNA polymerase (pol II), because all animal pol II's that have been tested to date are strongly inhibited by amanitin. This includes not just mammals but also insects, nematodes, and molluscs. Oral insensitivity must therefore be due other causes, for example, chemical or enzymatic degradation of the toxins in the digestive tract. This could be especially significant for ruminants such as sheep or cows, which probably frequently accidentally consume mushrooms while grazing. Another possible mechanism of resistance is failure to absorb the amatoxins from the digestive tract into the bloodstream. This is the most widely assumed explanation for the resistance of mice and rats to oral poisoning (Wieland 1986).

6.1.3.4 Alternate Theories for the Sensitivity of Mammals to Amatoxins

Clearly, some animals are very sensitive to the *Amanita* cyclic peptide toxins delivered orally, especially the amatoxins, but this does not necessarily signify the reason the mushrooms make the toxins. Many people have pointed out the paradox that, despite their high toxicity, amatoxins should be poor at discouraging fungivory by large, fast-moving animals. The toxins have no taste (at least to humans), and they are "slow acting," i.e., animals die only after many days. These properties do not make an effective deterrent – the mushroom can be entirely consumed before the animal gets sick, and the long delay lessens the possibility for any learned adverse reaction, assuming the animal even survives. Bitter tasting, fast-acting emetics are much more effective feeding deterrents. By this logic, the amatoxins are more likely to protect against organisms that eat or colonize mushrooms slowly, such as gastropods or insect larvae.

Nonetheless, the theory that amatoxins deter mycophagy by mammals should not yet be completely abandoned. In one of the few experimental studies on mammalian poisoning by any mushroom, Camazine (1983) showed that opossums (*Didelphis virginiana*), which are marsupials native to North America, learned to avoid otherwise edible mushrooms if the mushrooms had been treated with muscimol, a nonpeptide emetic toxin made by *Amanita muscaria* and other mushrooms (Fig. 2.11). Even after being made sick by *A. muscaria* (which does not contain amatoxins or any other cycloamanide) or by nonpoisonous mushrooms treated with muscimol, the opossums continued to eat other nontoxic control mushrooms. That is, they could distinguish between mushroom species by sight and/or by smell. Opossums became sick within 30–75 min after eating *A. muscaria* or a harmless puffball (*Calvatia gigantea*) treated with muscimol.

Based on the opossum/muscimol model, the short-term gastrointestinal distress caused by the amatoxins *might* be sufficient to induce a learned food aversion. If an obligate or facultative mycophagous animal ate enough to get sick but not enough to die, and was capable of distinguishing one mushroom species from another, then the animal would presumably thereafter avoid the poisonous mushrooms but still be able to recognize and benefit from the consumption of nonpoisonous ones.

A major question raised by this theory is whether a learned food aversion response to amatoxins would be adequate in timing, duration, and strength. The animal would have to be able to mentally associate the disagreeable reaction with the specific mushroom species, i.e., remember what it had eaten before getting sick. The gastrointestinal phase in response to amatoxins in humans starts approximately 8–12 h after consumption with a range of 6–24 h (Benjamin 1995; Bresinsky and Besl 1990). Is this a sufficiently short time between exposure and response to effectively link the two phenomena in the mind of an animal such as an opossum or a squirrel?

In regard to whether the strength of the gastrointestinal response to amatoxins is compatible with a learned food aversion, it is important to know the dose relationship between an unpleasant experience and death. Is there an amount of amanitin that causes gastrointestinal distress without fatal liver damage? If the doses triggering the two responses were strongly separated (say, by tenfold or more), then there is a reasonable possibility of an animal being exposed to enough to get sick but not die. On the other hand, if the doses causing the two responses were almost equal, there is little chance of developing a learned aversion, on the principle that knowledge is wasted on the nonliving. (Humans are not a good test species for this hypothesis because medical intervention is more common than not).

The possibility of a learned food aversion response to amatoxins is not necessarily restricted to mammals. Gelperin (1975) showed that a single exposure to a noxious chemical was sufficient to deter slugs (*Limax maximus*) from eating a preferred food source (mushrooms, as it turns out). The slugs still avoided the noxious food 91 days later (opossums remembered for >70 days; Camazine 1983).

Lastly, we must consider the possibility that the human sensitivity to amatoxins is ecologically meaningless. Mammals might just be collateral damage in the age-old struggle between amatoxin-producing fungi and their ecologically relevant antagonists. Such accidents of nature must occur frequently. If one estimates that the average fungus makes ~10 secondary metabolites (and some fungi are known to make many more than that), that there are ~3 million species of fungi, and that the typical secondary metabolite is found in an average of three different species, there might be as many as ~10 million compounds in the Kingdom Fungi. When one considers that most of these are under some sort of natural selection to have *some* type of biological activity against *something*, then it seems not improbable that at *some* time in evolution, *some* fungal compound arose that is a potent inhibitor of just about every possible cellular target. Even though humans exploit thousands of microbial secondary metabolites for a wide range of uses, little is known about the ecological functions of most of them. Why would a soil fungus make a mammalian immunosuppressant such as cyclosporin? Why does *Aspergillus terreus* make the cholesterol-lowering drug lovastatin? Why do opium poppies make the human analgesic morphine? The high human toxicity of amatoxins might be an equally fortuitous accident of nature.

6.1.4 Possible Fitness Contributions of the Phallotoxins and Extended Cycloamanide Family

The phallotoxins are not toxic to mammals when ingested orally. Therefore, there is no scope to speculate about a role for phallotoxins in defense against mammals. Furthermore, it is not even known if phallotoxins are poisonous orally to *any* animal including slugs, nematodes, or insects, so at this point we can say nothing about what defensive role, if any, phallotoxins might play against mycophagous organisms. Possibly the phallotoxins are toxic by absorption through the epidermis (cuticle or skin), which might be a plausible route by which insect larvae or nematodes that are in close contact with mushroom tissues might be physiologically exposed to phallotoxins.

As discussed in Chap. 4, fungi in sect. *Phalloideae* contain the genetic capacity to make many additional cyclic peptides besides the amatoxins and phallotoxins, known collectively as the cycloamanides (Pulman et al. 2016; Wieland 1986; unpublished results from the author's laboratory). Most of the cycloamanides are unmodified by cross-bridging or hydroxylations. The chemically characterized cycloamanides from species of *Amanita* include CylA through CylF and antamanide (Table 2.2). *Lepiota subincarnata* has a small family (five members) of cycloamanide genes (Fig. 4.12), several of which are translated into cyclic products, but *Galerina marginata* has genes only for α-amanitin (Luo et al. 2012; Riley et al. 2014). Little is known about the biological activities of the cycloamanides, and what effects that have been described (e.g., immunosuppression; Chap. 5) are probably not relevant to their functional targets, if any, in nature. The discovery that some species of *Amanita* have dozens of cycloamanide genes that are species-unique raises interesting evolutionary questions that are discussed below.

6.1.5 Why Do the Mushrooms Make the Toxins, Revisited

We still do not know the answer to this question with any certainty, but at least for the amatoxins, the simplest explanation is that they provide defense against mycophagous insects and perhaps gastropods. Insects and slugs probably exert stronger evolutionary pressure on mushrooms than mammals and reptiles simply because they are so much more numerous. Simply by killing or injuring the mycophagous organisms that ingest the toxins inadvertently while consuming the flesh of the mushrooms, the toxins prevent further mycophagy and hence enhance the survival of the toxin-containing mushrooms. An alternate but not mutually exclusive hypothesis is that the toxins might induce a learned food aversion by causing gastrointestinal distress. Because mammals are presumed to be better learners than invertebrates, this hypothesis seems more applicable to mammals.

In arriving at the conclusion that protection against mycophagous insects is the evolutionary rationale for the amatoxins, the research on natural resistance in the genus *Drosophila* is particularly persuasive, because evolved resistance is the expected outcome of an arms race between a food source and its consumer. That is, insofar as toxin resistance is a recently derived trait within *Drosophila*, it probably arose in response to selection pressure exerted by the toxins on the flies. This in turn implies that the original adaptive function of the toxins was to protect against mycophagous insects.

A role in defense against invertebrates raises some testable hypotheses. The mycophagy hypothesis could be tested by feeding preference experiments in which insects or gastropods were presented with specimens of related mushrooms that did or did not contain amatoxins and/or phallotoxins. Do mycophagous insects or gastropods avoid toxin-containing mushrooms either because of learned behavior, an ability to sense the toxins, or genetically programmed instinct? In some species, like *A. bisporigera*, it is not rare to find individual fruiting bodies lacking the toxins, which would make excellent controls for these kinds of experiments. Feeding behavior and mortality would have to be tracked over an extended period of time (say, 2 weeks). It could be feasible to do such experiments in the field, using the Wieland-Meixner test (Chap. 2) to distinguish in situ between mushroom specimens that contained amatoxins and those that did not. Simply observing what creatures were present naturally on toxin positive/negative mushrooms in the field might be highly informative. As discussed above, it would also be feasible to test whether any mycophagous flies accumulate amatoxins inside their bodies in order to protect themselves against predators or parasites. Insects can be trapped using sticky tape, and modern analytical methods can detect μg levels of α-amanitin (Oliveira et al. 2015) (Chap. 2).

In regard to mechanisms of resistance among insects, intrinsic pol II insensitivity has only been found under conditions of artificial selection in the laboratory (Chap. 5). The best hypothesis for natural resistance is some type of detoxification, including chemical modification and sequestration, with the caveats discussed above.

6.1.6 Amatoxins and the Mycorrhizal Symbiosis

So far in this chapter we have considered only the possible defensive roles of the cyclic peptide toxins against fungivores. However, of the many inter-organismal relationships in which cyclic peptide toxin-producing species of *Amanita* participate, the most important to overall global ecology is their mycorrhizal association with higher plants. Plant pol II is sensitive to α-amanitin, and phalloidin binds to plant actin (Chap. 5), so the toxins have the potential to affect the hosts of ectomycorrhizal fungi. For example, the amatoxins and phallotoxins might directly influence root growth or participate in a tritrophic interaction that indirectly affects plant growth. Additionally, it is not impossible (although unsupported by any evidence to date) that the peptide toxins from decaying fruiting bodies might influence the microbial community in the rhizosphere.

In considering the relationship between the mycorrhizal niche and the cyclic peptide toxins, it is first important to note that some of the toxin-producing fungi, such as *Galerina* and *Lepiota*, are saprobic, i.e., they obtain their energy from decomposition of dead plant material. (*Saprophytic* and *saprotrophic* are synonymous with *saprobic*.) A hallmark of the saprobic lifestyle is the biosynthesis and secretion of enzymes capable of attacking the more than 100 chemical linkages found in plant polymers such as polysaccharides, proteins, and nucleic acids. Fungi are especially well-endowed with enzymes that attack the plant cell wall, such as cellulases, hemicellulases, proteases, lytic polysaccharide monooxygenases, lyases, and polysaccharide esterases. There are more than 80 families of fungal glycosyl hydrolases alone (www.cazy.org).

In contrast to the saprobic niche of *Galerina* and *Lepiota*, most species of *Amanita* are mycorrhizal, that is, they form obligatory symbiotic relationships on the roots of host plants. The mycorrhizal habit apparently evolved once from a saprobic ancestor within the genus *Amanita*, i.e., the trait is monophyletic within

Amanita (Wolfe et al. 2012). The toxin trait is also probably monophyletic in *Amanita*, but it evolved after the evolution of the mycorrhizal trait later because all toxin-producing species are mycorrhizal, but not all mycorrhizal species are toxicogenic (Cai et al. 2014; Wolfe et al. 2012).

More than 80% of green plants are mycorrhizal (http://en.wikipedia.org/wiki/Mycorrhiza). There are two main types of mycorrhizae, ectomycorrhizae (abbreviated EM or ECM), and arbuscular mycorrhizae (AM or VAM) (Smith and Read 2008). EM fungi, which include many species of *Amanita*, form a sheath of fungal mycelial tissue around the root, which penetrates between the epidermal and cortical cells and extends into the soil (Figs. 6.3 and 6.4). This network greatly increases the

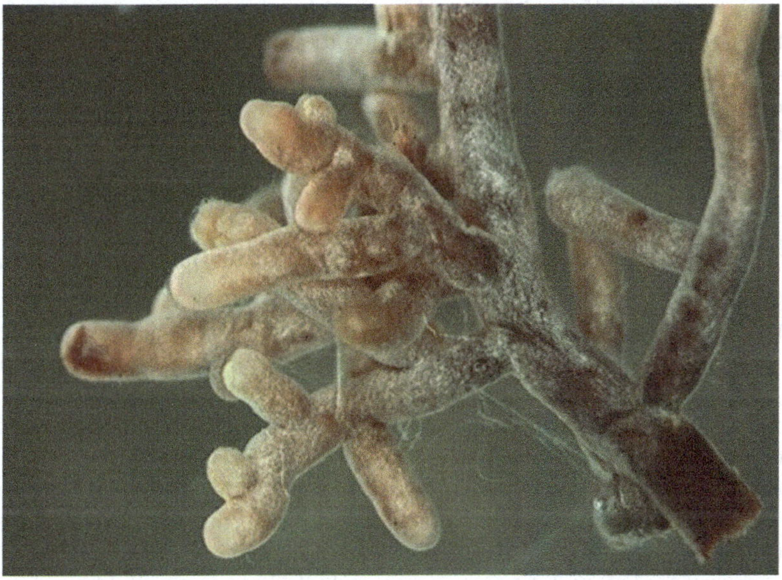

Fig. 6.3 A root colonized by an ectomycorrhizal species of *Amanita*. (From Nilsson et al. 2005. Photo credit: Ellen Larsson, University of Gothenburg. Creative Commons Attribution license)

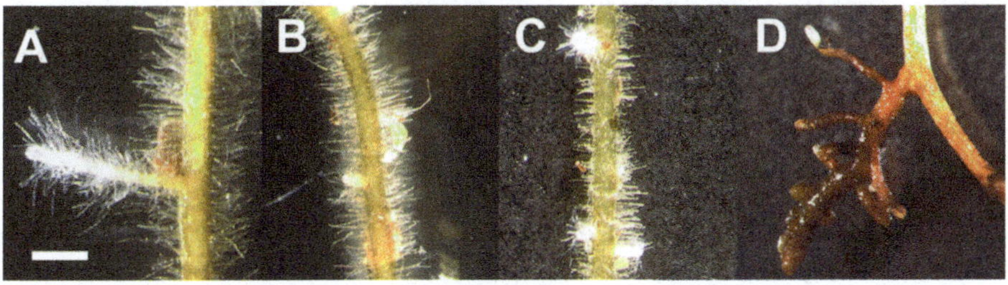

Fig. 6.4 Appearance of roots of *Picea abies* (Norway spruce) inoculated with various *Amanita* species. (**a**) Uninoculated, (**b**) inoculated with *A. thiersii* (saprobic), (**c**) inoculated with *A. inopinata* (saprobic), and (**d**) inoculated with *A. muscaria* (ectomycorrhizal). (Photo credit: Benjamin Wolfe, Tufts University. © 2012 Wolfe et al. (2012). Creative Commons Attribution license)

effective surface area of the host's roots and facilitates the uptake and transfer of nutrients from the soil into the host. *Amanita* forms EM associations with many host plants, including beech, oaks, pines, and eucalyptus. The EM symbiosis is important for supplying the host plants with soil nutrients, especially phosphorus and nitrogen, but likely also K^+, Mg^{++}, and Ca^{++} when they are limiting (Smith and Read 2008). In exchange, the fungal partner obtains fixed carbon (in the form of sugars) from the host. Mature forest trees can exchange significant amounts of carbon among themselves through the ectomycorrhizal network (Klein et al. 2016). Furthermore, EM inhibit root damage from antagonists such as pathogenic bacteria and fungi, thereby influencing temperate forest population structure (Bennett et al. 2017).

During the mycorrhizal symbiosis, sucrose from the host is exported to the plant apoplast, converted to free glucose and fructose by plant invertase, and the glucose then taken up by specialized transporters into the fungal cytoplasm. *Amanita muscaria*, although not a producer of the cyclic peptide toxins, is one of the better studied ectomycorrhizal species at the molecular level. It has at least two membrane-localized transporters, AmMst1 and AmMst2, which are responsible for transporting glucose across the fungal plasma membrane from the apoplast of the host into the fungal cytoplasm (Nehls et al. 2010). These and other molecular aspects of the *A. muscaria* mycorrhizal interaction most likely pertain to toxin-producing mycorrhizal species of *Amanita* as well (Buscot et al. 2000). As one would expect, *A. phalloides* and *A. bisporigera* have strong homologs of AmMst1 and AmMst2 (Pulman et al. 2016).

A hallmark of ectomycorrhizal fungi is their loss of the genes encoding many or most secreted cell wall-degrading enzymes, which saprobic fungi utilize to scavenge carbon from dead plant material (Martin et al. 2008; Nagendran et al. 2009; Wolfe et al. 2012). A plausible adaptive rationale for this loss of extracellular cell wall-degrading enzymes is twofold. First, because the mycorrhizal fungi obtain their carbon from their living host plants instead of from dead plant material, the degradative enzymes are not needed. Second, many cell wall-degrading enzymes are triggers of host plant defense responses, which, if present, would interfere with development of the symbiotic relationship (Walton 1994).

As a result of their obligate symbiotic relationships, species of *Amanita* are difficult or impossible to culture in the laboratory on any synthetic or complex medium that anyone has ever tried. This limits our ability to experimentally manipulate them. Some species have been cultivated with limited success (e.g., the toxin-nonproducing species *A. muscaria* and the toxin-producing species *A. exitialis*; Zhang et al. 2005a, b). (A report of rapid in vitro growth of *A. hygroscopica* is probably spurious; Chap. 3.)

Do the toxins play any role in the mycorrhizal symbiotic interaction? This seems unlikely for several reasons. First, α-amanitin is also produced by the saprobic fungi *Galerina* and *Lepiota*, and therefore if it contributes to the initiation or maintenance of the symbiotic relationship, it must have a different function in these other two fungi, i.e., among the agarics the toxins are apparently not an adaptation to a particular nutritional niche. Second, whereas the cyclic peptide toxins are made at some level in the mycelium, as shown for *G. marginata* and *A. exitialis*, they are present at much higher levels in fruiting bodies, which do not come into direct contact with the host plant. Whether the mycelium in nature can produce the cyclic peptide toxins is not known but, if so, this could be an important clue to the ecological functions of the toxins. For example, toxins in the mycelium could deter mycophagy or influence the rhizosphere. Third, insofar as the toxins contribute to the survival of the fungi per se, for which the symbiotic interaction is essential, then one would have to propose a positive or at least neutral effect of the amatoxins and phallotoxins on the host plant. However, the toxins are known only to cause deleterious effects on the metabolism of eukaryotic cells including plants, and no has ever observed or even postulated a way in which they could have a *direct* positive effect on any organism. Fourth, many species of *Amanita*, such as *A. muscaria*, that do not produce the toxins are still able to form robust ectomycorrhizal associations. Thus, the toxins are neither restricted to mycorrhizal fungi nor necessary for the symbiotic interaction. Taken together, the evidence does not support a direct or indirect role for the *Amanita* cyclic peptide toxins in the mycorrhizal symbiosis.

6.1.7 **Amanita phalloides** *as an Invasive Species in North America*

A. phalloides is native to Europe, but sometime in the first half of the twentieth century, it was introduced into North America. It has also been introduced into Africa and Australia. In North America there are two disjunct populations, one along the East coast from Maryland to Maine including New Jersey, New York, and New Hampshire, and one on the West coast from Southern California to British Columbia including inland to the foothills of the Sierra Nevada mountains (Wolfe et al. 2010). A recent introduction into the West coast is supported by chemical and genomic profiling of European and California populations (Pringle et al. 2009; Pulman et al. 2016; Sgambelluri et al. 2014).

A. phalloides is increasingly abundant in the fall and winter months in the San Francisco Bay region of California. Based on molecular analysis of herbarium specimens, *A. phalloides* was probably introduced into California in the 1930s on the roots of oak trees such as *Quercus suber* (cork oak) (Pringle et al. 2009). It is recently becoming widely established in British Columbia including on city street trees (Bazzicalupo et al. 2017).

Like other ectomycorrhizal fungi, *A. phalloides* can associate with many tree species. In Europe it grows mainly on beech, oak, pine, and chestnut, but in North America it appears to be found mainly on native oaks such as coast live oak (*Q. agrifolia*), pine (*Pinus* species), and hemlock (*Tsuga* species). In Australia *A. phalloides* grows on native *Eucalyptus* species, in Tanzania on leguminous trees, in South America on various native trees, and in New Zealand on native *Leptospermum* species (R. Tulloss, www.amanitaceae.org; https://en.wikipedia.org/wiki/Amanita_phalloides). Thus, it clearly has no trouble jumping to new endemic hosts. In the Pt. Reyes peninsula of central California, *A. phalloides* can be the dominant ectomycorrhizal fungus and might thereby be altering the local fungal and plant communities (Wolfe et al. 2010). Amateur mycologists in the San Francisco Bay Area have reported that *A. phalloides* is displacing another ectomycorrhizal fungus, the native chanterelle (*Cantharellus* species) (Randy Garrett, personal communication). The possible importance of the cyclic peptide toxins to the success of *A. phalloides* as an invasive species in North America has not been studied.

6.2 Evolution of the *Amanita* Cyclic Peptide Toxins

In this section we will first consider the *Amanita* peptide toxins in relation to other natural products, especially other ribosomally synthesized peptides (RiPPs). It is clear that despite some common themes between the *Amanita* RiPPs and RiPPs from nonfungal sources, there is no evolutionary relationship, i.e., the commonalities are due to convergent evolution and universal biochemical constraints and not to descent with modification from a common ancestor (i.e., true homology). We will then turn to the situation within the agarics, comparing the biosynthetic routes for the cyclic peptides among *Amanita*, *Galerina*, and *Lepiota*, all of which make α-amanitin, and examine whether this might be an example of true homology or convergent evolution. Finally, we consider some speculative theories to explain the origin, amplification, and species diversity of the cycloamanide gene family.

6.2.1 Amanita *Cyclic Peptides in Relation to Other Natural Products*

Cyclic peptides are among the ecologically and pharmaceutically most important classes of secondary metabolites. Cyclic peptides can be cyclized by all head-to-tail peptide bonds (*homodetic* cyclic peptides), by a combination of peptide and ester bonds (*depsipeptides*), between one of the ends and an internal amino acid (*isopeptides*), and/or through internal disulfide bonds between pairs of cysteine residues. Some cyclic peptides such as the cyclotides, amatoxins, and phallotoxins, are multicyclic due to both head-to-tail peptide bonds as well as internal bonds (Fig. 4.15). The chemical properties of cyclic peptides that contribute to their propensity for potent biological activity are considered in more detail in Chap. 7.

Cyclic peptides have evolved multiple times in all branches of life, including animals, plants, bacteria, and fungi (Craik and Allewell 2012). Their potential diversity is huge; there are 2.6×10^{10} possible octapeptides even with just the canonical 20 amino acids. Some small peptides, such as the disulfide-cross-linked nonapeptides vasopressin and oxytocin, are animal hormones and can therefore be considered to be primary rather than secondary metabolites; here we will consider only those small peptides that fall into the broad category of secondary metabolites.

Small cyclic peptides are biosynthesized either by nonribosomal peptide synthetases (NRPSs) or on ribosomes (Chap. 4). In the latter case, they are called RiPPs for ribosomally synthesized and posttranslationally modified peptides (Arnison et al. 2013). Prior to the discovery of the *Amanita* toxin genes, all fungal cyclic peptides were believed to be biosynthesized by NRPSs (Hallen et al. 2007). RiPPs are now also known in another basidiomycete (agaric) (Ramm et al. 2017; Van der Velden et al. 2017) and in several ascomycetous fungi (Ding et al. 2016; Nagano et al. 2016; Umemura et al. 2014) (Chap. 4). RiPPs might be much more widespread in fungi than currently appreciated, but at the moment we lack the bioinformatics tools to identify RiPPs de novo from fungal genome sequences.

Other classes of RiPPs and their biosynthetic similarities and differences to the *Amanita* toxins are considered in Chap. 4. Here we revisit one class, the cone snail toxins (conotoxins), to emphasize how convergent evolution can lead to interesting parallels. A single species of cone snail can have more than 150 conotoxin genes, each gene producing a different toxin from its core region, and different species of cone snails make different conotoxins (Arnison et al. 2013; Duda and Palumbi 1999). Similarly, one species of *Amanita* can have ~30 members of the cycloamanide family, and the core regions in different species of *Amanita* are largely non-overlapping (Pulman et al. 2016). Both the conotoxins and the *Amanita* toxins are translated as larger precursor peptides with a conserved region common to all members of the family in a single species and a variable region that comprises the amino acids found in the mature toxins. The conotoxins lack a conserved follower sequence but have a classic signal peptide for secretion into the venom duct (Olivera et al. 2012; Robinson and Norton 2014; Verdes et al. 2016). The *Amanita* toxin precursor peptides lack a signal peptide. The conotoxins therefore have the overall structure: signal sequence-conserved leader-variable core region (Fig. 4.14), whereas the *Amanita* toxins have the structure: conserved leader-variable core region-conserved follower (Fig. 4.12). The venom genes of funnel web spiders, in which the mature toxins are 36–37 amino acids with 3 disulfide bridges, have the same structural pattern as the cone snail toxins (Pineda et al. 2014). Another parallel between the *Amanita* cyclic peptide toxins and the cone snail toxins is that both can undergo the same posttranslational modifications, including Pro hydroxylation and α-carbon epimerization (Buczek et al. 2005). Pro hydroxylation in the conotoxins is catalyzed by a dioxygenase, like collagen in animals and extensins in plants (Gorres and Raines 2010), but the Pro hydroxylase for the *Amanita* cyclic peptides is not known (Chap. 4). The enzyme responsible for amino acid epimerization is unknown for either class of peptide.

6.2.2 *Origins of the* Amanita *Peptides*

The structure of the genes encoding the peptide toxins in *Amanita* and *Galerina* (Chap. 4) impose several constraints on possible theories of the origins and continuing evolution of the *Amanita* cyclic peptides. Salient points are, first, fungi that do not make the *Amanita* cyclic peptides completely lack the encoding genes. That is, the absence of amatoxins and phallotoxins in most agarics is not because of small mutations or secondary biochemical effects such as a block in toxin transport (Hallen et al. 2007). Second, *Galerina marginata* and *Lepiota subincarnata* also make α-amanitin as 34- or 35-amino acid precursor peptides of similar overall structure (i.e., a 9- or 10-amino acid leader, an 8-amino acid core region, and a 17-amino acid follower). In the three amanitin genes from the three genera, the amino acid leader peptides have in common a sequence of five amino acids immediately upstream of the core region (i.e., Asn-Ala-Thr-Arg-Leu) (Fig. 4.4). This could indicate either a common evolutionary origin or an obligatory constraint for processing of the precursor peptide by the cognate POPB enzymes. In the three genera, the follower peptide has low primary amino acid conservation but conserved secondary structure (Fig. 4.8).

Within the genus *Amanita*, amatoxin production is probably *monophyletic*, i.e., it originated evolutionarily only once in an ancestral species in sect. *Phalloideae* and is therefore a *synapomorphy*. A monophyletic origin is consistent with the phylogenetic trees of Weiß et al. (1998), Drehmel et al. (1999), and Cai et al. (2014). Cai et al. (2014) reported that *A. areolata*, *A. hesleri*, and *A. zangii* belong to sect. *Phalloideae* but do not produce amatoxins or phallotoxins. If true, then these three fungi constitute a basal lineage of sect. *Phalloideae*, i.e., toxin production evolved after these species split off from the toxin-containing clade of *Phalloideae*. This interpretation must be considered tentative because only one or a few specimens of these three species have been tested for the presence of toxins, and the toxin trait can be variable in some *Amanita* species (Chap. 3). A more definitive test of the genetic potential of these basal lineages to make amatoxins and/or phallotoxins would be a molecular test for the presence or absence of the cycloamanide gene family. A different phylogenetic analysis, whose focus was on the evolution of the mycorrhizal habit in *Amanita* and not of cyclic peptide toxin production per se, showed some intermingling of toxin producers and nonproducers (Wolfe et al. 2012). However, overall toxin production is probably monophyletic within sect. *Phalloideae*, but a definitive answer will require broader phylogenetic sampling within *Amanita* (including multiple specimens of each species), accompanied by a concurrent genetic analysis of the cycloamanide family through genome sequencing and chemical profiling.

Within the genus *Galerina*, only some species have the genes for *AMA1* and *POPB* and make amatoxins (Luo et al. 2012). The most recent phylogenetic analysis indicates that amatoxin biosynthesis is also monophyletic in this genus (B. Landry and M. Berbee, University of British Columbia, unpublished results).

6.2.3 Discontinuous Distribution of the Cyclic Peptide Toxins and Horizontal Gene Transfer

Amanita, *Galerina*, and *Lepiota* are not closely related among the agarics, belonging to different families (Fig. 6.5). Amatoxin biosynthesis is thus not monophyletic within the order Agaricales. It is an example of a *symplesiomorphy*, a trait found in

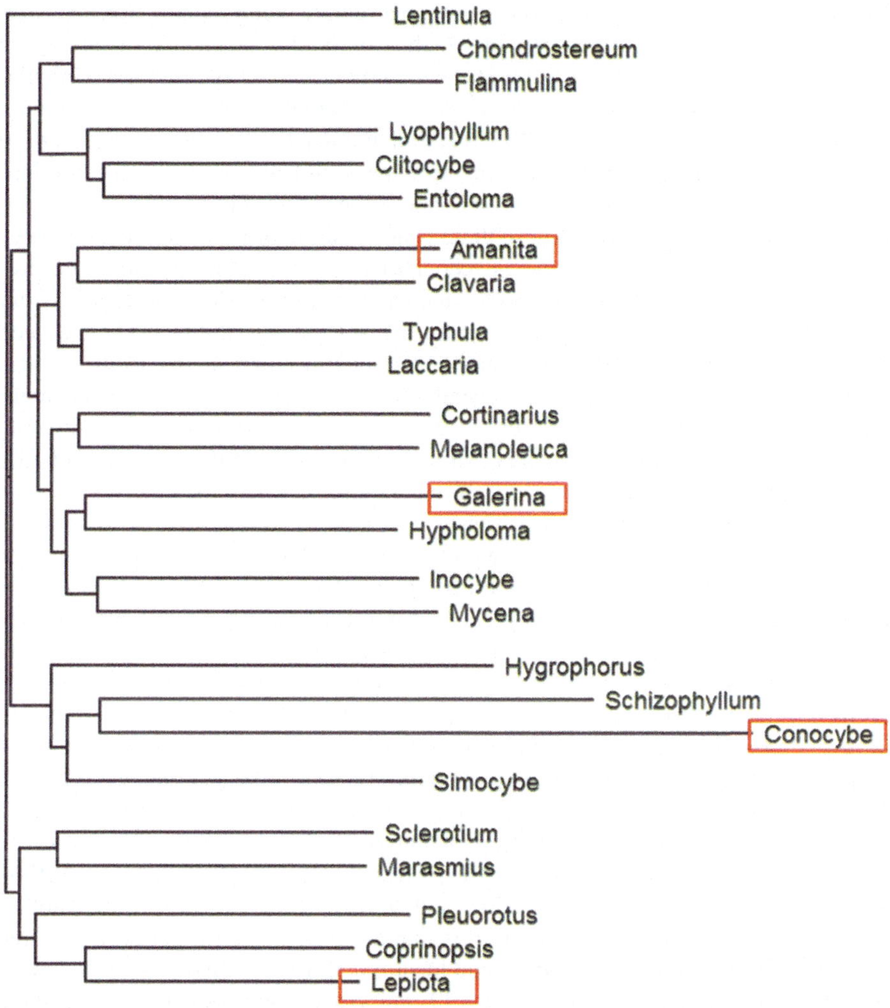

Fig. 6.5 Phylogenetic tree showing the relatedness of toxin-producing mushrooms to each other and to other agarics. The tree is based on a single representative *rpb2* amino acid sequence from each of the major families of agarics and is meant to show relative relatedness and not a rigorous multilocus phylogeny (Matheny et al. 2006).The known or suspected toxin-producing genera (names in red boxes) are *Lepiota* (Family Agaricaceae), *Galerina* (Hymenogastraceae or Strophariaceae), *Amanita* (Amanitaceae), and perhaps *Conocybe* (Chap. 3) (Bolbitaceae). GenBank accession numbers for the sequences are: *Lentinula*, AY218492; *Chondrostereum* AY218477; *Flammulina* AY786055; *Lyophyllum* DQ367433; *Clitocybe* AY780942; *Entoloma* DQ385883; *Amanita* AY485609; *Clavaria* AY780940; *Typhula* AY218525; *Laccaria* DQ472731; *Cortinarius* DQ083920; *Melanoleuca* DQ474119; *Galerina* AY337357; *Hypholoma* AY337413; *Inocybe* AY337409; *Mycena* DQ474121; *Hygrophorus* DQ472720; *Schizophyllum* AY218515; *Conocybe* DQ470834; *Simocybe* DQ484053; *Sclerotium* AY641027; *Marasmius* DQ474118; *Pleurotus* AY786062; *Coprinopsis* AY780934; *Lepiota* JN993699

two or more taxa that are related only through a remote common ancestor. This pattern of discontinuous distribution, in which a particular secondary metabolite is found in widely unrelated taxa yet at the same time not in all species of a particular genus, has been documented on multiple occasions for natural products in both fungi and bacteria (e.g., Khaldi and Wolfe 2011; Slot and Rokas 2011; Wight et al. 2013).

What evolutionary mechanism or mechanisms could account for the discontinuous distribution of the *Amanita* cyclic peptides? We cannot say anything significant about the *origin* of the trait, because there is no fossil record buried in the genomes of any known organism that allows reconstruction of the route from some pre-existing gene family to the extant cycloamanide gene family. In regard to its current existence in three unrelated agarics, there are several possibilities: (1) the deep ancestor of all three genera made amanitin, and the trait has been lost from all intervening taxa, (2) the three genera independently evolved the capacity to make amanitin (i.e., convergent evolution), or (3) the genes moved from one fungus to the others (i.e., horizontal gene transfer). All three theories share the property of being highly improbable, and none appears more parsimonious than the others.

Loss from all intervening taxa cannot be excluded but one can hardly imagine what such a deep ancestor might have looked like or what niche it occupied. Insofar as all agarics face many of the same ecological challenges, including insect herbivory, why would the presumably useful toxin trait have been lost in most lineages? Convergent evolution seems unlikely considering the strong similarity of the amanitin biosynthetic pathway in *Amanita*, *Galerina*, and *Lepiota* (Chap. 4). However, there are precedents for a complex metabolite evolving multiple times in unrelated organisms by convergent evolution, evidenced by key biosynthetic steps being catalyzed by different classes of enzymes (e.g., Bömke and Tudzynski 2009). If the amanitin trait evolved in the three genera by convergent evolution, it seems most likely that at least one of the fungi would have evolved a pathway to amanitin that involved an NRPS to make the cyclic intermediate, cyclo(IWGIGCNP). NRPSs are more common in fungi than RiPPs and show tremendous versatility. Another argument against convergent evolution is that all three amanitin genes in all three fungi have an intron in exactly the same location, all the precursor peptides have a 9- or 10-amino acid leader peptide and a 17-amino acid follower peptide, and all are predicted to be cyclized by a dedicated prolyl oligopeptidase (POP). These facts support a common ancestor of the three amanitin genes, i.e., they are true homologs. This conclusion favors horizontal gene transfer (HGT) as the most parsimonious explanation for their extant taxonomic distribution.

HGT is accepted as common in bacteria, but well-documented examples are still rare in fungi. Based largely on statistical bioinformatics analyses, it is now frequently deduced to be the most likely explanation to account for the presence of certain secondary metabolites and their corresponding biosynthetic gene clusters in unrelated fungi (e.g., Wisecaver and Rokas 2015). However, the *mechanisms* of HGT in fungi are still largely unresolved (Friesen et al. 2006; Rosewich and Kistler 2000; Van der Does and Rep 2012; Walton 2000; but see He et al. 1998).

If HGT is the explanation for the presence of the cyclic peptide toxins in three unrelated genera, then what might have been the direction of transfer – i.e., in which fungus did it first evolve? Two aspects are probably relevant. One is that there is a biosynthetic gene cluster (albeit probably partial) in *Galerina* but not in *Lepiota* or *Amanita*, and the other is that the cycloamanide gene family is much larger in *Amanita* than in *Galerina* and of intermediate size in *Lepiota* (Chap. 4). Assuming that *AMA1* (encoding the α-amanitin core peptide) and *POPB* (encoding the first essential processing step of the core precursor peptide) must be transferred together to be a functional unit, then the most plausible direction of horizontal movement would be from *Galerina* to *Amanita* and/or to *Lepiota*, rather than from *Lepiota* or *Amanita* to *Galerina*. This conclusion is based on the assumption that HGT typically involves the transfer of subgenomic fragments (anywhere from 50 kb to a whole chromosome) and that a gene cluster for a biosynthetic pathway is more likely to be transferred by HGT if the genes are in close genomic proximity. If the genes are scattered in the genome (as apparently *POPB* and *AMA1* are in *Amanita* and *Lepiota*; Pulman et al. 2016, and unpublished results from the author's laboratory), they would be much less likely to be transferred together (Walton 2000; Wisecaver and Rokas 2015). Two corollaries of the "*Galerina* first" hypothesis are that genomic separation of *POPB* and *AMA1* and expansion of the cycloamanide family in *Amanita* and *Lepiota* occurred after transfer, plausibly by well-documented processes such as chromosomal translocations, inversions, gene duplications, etc. A post-transfer expansion is supported by the much higher amino acid identity of the conserved regions (leader and follower) of the cycloamanide family members within a genus than between genera (Figs. 4.4, 4.12, and 4.13). In regard to the evolutionary trajectory of *POPB*, presumably it evolved from *POPA*, which encodes the housekeeping POP, by the canonical processes of gene duplication and neo-functionalization (Luo et al. 2014). There is no indication where it first evolved into a peptide macrocyclase; the *Lepiota* POPB homolog has not been described as of this writing. The agaric *Omphalotus olearius* also has a presumptive macrocyclizing POP, but the evolutionary relationship between it and POPB of *Amanita* and *Galerina* has not been examined (Ramm et al. 2017).

6.2.4 Cycloamanide Gene Structure and Selection

The cycloamanide family in all three genera (*Amanita*, *Galerina*, and *Lepiota*) is characterized by a conserved leader (9 or 10 amino acids) and follower peptide (always 17 amino acids) flanking a variable core region of 6–10 amino acids. A gene/precursor peptide structure consisting of conserved and variable domains is also characteristic of animal venom precursor peptides such as those for the amphibian antimicrobial peptides and cone snail, snake, spider, centipede, and scorpion toxins (Arnison et al. 2013; Nicolas et al. 2003; Sollod et al. 2005; Tang et al. 2010). Such gene structures can be interpreted to be the result of negative (purifying) selection operating on the conserved regions (i.e., most mutations in these regions result in reduced cyclic peptide production and are therefore eliminated by natural selection) and positive (diversifying) selection operating on the variable core region (i.e., rapid change is advantageous due to rapid coevolution with the organism or organisms against which the toxins are targeted). There are several bioinformatic tests to distinguish if genes are under purifying or diversifying selection based on the dN/dS metric, i.e., the ratio of differences in nonsynonymous to synonymous codons (Hedrick 2011). Animal venom genes have been subjected to such tests numerous times (e.g., Conticello et al. 2000, 2001; Sollod et al. 2005; Sunagar and Moran 2015; Tang et al. 2010). Whereas some of these studies have found evidence for positive selection (e.g., Casewell et al. 2013), in other cases researchers found evidence for negative selection, no statistically significant selection in either direction or a combination of positive and negative selection at different stages of gene family evolution (Sollod et al. 2005; Sunagar and Moran 2015).

Li et al. (2014) concluded that the *AMA* and *PHA* genes in six species of *Amanita* were under purifying selection. However, a different evolutionary trajectory might have been deduced if the test were performed on just the core toxin regions rather than on the whole precursor peptides (i.e., cores plus flanking conserved domains). It is possible that the conserved domains and core domains are under different types of selection, as they are in the signal peptide, propeptide, and core toxin regions in some of the animal toxins (Sollod et al. 2005).

In cone snail toxin genes, the fact that introns separate the three domains has been suggested to be related to the apparent different rates of evolution of the domains (Olivera et al. 1999). This cannot be relevant to the cycloamanide genes, which have only a single intron in the precursor peptide coding region. This intron occurs near the carboxyl terminus of the precursor peptide and not between any of the domains (Fig. 4.1). However, it is striking that in all of the genes in all three genera of cycloamanide-producing fungi, the intron is in exactly the same location, namely, interrupting the fourth from the last codon (Luo et al. 2012; Pulman et al. 2016; unpublished results from the author's lab). Insofar as intron presence and location have no functional significance and are therefore not under selective pressure (a controversial subject for introns in general), this fact could be used as an argument in favor of a common origin for the α-amanitin genes in all three genera, i.e., in support of HGT or descent from a common ancestor.

6.2.5 Evolution of the Broader Cycloamanide Family

Although the amatoxins share the same biosynthetic pathway with the traditional cycloamanides sensu Wieland (1986), from the ecological and evolutionary perspective, they have important distinctions. Amanitin is the most widespread and abundant of the cycloamanides, and it is highly biologically active. These are all traits expected for a molecule conferring a strong selective advantage on the fungi that make it. In contrast, most of the members of the larger cycloamanide family are made at lower levels, if at all, they have no known biological activities that make ecological sense, such as toxicity (Chap. 5), and they are highly variable between species of *Amanita*. This last point is the most intriguing from an evolutionary point of view. *A. bisporigera* and *A. phalloides* each contain ~30 cycloamanide genes, only three of which are in common (Pulman et al. 2016). Even between two closely related species within the *A. bisporigera* species complex, there is almost no overlap in the cycloamanide gene complement (Pulman et al. 2016). It is important to emphasize that at least half of the cycloamanide gene family are expressed at the RNA level in *A. bisporigera* and at least ten of them at the chemical level in *Amanita* and *Lepiota* (Chap. 4). Furthermore, none of the predicted cycloamanide genes contain stop codons or frame shifts, so they are probably not pseudogenes, i.e., nonfunctional evolutionary relics.

6.2.5.1 Evolutionary Scenarios for the Cycloamanides: The Strict Adaptationist Hypothesis

In regard to the expanded cycloamanide family, a strict adaptationist stance would argue that every one of the cycloamanides confers functional advantage to the producing fungus. That is, each of the 30+ cycloamanides in *A. phalloides* or *A. bisporigera* has a particular target in some organism with which these *Amanita* species have an antagonistic relationship. Such a scenario could conceivably emerge from a protracted arms race with a rapidly evolving antagonist such as a mycophagous insect, although at this stage we do not know what that antagonist is, much less what its complementary genome evolution looks like. An advanced stage of an arms race has been proposed to account for why the average cone snail makes 50–200 toxic peptides, but it is not clear what all of those hypothetical cone snail antagonists are. In the case of cone snails, antagonists could include not just pathogens and predators but also competitive snails and prey such as worms, fish, and crustaceans. On the one hand, diversification of the cone snail toxins could "enable rapid access to new prey" (Olivera et al. 2012), but on the other hand, why would a snail need 150+ new compounds in order to prey on one or two new groups of organisms, e.g., to move from hunting worms to hunting fish? *Amanita* has pathogens, predators, and competitors, but there is no evidence that it is itself a predator (although some mushroom-forming fungi do attack and consume nematodes; see above). While one could conceive that any particular species of *Amanita* might have 30+ antagonists, it seems unlikely that small molecules such as cycloamanides would be effective deterrents against all types of potential enemies.

A critical point that needs to be factored into any evolutionary and ecological scenario for *Amanita* is the fact that the cycloamanide gene families in *A. phalloides* and *A. bisporigera* (and even within the *A. bisporigera* complex) are largely different (Pulman et al. 2016). This implies that the different species of *Amanita* are being driven in different directions by different selective pressures. For example, *A. phalloides* makes CylA (Table 2.2) but not amanexitide, whereas the opposite holds true for *A. exitialis*. To the strict adaptationist, does this mean that the antagonist against which CylA evolved has never been a threat to *A. exitialis* and vice versa? This seems improbable because no one has ever documented that any mushroom antagonist such as a mycophagous fly specializes on different species of *Amanita*.

Perhaps the arms race partners of the cone snail toxin and cycloamanide families are not *species* of antagonists but *variants* of specific molecular targets. For example, if the gene for an ion channel that is inhibited by a particular cone snail toxin were to mutate to resistance, that would drive evolution of the snail toxin gene to encode a modified toxin that can now inhibit the formerly resistant ion channel, leading in turn to another mutation in the ion channel gene, etc. These target variants could exist as multiple alleles within the population of a single antagonist species of, e.g., prey fish. In order for a particular cone snail to successfully hunt that species of fish, it would need to make multiple variants of a particular toxin. A molecular arms race within the cycloamanide gene family might be the reason why some cycloamanides, e.g., CylC and CylD, and CylB and CylE (Table 2.2), have nearly identical structures, which would be expected if they inhibit variants of the same target. In regard to the *Amanita* cycloamanides, we will not be able to speculate about a possible molecular arms race until we have a clearer idea of what the major *Amanita* mushroom coevolutionary antagonists are, as well as knowing more about the modes of action of the cycloamanides other than amatoxins and phallotoxins.

Another potentiality that would complicate the rather simplistic evolutionary scenarios so far discussed is that the cycloamanides might not be active individually but rather in combinations. There is at least one example of a cone snail toxin pair that acts cooperatively (Wong and Belov 2012). It is easy to imagine how mixtures of toxic compounds might be more effective at stopping an antagonist than individual compounds with specific modes of action, no matter how toxic. Certainly, the venoms of many animals besides cone snails are complex "cocktails" of many components, usually peptidic, including hydrolytic enzymes, hemotoxins, cytotoxins, and inhibitors of ion channels and neurotransmitter receptors (https://en.wikipedia.org/wiki/Snake_venom). Perhaps against some antagonists, a combination of amatoxins, phallotoxins, and/or cycloamanides are more effective than any one class or molecular species alone. In this case, the individual cycloamanide genes would not be evolving independently but rather influencing each other's adaptive advantage.

6.2.5.2 Evolutionary Scenarios for the Cycloamanides: The Nonadaptationist Hypothesis

At the other extreme, we need to consider the possibility that the cycloamanide diversity found within and between species of *Amanita* originated partly or completely as an accidental by-product of random mutation and gene amplification. Instead of each cycloamanide conferring a selective advantage to the fungus, could at least some of the cycloamanide genes be simply evolutionary relics or "genomic noise"? This could be the case even if any one cycloamanide, which would most likely be amanitin, contributed to fitness of any *Amanita* species. Many evolutionary biologists have argued that not all organismal traits are necessarily the result of adaptive selective pressure (i.e., they could be spandrels, sensu Gould and Lewontin 1979).

There are several arguments against the genomic noise hypothesis. First, evolution tends to be intolerant of truly useless genes, so they would not likely persist on any realistic evolutionary time scale as clean open reading frames. On the contrary, there is no evidence from the genomes of *A. bisporigera* and *A. phalloides* that any of the cycloamanide genes are pseudogenes. Second, the cycloamanide family in *Amanita* and *Lepiota* have highly conserved leader and follower peptides, indicating that these portions of the cycloamanide genes at least are under negative (purifying) selection. If they were not useful, i.e., not under selection, the follower and leader peptides would be expected to be more variable than they are, probably to the point of not being able to recognize them any longer by bioinformatic approaches. Third, even within the hypervariable core regions, the amino acid distribution is not random. Some codons (e.g., for Pro, Phe, and Ile) are overrepresented, and others (e.g., stop, Ser, and Arg) are underrepresented (Pulman et al. 2016). This implies that the core regions are not pseudogenes, are under selective pressure, and are therefore functional.

6.2.5.3 Evolutionary Scenarios for the Cycloamanides: The "Soft Adaptationist" Hypothesis

This hypothesis is the most subtle and has gradations. Perhaps the cycloamanides do provide a low level of selective advantage to the fungi, but the precise chemical structures and quantities are not very important. The genes arise by random mutation and persist for many generations despite a marginal contribution to fitness, because the cost of making the compounds is very low. Each gene is small, only ~105 base pairs (35 amino acids × 3 base pairs/amino acid). Translating those genes into the actual cyclic peptides also seems like a small metabolic burden. Because they are not involved in primary metabolism, mutations in the cycloamanide genes are less likely to be strongly deleterious and hence less likely to be rapidly eliminated (Conticello et al. 2001; Olivera et al. 2012).

(It is instructive to consider an alternate universe in which the amatoxins were biosynthesized by a nonribosomal peptide synthetase (NRPS) (Chap. 4). The theoretical amanitin NRPS would have to be ~1.0 MDa in size and require ~30 kb of encoding DNA (Walton et al. 2004). This is about 0.05% of the total genome (~60 MB), maintenance of which would amount to a much higher metabolic burden than the actual amanitin RiPP (i.e., requiring ~300-times more amino acids). Furthermore, NRPSs are less enzymatically flexible than RiPPs, so phalloidin and amanitin would require their own NRPS. Although NRPSs have looser amino acid substrate specificity than proteins made on ribosomes, such that some NRPSs are able to make small families of cyclic peptides with conserved amino acid replacements, the overall evolvability, and hence versatility, of NRPSs is much lower than that of RiPP pathways).

However, if any of the minor cycloamanides should ever become slightly, or slightly more, useful to its producer (i.e., enhance its fitness), then natural selection could quickly drive up its production levels (e.g., by promoter mutations) as well as biological activity (e.g., by additional posttranslational modifications). In a situation where the cycloamanides were being made but did not (yet) have a function, they could be considered preadaptations or "exaptations" (Gould and Vrba 1982).

Imagine, for example, that a primitive cycloamanide of structure cyclo(IWGIGCNP) had originally some interesting property that contributed, however modestly, to the fitness of some ancestral species of *Amanita*. Subsequent random mutation (e.g., of its promoter and/or existing CYP450 or FMO genes) and natural selection led to the production of progressively higher levels and more active and/or more stable variants until the fungi were making the high levels of α-amanitin found today in the basidiocarps of *Amanita phalloides*. Mushrooms cannot, of course, anticipate their future ecological needs, but it does not seem anti-Darwinian to hypothesize selective pressure to maintain and amplify the cycloamanide pathway in general, without having to propose the inevitable evolution of any particular structures or compounds with particular biological activities.

Diversity for its own sake might have a selective advantage. Selection for diversity as opposed to selection on individual members of a complex family occurs in some other biological contexts, such as the adaptive immune system of mammals. Panaccione (2005) has argued that the apparent metabolic sloppiness of ergot alkaloid biosynthesis in fungi might be an adaptive advantage because in this way the same pathway generates more compounds with greater potential for interesting new and therefore useful biological activities than a more finely tuned pathway would. Olivera et al. (2012) proposed that the size and diversity of the cone snail toxin gene families are a means by which cone snails can adapt quickly to new niches. This might be true for *Amanita* as well, although it begs the question of what those niches might be. Selection for diversity would require molecular genetic mechanisms that generate new cycloamanide gene members faster than they disappear through random mutation (see below).

Another speculative possibility in considering the evolutionary pressures acting on the cycloamanide gene family is that the genes for known toxic compounds that are found throughout *Amanita* sect. *Phalloideae*, notably α-amanitin and to a lesser extent one or more of the phallotoxins, might be under different evolutionary constraints than the other cycloamanide genes, which are mainly restricted to a single species (Cai et al. 2014; Pulman et al. 2016). In particular, α-amanitin has the strongest case for contributing to the fitness of the fungi that make it because of its widespread distribution, high levels, and high toxicity to potential fungivores (insects and gastropods). α-Amanitin might be the "keystone" cyclic peptide conferring strong selective advantage, but the existence of the biochemical machinery to process it enables the existence and persistence of the rapidly evolving, highly divergent, and weakly advantageous pool of cycloamanides.

6.2.6 Focal Hypermutation in the Cycloamanide Family?

There has been speculation that gene families that show high diversity, such as animal venoms, might be subject to specific mechanisms that actively promote mutation and hence functional hypervariability. The active process of generating exceptional diversity in narrow regions of the genome has been called *focal* or *targeted hypermutation*. For example, cone snail toxins have highly conserved Cys residues embedded in their hypervariable regions (Fig. 4.14), and it has been hypothesized that there is a mechanism that accelerates mutation of the hypervariable regions and simultaneously protects these essential residues against mutation (Conticello et al. 2001). That such a process might be contributing to the evolution of the *Amanita* cyclic peptide toxins is compatible with the high but intragenically selective divergence of the cycloamanide gene family in the *A. bisporigera* species complex (Pulman et al. 2016).

Several tests have been designed to detect focal hypermutation, for example, site-specific codon bias and transversion/transition ratios (Conticello et al. 2000, 2001; Nicolas et al. 2003). Mechanistically, error-prone DNA polymerase V has been implicated in hypermutation in bacteria, but to the best of the author's knowledge an equivalent polymerase is not known in any eukaryote. Somatic hypermutation in the mammalian immune system, one of the mechanisms by which antibody diversity is generated, involves cytosine deamination followed by mutagenic DNA repair (Di Noia and Neuberger 2007). Again, however, there is no evidence that an analogous system capable of generating hypervariability might be functioning in *Amanita* or in any other organism making complex families of small peptides.

The hypothesis of focal hypermutation includes not only a mechanism of generating hypervariability but also a counteracting force that protects the flanking conserved regions (Conticello et al. 2001). However, it could also be that the observed conservation of the leader and follower peptides of the *Amanita* cyclic peptides is not due to any special mechanism to protect those regions from mutation but rather just because they are evolutionarily young and have not had time to accumulate random mutations, i.e., the core regions are "prematurely aged" relative to the flanking conserved regions because for them evolutionary time has speeded up due to hypermutation. This might apply also to the cone snail toxins. The apparent high conservation of the cone snail signal peptides (Fig. 4.14) might simply reflect a relatively short time having elapsed since the divergence of the gene family members (Olivera et al. 2012).

The possibility of the evolution of evolvability, in particular during periods of stress, has been proposed and discussed many times (Pigliucci 2008). However, conventional natural selection is probably sufficient to account for the observed array of cycloamanide gene family in agarics. That is, the cycloamanide genes mutate at the same rate as the other genes in the genome, and purifying selection eliminates most mutations in the leader and follower because without these not even traces of the cyclic peptide encoded in the core region can be biosynthesized. Whenever the core regions mutate to a novel amino acid sequence that contributes to the organism's fitness, they, too, then become subject to purifying selection. Otherwise, the novel cyclic peptides persist to a stochastically determined greater or lesser extent with no strong selection for or against. The different species of *Amanita* sect. *Phalloideae* extant today have arrived at their unique sets of cycloamanides mainly through genetic drift.

References

Arnison PG, Bibb MJ, Bierbaum G, Bowers AA, Bugni TS, Bulaj G, Camarero JA, Campopiano DJ, Challis GL, Clardy J, Cotter PD, Craik DJ, Dawson M, Dittmann E, Donadio S, Dorrestein PC, Entian KD, Fischbach MA, Garavelli JS, Göransson U, Gruber CW, Haft DH, Hemscheidt TK, Hertweck C, Hill C, Horswill AR, Jaspars M, Kelly WL, Klinman JP, Kuipers OP, Link AJ, Liu W, Marahiel MA, Mitchell DA, Moll GN, Moore BS, Müller R, Nair SK, Nes IF, Norris GE, Olivera BM, Onaka H, Patchett ML, Piel J, Reaney MJ, Rebuffat S, Ross RP, Sahl HG, Schmidt EW, Selsted ME, Severinov K, Shen B, Sivonen K, Smith L, Stein T, Süssmuth RD, Tagg JR, Tang GL, Truman AW, Vederas JC, Walsh CT, Walton JD, Wenzel SC, Willey JM, van der Donk WA (2013) Ribosomally synthesized and post-translationally modified peptide natural products: overview and recommendations for a universal nomenclature. Nat Prod Rep 30:108–160. https://doi.org/10.1039/c2np20085f

Arnold GRW (1976) Internationale Bibliographie der Hypomycetaceae (Mycophyta, Ascomycotina). Bibliographische Mitteilungen der Universitätsbibliothek, Jena

Ballou WH (1927) Squirrels as mushroom eaters. J Mammal 8:57–58

Bazzicalupo A, Berch S, Callan B, Ceska O, Kroeger P, Li R, McIntyre L, Miao V, Stanwick R, Trudell S, Vellinga EC, Walton JD, Arevalo W, Asay A, Auxier B, Ceska A, Dee J, Gamiet S, Landry B, Le Renard L, Schwarz C, Super L, van der Meer B, Berbee M (2017) White paper on strategies to reduce risks and expand appreciation of foraged wild mushrooms. McIlvanea 26: http://www.namyco.org/docs/EdiblePoisonousReport20170914.pdf

Begun DJ, Whitley P (2000) Genetics of α-amanitin resistance in a natural population of *Drosophila melanogaster*. Heredity 85:184–190

Benjamin DR (1995) Mushrooms: poisons and panaceas. A handbook for naturalists, mycologists, and physicians. WH Freeman, New York

Bennett JA, Maherali H, Reinhart KO, Lekberg Y, Hart MM, Klironomos J (2017) Plant-soil feedbacks and mycorrhizal type influence temperate forest population dynamics. Science 355:181–184. https://doi.org/10.1126/science.aai8212

Beug MW, Shaw M, Cochran KW (2006) Thirty-plus years of mushroom poisoning: summary of the approximately 2000 reports in the NAMA case registry. McIlvainea 16:47–68

Bock KW (2016) The UDP-glycosyltransferase (UGT) superfamily expressed in humans, insects, and plants: animal-plant arms-race and co-evolution. Biochem Pharmacol 99:11–17. https://doi.org/10.1016/j.bcp.2015.10.001

Bömke C, Tudzynski B (2009) Diversity, regulation, and evolution of the gibberellin biosynthetic pathway in fungi compared to plants and bacteria. Phytochemistry 70:1876–1893. https://doi.org/10.1016/j.phytochem.2009.05.020

Brakhage AA (2013) Regulation of fungal secondary metabolism. Nat Rev Microbiol 11:21–32. https://doi.org/10.1038/nrmicro2916

Bresinsky A, Besl H (1990) A colour atlas of poisonous fungi: a handbook for pharmacists, doctors, and biologists. Wolfe Publishing, London

Brower LP, Fink LS (1985) A natural toxic defense system: cardenolides in butterflies versus birds. Ann N Y Acad Sci 443:171–188

Buczek O, Bulaj G, Olivera BM (2005) Conotoxins and the posttranslational modification of secreted gene products. Cell Mol Life Sci 62:3067–3079. https://doi.org/10.1007/s00018-005-5283-0

Buscot F, Munch JC, Charcosset JY, Gardes M, Nehls U, Hampp R (2000) Recent advances in exploring physiology and biodiversity of ectomycorrhizas highlight the functioning of these symbioses in ecosystems. FEMS Microbiol Rev 24:601–614

Cai Q, Tulloss RE, Tang LP, Tolgor B, Zhang P, Chen ZH, Yang ZL (2014) Multi-locus phylogeny of lethal amanitas: implications for species diversity and historical biogeography. BMC Evol Biol 14:143. https://doi.org/10.1186/1471-2148-14-143

Camazine S (1983) Mushroom chemical defense: food aversion learning induced by hallucinogenic toxin, muscimol. J Chem Ecol 9:1473–1481. https://doi.org/10.1007/BF00988513

Casewell NR, Wuster W, Vonk FJ, Harrison RA, Fry BG (2013) Complex cocktails: the evolutionary novelty of venoms. Trends Ecol Evol 28:219–229. https://doi.org/10.1016/j.tree.2012.10.020

Conticello SG, Pilpel Y, Glusman G, Fainzilber M (2000) Position-specific codon conservation in hypervariable gene families. Trends Genet 16:57–59. https://doi.org/10.1016/S0168-9525(99)01956-3

Conticello SG, Gilad Y, Avidan N, Ben-Asher E, Levy Z, Fainzilber M (2001) Mechanisms for evolving hypervariability: the case of conopeptides. Mol Biol Evol 18:120–131

Craik DJ, Allewell NM (2012) Thematic minireview series on circular proteins. J Biol Chem 287:26999–27000. https://doi.org/10.1074/jbc.R112.390344

Dawkins R (1986) The blind watchmaker. WW Norton & Co, New York

de Roode JC, Lefèvre T, Hunter MD (2013) Self-medication in animals. Science 340:150–151. https://doi.org/10.1126/science.1235824

Degenkolb T, Vilsinskas A (2016) Metabolites from nematophagous fungi and nematicidal natural products from fungi as alternatives for biological control. Part II: metabolites from nematophagous basidiomycetes and non-nematophagous fungi. Appl Microbiol Biotechnol 100:3813–3824. https://doi.org/10.1007/s00253-015-7234-5

Di Noia JM, Neuberger MS (2007) Molecular mechanisms of antibody somatic hypermutation. Annu Rev Biochem 76:1–22. https://doi.org/10.1146/annurev.biochem.76.061705.090740

Ding W, Liu WQ, Jia Y, Li Y, van der Donk WA, Zhang Q (2016) Biosynthetic investigation of phomopsins reveals a widespread pathway for ribosomal natural products in Ascomycetes. Proc Natl Acad Sci U S A 113:3521–3526. https://doi.org/10.1073/pnas.1522907113

Drehmel D, Moncalvo JM, Vigalys R (1999) Molecular phylogeny of *Amanita* based on large-subunit ribosomal DNA sequences: implications for taxonomy and character evolution. Mycologia 91:610–618. https://doi.org/10.1006/mpev.2000.0782

Duda TF Jr, Palumbi SR (1999) Molecular genetics of ecological diversification: duplication and rapid evolution of toxin genes of the venomous gastropod *Conus*. Proc Natl Acad Sci U S A 96:6820–6823

Fogel R, Trappe JM (1978) Fungus consumption (mycophagy) by small animals. Northwest Sci 52:1–31

Friesen TL, Stukenbrock EH, Liu Z, Meinhardt S, Ling H, Faris JD, Rasmussen JB, Solomon PS, McDonald BA, Oliver RP (2006) Emergence of a new disease as a result of interspecific virulence gene transfer. Nat Genet 38:953–956. https://doi.org/10.1038/ng1839

Gelperin A (1975) Rapid food-aversion learning by a terrestrial mollusk. Science 189:567–570

Gorres KL, Raines RT (2010) Prolyl 4-hydroxylase. Crit Rev Biochem Mol Biol 45:106–124. https://doi.org/10.3109/10409231003627991

Gould SJ, Lewontin RC (1979) The spandrels of San Marco and the Panglossian paradigm: a critique of the adaptationist programme. Proc R Soc Lond B 205:581–598

Gould SJ, Vrba ES (1982) Exaptation – a missing term in the science of form. Paleobiology 8:4–15

Greenleaf AL, Borsett LM, Jiamachello PF, Coulter DE (1979) α-Amanitin-resistant D. melanogaster with an altered RNA polymerase II. Cell 18:613–622

Hallen HE, Luo H, Scott-Craig JS, Walton JD (2007) Gene family encoding the major toxins of lethal *Amanita* mushrooms. Proc Natl Acad Sci U S A 104:19097–19101. https://doi.org/10.1073/pnas.0707340104

Harder A, Schmitt-Wrede HP, Krücken J, Marinovski P, Wunderlich F, Willson J, Amliwala K, Holden-Dye L, Walker R (2003) Cyclooctadepsipeptides--an anthelmintically active class of compounds exhibiting a novel mode of action. Int J Antimicrob Agents 22:318–331

Hastings S, Mottram JC (1916) Observations upon the edibility of fungi for rodents. Trans Br Mycol Soc 5:364–378

Hatt RT (1929) The red squirrel: its life history and habits. Roosevelt Wild Life Annal 2:7–146

He C, Rusu AG, Poplawski AM, Irwin JA, Manners JM (1998) Transfer of a supernumerary chromosome between vegetatively incompatible biotypes of the fungus *Colletotrichum gloeosporioides*. Genetics 150:1459–1466

Hedrick PW (2011) Genetics of populations, 4th edn. Jones & Bartlett, Sudbury

Hess W (2014) Squirrel doing mushrooms. https://www.youtube.com/watch?v=opCZ3b5ePd4

Hoffmeister D, Keller NP (2007) Natural products of filamentous fungi: enzymes, genes, and their regulation. Nat Prod Rep 24:393–416. https://doi.org/10.1039/b603084j

Hsueh YP, Mahanti P, Schroeder FC, Sternberg PW (2013) Nematode-trapping fungi eavesdrop on nematode pheromones. Curr Biol 23:83–86. https://doi.org/10.1016/j.cub.2012.11.035

Hutchison LH, Mazdia SE, Barron GL (1996) The presence and antifeedant function of toxin-producing secretory cells on hyphae of the lawn-inhabiting agaric *Conocybe lactea*. Can J Bot 74:431–434

Jaenike J (1985) Parasite pressure and the evolution of amanitin tolerance in *Drosophila*. Evolution 39:1295–1301. https://doi.org/10.1111/j.1558-5646.1985.tb05695.x

Jaenike J, Grimaldi DA, Sluder AE, Greenleaf AL (1983) α-Amanitin tolerance in mycophagous *Drosophila*. Science 221:165–167. https://doi.org/10.1126/science.221.4606.165

Jakoby WB, Ziegler DM (1990) The enzymes of detoxification. J Biol Chem 265:20715–20718

Kacsoh B, Lynch ZR, Mortimer NT, Schlenke TA (2013) Fruit flies medicate offspring after seeing parasites. Science 339:947–950. https://doi.org/10.1126/science.1229625

Khaldi N, Wolfe KH (2011) Evolutionary origins of the fumonisin secondary metabolite gene cluster in *Fusarium verticillioides* and *Aspergillus niger*. Int J Evol Biol 2011:423821. https://doi.org/10.4061/2011/423821

Klein T, Siegwolf RT, Körner C (2016) Belowground carbon trade among tall trees in a temperate forest. Science 352:342–344. https://doi.org/10.1126/science.aad6188

Klugh AB (1927) Ecology of the red squirrel. J Mammal 8:1–32. https://doi.org/10.2307/1373393

Lewis DF, Dickins M (2004) Compound lipophilicity for substrate binding to human P450s in drug metabolism. Drug Discov Today 15:530–537. https://doi.org/10.1016/S1359-6446(04)03115-0

Li P, Deng W, Li T (2014) The molecular diversity of toxin gene families in lethal *Amanita* mushrooms. Toxicon 83:59–68

Limousin H, Petit G (1932) Essai thérapeutique au cours des intoxications par l'amanite phalloïde. Bull Acad Nat Méd 1932:698–701

Luo H, Hallen-Adams HE, Scott-Craig JS, Walton JD (2012) Ribosomal biosynthesis of α-amanitin in *Galerina marginata*. Fungal Genet Biol 49:123–129. https://doi.org/10.1016/j.fgb.2011.12.005

Luo H, Hong SY, Sgambelluri RM, Angelos E, Li X, Walton JD (2014) Peptide macrocyclization catalyzed by a prolyl oligopeptidase involved in α-amanitin biosynthesis. Chem Biol 21:1610–1617. https://doi.org/10.1016/j.chembiol.2014.10.015

Martin F, Aerts A, Ahrén D, Brun A, Danchin EG, Duchaussoy F, Gibon J, Kohler A, Lindquist E, Pereda V, Salamov A, Shapiro HJ, Wuyts J, Blaudez D, Buée M, Brokstein P, Canbäck B, Cohen D, Courty PE, Coutinho PM, Delaruelle C, Detter JC, Deveau A, DiFazio S, Duplessis S, Fraissinet-Tachet L, Lucic E, Frey-Klett P, Fourrey C, Feussner I, Gay G, Grimwood J, Hoegger PJ, Jain P, Kilaru S, Labbé J, Lin YC, Legué V, Le Tacon F, Marmeisse R, Melayah D, Montanini B, Muratet M, Nehls U, Niculita-Hirzel H, Oudot-Le Secq MP, Peter M, Quesneville H, Rajashekar B, Reich M, Rouhier N, Schmutz J, Yin T, Chalot M, Henrissat B, Kües U, Lucas S, Van de Peer Y, Podila GK, Polle A, Pukkila PJ, Richardson PM, Rouzé P, Sanders IR, Stajich JE, Tunlid A, Tuskan G, Grigoriev IV (2008) The genome of *Laccaria bicolor* provides insights into mycorrhizal symbiosis. Nature 452:88–92. https://doi.org/10.1038/nature06556

Matheny PB, Curtis JM, Hofstetter V, Aime MC, Moncalvo JM, Ge ZW, Slot JC, Ammirati JF, Baroni TJ, Bougher NL, Hughes KW, Lodge DJ, Kerrigan RW, Seidl MT, Aanen DK, DeNitis M, Daniele GM, Desjardin DE, Kropp BR, Norvell LL, Parker A, Vellinga EC, Vilgalys R, Hibbett DS (2006) Major clades of Agaricales: a multilocus phylogenetic overview. Mycologia 98:982–995

Maunder JE, Voitk AJ (2010) What we don't know about slugs and mushrooms! Fungi 3:36–44

Mayer A, Kilian M, Hoster B, Sterner O, Ankel H (1999) In-vitro and in-vivo nematicidal activities of the cyclic dodecapeptide omphalotin A. Pestic Sci 55:27–30. https://doi.org/10.1002/(SICI)1096-9063(199901)55:1<27::AID-PS854>3.0.CO;2-K

Mitchell CL, Saul MC, Lei L, Wei H, Werner T (2014) The mechanisms underlying α-amanitin resistance in *Drosophila melanogaster*: a microarray analysis. PLoS One 9:e93489. https://doi.org/10.1371/journal.pone.0093489

Montarolo PG, Goelet P, Castelluci VF, Morgan J, Kandel ER, Schacher S (1986) A critical period for macromolecular synthesis in long-term synaptic facilitation in *Aplysia*. Science 234:1249–1254

Mullenax CH, Mullenax PB (1962) Mushroom poisoning in cats – two possible cases. Mod Vet Pract 43:61

Nagano N, Umemura M, Izumikawa M, Kawano J, Ishii T, Kikuchi M, Tomii K, Kumagai T, Yoshimi A, Machida M, Abe K, Shin-Ya K, Asai K (2016) Class of cyclic ribosomal peptide synthetic genes in filamentous fungi. Fungal Genet Biol 86:58–70. https://doi.org/10.1016/j.fgb.2015.12.010

Nagendran S, Hallen-Adams HE, Paper JM, Aslam N, Walton JD (2009) Reduced genomic potential for secreted plant cell-wall-degrading enzymes in the ectomycorrhizal fungus *Amanita bisporigera*, based on the secretome of *Trichoderma reesei*. Fungal Genet Biol 46:427–435. https://doi.org/10.1016/j.fgb.2009.02.001

Nakamori T, Suzuki A (2007) Defensive role of cystidia against *Collembola* in the basidiomycetes *Russula bella* and *Strobilurus ohshimae*. Mycol Res 111:1345–1351. https://doi.org/10.1016/j.mycres.2007.08.013

Nehls U, Göhringer F, Wittulsky S, Dietz S (2010) Fungal carbohydrate support in the ectomycorrhizal symbiosis: a review. Plant Biol 12:292–301. https://doi.org/10.1111/j.1438-8677.2009.00312.x

Nicolas P, Vanhoye D, Amiche M (2003) Molecular strategies in biological evolution of antimicrobial peptides. Peptides 24:1669–1680. https://doi.org/10.1016/j.peptides.2003.08.017

Nilsson RH, Kristiansson E, Ryberg M, Larsson KH (2005) Approaching the taxonomic affiliation of unidentified sequences in public databases–an example from the mycorrhizal fungi. BMC Bioinforma 6:178. https://doi.org/10.1186/1471-2105-6-178

Nützmann HW, Osbourn A (2014) Gene clustering in plant specialized metabolism. Curr Opin Biotechnol 26:91–99. https://doi.org/10.1016/j.copbio.2013.10.009

Oliveira AG, Stevani CV, Waldenmaier HE, Viviani V, Emerson JM, Loros JJ, Dunlap JC (2015) Circadian control sheds light on fungal bioluminescence. Curr Biol 25:964–968. https://doi.org/10.1016/j.cub.2015.02.021

Olivera BM, Walker C, Cartier GE, Hooper D, Santos AD, Schoenfeld R, Shetty R, Watkins M, Bandyopadhyay P, Hillyard DR (1999) Speciation of cone snails and interspecific hyperdivergence of their venom peptides: potential evolutionary significance of introns. Ann N Y Acad Sci 870:223–237

Olivera BM, Watkins M, Bandyopadhyay P, Imperial JS, de la Cotera EP, Aguilar MB, Vera EL, Concepcion GP, Lluisma A (2012) Adaptive radiation of venomous marine snail lineages and the accelerated evolution of venom peptide genes. Ann N Y Acad Sci 1267:61–70. https://doi.org/10.1111/j.1749-6632.2012.06603.x

Panaccione DG (2005) Origins and significance of ergot alkaloid diversity in fungi. FEMS Microbiol Lett 251:9–17. https://doi.org/10.1016/j.femsle.2005.07.039

Panaccione DG, Scott-Craig JS, Pocard JA, Walton JD (1992) A cyclic peptide synthetase gene required for pathogenicity of the fungus *Cochliobolus carbonum* on maize. Proc Natl Acad Sci U S A 89:6590–6594

Phillips JP, Willms J, Pitt A (1982) α-Amanitin resistance in three wild strains of *Drosophila melanogaster*. Can J Genet Cytol 24:151–162

Piercy PL, Hargis G, Brown CA (1944) Mushroom poisoning in cattle. J Am Vet Med Assoc 105:206–208

Pigliucci M (2008) Is evolvability evolvable? Nat Rev Genet 9:75–82. https://doi.org/10.1038/nrg2278

Pilát A, Ušák O (1955) Mushrooms. Spring Books, London

Pineda SS, Sollod BL, Wilson D, Darling A, Sunagar K, Undheim EA, Kely L, Antunes A, Fry BG, King GF (2014) Diversification of a single ancestral gene into a successful toxin superfamily in highly venomous Australian funnel-web spiders. BMC Genomics 15:177. https://doi.org/10.1186/1471-2164-15-177

Preston JF, Johnson BEC, Little M, Romeo T, Stark HJ, Mullersman JE (1982) Investigations on the function of amatoxins in *Amanita* species: a case for amatoxins as potential regulators of transcription. In: Kleinkauf H, von Döhren H (eds) Peptide antibiotics: biosynthesis and functions. Walter de Gruyter, Berlin

Pringle A, Adams RI, Cross HB, Bruns TD (2009) The ectomycorrhizal fungus *Amanita phalloides* was introduced and is expanding its range on the west coast of North America. Mol Ecol 18:817–833. https://doi.org/10.1111/j.1365-294X.2008.04030.x

Pulman JA, Childs KL, Sgambelluri RM, Walton JD (2016) Expansion and diversification of the MSDIN family of cyclic peptide genes in the poisonous agarics *Amanita phalloides* and *A. bisporigera*. BMC Genomics 17:1038. https://doi.org/10.1186/s12864-016-3378-7

Ramm S, Krawczyk B, Mühlenweg A, Poch A, Mösker E, Süssmuth RD (2017) A self-sacrificing N-methyltransferase is the precursor of the fungal natural product omphalotin. Angew Chem Int Ed Eng 56:9994–9997. https://doi.org/10.1002/anie.201703488

Riley R, Salamov AA, Brown DW, Nagy LG, Floudas D, Held BW, Levasseur A, Lombard V, Morin E, Otillar R, Lindquist EA, Sun H, LaButti KM, Schmutz J, Jabbour D, Luo H, Baker SE, Pisabarro AG, Walton JD, Blanchette RA, Henrissat B, Martin F, Cullen D, Hibbett DS, Grigoriev IV (2014) Extensive sampling of basidiomycete genomes demonstrates inadequacy of the white-rot/brown-rot paradigm for wood decay fungi. Proc Natl Acad Sci U S A 111:9923–9928. https://doi.org/10.1073/pnas.1400592111

Robinson SD, Norton RS (2014) Conotoxin gene superfamilies. Mar Drugs 12:6058–6101. https://doi.org/10.3390/md12126058

Roderson CT, Samuels GJ (1994) Agaricicolous species of Hypomyces. Mycologia 86:839–866

Rosewich UL, Kistler HC (2000) Role of horizontal gene transfer in the evolution of fungi. Annu Rev Phytopathol 38:325–363. https://doi.org/10.1146/annurev.phyto.38.1.325

Seymour FR (1932) Therapeutic experiments on the manner of intoxication by *Amanita phalloides*. Brit Med J 2:220–221

Sgambelluri RM, Epis S, Sassera D, Luo H, Angelos ER, Walton JD (2014) Profiling of amatoxins and phallotoxins in the genus *Lepiota* by liquid chromatography combined with UV absorbance and mass spectrometry. Toxins 6:2336–2347. https://doi.org/10.3390/toxins6082336

Slot JC, Rokas A (2011) Horizontal transfer of a large and highly toxic secondary metabolic gene cluster between fungi. Curr Biol 21:134–139. https://doi.org/10.1016/j.cub.2010.12.020

Smith SE, Read DJ (2008) Mycorrhizal symbiosis, 3rd edn. Academic Press, New York

Sollod BL, Wilson D, Zhaxybayeva O, Gogarten JP, Drinkwater R, King GF (2005) Were arachnids the first to use combinatorial peptide libraries? Peptides 26:131–139. https://doi.org/10.1016/j.peptides.2004.07.016

Spicer GS, Jaenike J (1996) Phylogenetic analysis of breeding site use and α-amanitin tolerance within the *Drosophila quinaria* species group. Evolution 50:2328–2337. https://doi.org/10.1111/j.1558-5646.1996.tb03620.x

Steppuhn A, Gase K, Krock B, Halitschke R, Baldwin IT (2004) Nicotine's defensive function in nature. PLoS Biol 2:e217. https://doi.org/10.1371/journal.pbio.0020217

Stump AD, Jablonski SE, Bouton L, Wilder JA (2011) Distribution and mechanism of α-amanitin tolerance in mycophagous *Drosophila* (Diptera:Drosophilidae). Environ Entomol 40: 1604–1612. https://doi.org/10.1603/EN11136

Sunagar K, Moran Y (2015) The rise and fall of an evolutionary innovation: contrasting strategies of venom evolution in ancient and young animals. PLoS Genet 11:e1005596. https://doi.org/10.1371/journal.pgen.1005596

Tang X, Zhang Y, Hu W, Xu D, Tao H, Yang X, Li Y, Jiang L, Liang S (2010) Molecular diversification of peptide toxins from the tarantula *Haplopelma hainanum* (*Ornithoctonus hainana*) venom based on transcriptomic, peptidomic, and genomic analyses. J Proteome Res 9:2550–2564. https://doi.org/10.1021/pr1000016

Thorn RG, Barron GL (1984) Carnivorous mushrooms. Science 224:76–78. https://doi.org/10.1126/science.224.4644.76

Tokarz D, Poppenga R, Kaae J, Filigenzi M, Lowenstine LJ, Pesavento P (2012) Amanitin toxicosis in two cats with acute hepatic and renal failure. Vet Pathol 49:1032–1035. https://doi.org/10.1177/0300985811429307

Tuno N, Takahashi KH, Yamashita H, Osawa N, Tanaka C (2007) Tolerance of *Drosophila* flies to ibotenic acid poisons in mushrooms. J Chem Ecol 33:311–317. https://doi.org/10.1007/s10886-006-9228-3

Umemura M, Nagano N, Koike H, Kawano J, Ishii T, Miyamura Y, Kikuchi M, Tamano K, Yu J, Shin-ya K, Machida M (2014) Characterization of the biosynthetic gene cluster for the ribosomally synthesized cyclic peptide ustiloxin B in *Aspergillus flavus*. Fungal Genet Biol 68: 23–30. https://doi.org/10.1016/j.fgb.2014.04.011

Van der Does HC, Rep M (2012) Horizontal gene transfer of supernumerary chromosomes in fungi. Methods Mol Biol 835:427–437. https://doi.org/10.1007/978-1-61779-501-5_26

Van der Velden NS, Kälin N, Helf MJ, Piel J, Freeman MF, Künzler M (2017) Autocatalytic backbone N-methylation in a family of ribosomal peptide natural products. Nat Chem Biol 13:833–835. https://doi.org/10.1038/nchembio.2393

Verdes A, Anand P, Gorson J, Jannetti S, Kelly P, Leffler A, Simpson D, Ramrattan G, Holford M (2016) From mollusks to medicine: a venomics approach for the discovery and characterization of therapeutics from Terebridae peptide toxins. Toxins 8:117. https://doi.org/10.3390/toxins8040117

Walton JD (1994) Deconstructing the cell wall. Plant Physiol 104:1113–1118

Walton JD (2000) Horizontal gene transfer and the evolution of secondary metabolite gene clusters in fungi: an hypothesis. Fungal Genet Biol 30:167–171. https://doi.org/10.1006/fgbi.2000.1224

Walton JD, Panaccione DG, Hallen H (2004) Peptide synthesis without ribosomes. In: Tkacz J, Lange L (eds) Advances in fungal biotechnology for industry, agriculture, and medicine. Kluwer Academic, New York, pp 127–162

Weiß M, Yang ZL, Oberwinkler F (1998) Molecular phylogenetic studies in the genus *Amanita*. Can J Bot 76:1170–1179

Wheeler Q, Blackwell M (eds) (1984) Fungus-insect relationships: perspectives in ecology and evolution. Columbia University, New York

Wieland T (1986) Peptides of poisonous *Amanita* mushrooms. Springer, Berlin

Wight WD, Labuda R, Walton JD (2013) Conservation of the genes for HC-toxin biosynthesis in *Alternaria jesenskae*. BMC Microbiol. 2013 13:165. https://doi.org/10.1186/1471-2180-13-165

Wisecaver JH, Rokas A (2015) Fungal metabolic gene clusters–caravans traveling across genomes and environments. Front Microbiol 6:1–10. https://doi.org/10.3389/fmicb.2015.00161

Wolfe BE, Richard F, Cross HB, Pringle A (2010) Distribution and abundance of the introduced ectomycorrhizal fungus *Amanita phalloides* in North America. New Phytol 185:803–816. https://doi.org/10.1111/j.1469-8137.2009.03097.x

Wolfe BE, Tulloss RE, Pringle A (2012) The irreversible loss of a decomposition pathway marks the single origin of an ectomycorrhizal symbiosis. PLoS One 7:e39597. https://doi.org/10.1371/journal.pone.0039597

Wong ES, Belov K (2012) Venom evolution through gene duplications. Gene 496:1–7. https://doi.org/10.1016/j.gene.2012.01.009

Wood WF, Archer CL, Largent DL (2001) 1-Octen-3-ol, a banana slug antifeedant from mushrooms. Biochem Syst Ecol 29:531–533

Wood WF, Clark TJ, Bradshaw DE, Foy BD, Largent DL, Thompson BL (2004) Clitolactone: a banana slug antifeedant from *Clitocybe flaccida*. Mycologia 96:23–25

Yee MM, Woods LW, Poppenga RH, Puschner B (2012) Amanitin intoxication in two beef calves in California. J Vet Diagn Investig 24:241–244. https://doi.org/10.1177/1040638711427947

Zhang C, Hampp R, Nehls U (2005a) Investigation of horizontal gene transfer in poplar/*Amanita muscaria* ectomycorrhizas. Environ Biosaf Res 4:235–242. https://doi.org/10.1051/ebr:2006004

Zhang P, Chen ZH, Hu JS, Wei BY, Zhang ZG, Hu WQ (2005b) Production and characterization of amanitin toxins from a pure culture of *Amanita exitialis*. FEMS Microbiol Lett 252:223–228. https://doi.org/10.1016/j.femsle.2005.08.049

Chapter 7
Medical and Biotechnological Aspects

> ***If you suspect mushroom poisoning, go to the nearest emergency room, or, in the United States, call the Poison Control Center (1-800-222-1222).*** *Keep a sample of the mushroom if it is available, in any shape – fresh, dried, cooked, from stomach contents, vomit, etc. Mushroom species can be identified from even small fragments of cooked mushrooms by microscopic examination or DNA sequencing. Correct identification of the consumed mushroom will facilitate effective treatment. If you suspect your pet has eaten a poisonous mushroom, take your animal to a veterinarian as soon as possible. For more information on human and animal mushroom poisonings, see the web links at the end of this chapter.*

7.1 Medical Aspects of *Amanita* Cyclic Peptide Poisoning

As discussed in Chap. 2, higher fungi including mushrooms make many bioactive compounds. In this chapter we restrict our discussion to the cycloamanide family of ribosomally biosynthesized cyclic peptides found, to date, only in some species of *Amanita*, *Lepiota*, and *Galerina*. Information on the taxonomy, chemistry, and clinical aspects of mushroom poisonings due to non-peptidic mushroom components, including orellanine, gyromitrin, muscarine, illudin, coprine, and muscimol, are covered in Bresinky and Besl (1990), Benjamin (1995), and other books and web sites cited at the end of this chapter. Within the cycloamanide family, from the medical point of view, it is appropriate that the emphasis be exclusively on the amatoxins. Poisonings of mammals are due solely to the amatoxins because the phallotoxins are not absorbed from the digestive tracts of mammals (or, to the best of our knowledge, any other animal), and none of the monocyclic cycloamanides are known to be poisonous to mammals by any mode of ingestion.

There is a large literature on human and veterinarian clinical aspects of mushroom poisoning. The coverage here is restricted to a summary of highlights of recent

J. Walton, *The Cyclic Peptide Toxins of Amanita and Other Poisonous Mushrooms*, https://doi.org/10.1007/978-3-319-76822-9_7

advances in our understanding of mushroom poisoning attributable to the cyclic peptide toxins. For descriptions of clinical aspects of mushroom poisoning, including detailed accounts of symptom progression, the reader is referred to Ammirati et al. (1985), Benjamin (1995), Beug (2016), Bresinksy and Besl (1990), Lincoff and Mitchel (1977), and Spoerke and Rumack (1994). For journal articles, see Garcia et al. (2015), Mengs et al. (2012), Olson et al. (1982), Santi et al. (2012), Smith and Davis (2016), and Ward et al. (2013).

7.1.1 Why Do So Many People Die from Eating Poisonous Mushrooms?

Poisons are widespread in nature, yet it seems that mushrooms account for a large percentage of the human fatalities from natural poisons. It is instructive to consider the many factors that contribute to this situation.

- **Poisonous mushrooms are large and attractive, and taste good.** Positive organoleptic properties make it more likely that people will pick and consume amatoxin-containing mushrooms compared to many other fungi and plants. Furthermore, some species such as *A. phalloides* can be locally abundant. While small, obscure toxicogenic mushrooms such as *Lepiota* and *Galerina* species are less likely to be attractive to the traditional five senses, they have been mistaken for mushrooms such as *Psilocybe* that contain the psychoactive compound psilocybin. Some poisonous species of *Lepiota* are common in gardens in Italy, where they are more likely to attract human attention (Anna Marini, Associazione Micologica Bresadola, personal communication).
- **Geographic distribution and geographic naiveté.** *A. phalloides* in particular has a wide distribution in its native European range and is now spreading aggressively to many areas around the world where people are not expecting to find it. In the current era of mass travel for recreation and migration, many mushroom poisonings are the result of collectors being unfamiliar with the mycoflora of a new geographical area (Bazzicalupo et al. 2018; Frank and Cummins 1987; Ward et al. 2013). Folk criteria for distinguishing poisonous from nonpoisonous mushrooms in a particular region are frequently invalid in another. For example, *A. phalloides* has been mistaken for the edible Asian species *Volvariella volvacea* and the edible cocorra (*Amanita calyptoroderma* or *A. lanei*, in sect. *Caesareae*), which is native to western North America.
- **Toxin levels in mushrooms are high.** As discussed in Chap. 3, mushrooms can contain 2–10 mg α-amanitin and other amatoxins per gram dry weight (Tang et al. 2016). Therefore, one medium-sized mushroom can deliver a fatal adult dose of 7–12 mg.
- **Amatoxins are chemically stable.** This is a general property of cyclic peptides (see below), and amatoxins in particular can withstand extremes of temperature and pH and all forms of cooking (baking, boiling, frying, acidic sauces, etc.).

They also survive the mammalian digestive tract and are not metabolically detoxified by the liver or kidneys.

- **Amatoxins are rapidly absorbed and poorly excreted.** Amatoxins are detectable in urine 90-120 min after ingestion. Of the amatoxins absorbed into the portal circulation, ~40% are cleared by the kidneys. The rest are taken up into liver cells by high affinity membrane transporters (Chap. 5). The toxins that are excreted from the liver through the bile duct reenter the duodenum and are thereby recirculated within the enterohepatic system, with ~60% being reabsorbed by the liver at each pass. The kidneys therefore have a limited opportunity to excrete them.
- **The symptoms are delayed.** The first delay is due to a latent period between ingestion and appearance of the first symptoms. A second delay occurs after an early bout (8–12 h, range 6–36 h) of gastrointestinal distress, when patients typically undergo a temporary remission of symptoms. However, hepatotoxicity continues to progress. Many patients seek medical care too late to be effective, or are released from medical care prematurely when the initial symptoms subside.
- **Amatoxins inhibit an essential enzyme.** The site of action of the amatoxins, DNA-dependent RNA polymerase II (pol II), is a critical enzyme without which cells cannot synthesize proteins (Chap. 5). The binding affinity of α-amanitin for mammalian pol II is high ($K_D \sim 10^{-9}$ M). In the absence of protein synthesis, liver cells eventually die. When the liver loses sufficient function, the patient dies.

7.1.2 Clinical Aspects of Amatoxin Poisoning

The detailed course of events following oral ingestion of amatoxin-containing mushrooms has been documented in numerous publications. There are typically three phases: the first, gastrointestinal distress; the second, apparent remission; and the third, acute liver failure ending in death (Santi et al. 2012; Smith and Davis 2016). In a clinical situation where time is of the essence, a history of wild mushroom consumption, delayed onset of symptoms, and evidence of liver injury as indicated by elevated serum transaminase and lactic dehydrogenase levels is sufficient to assume amatoxin poisoning and warrant immediate appropriate treatment. Further tests that permit a more definitive diagnosis include measurement of amatoxins in urine, blood, and/or bile by radioimmunoassay, ELISA, or liquid chromatography (HPLC) and mass spectrometry (MS) and identification of the consumed mushrooms by microscopic and/or molecular criteria such as PCR. However, chemical and biological analytical methods are usually of little use in a clinical setting because most medical institutions are not equipped to do such analyses, especially on short notice. Most physicians recommend that aggressive treatment begin immediately and not be postponed in the expectation of a more definitive diagnosis.

7.1.2.1 Do Amatoxins Cause All of the Symptoms of *Amanita* Mushroom Poisoning?

In the early days of amatoxin research, it was unclear if the early and late phases of *Amanita* mushroom poisoning are both due to the amatoxins, or whether other mushroom chemicals account for, or contribute to, the symptoms characteristic of one or the other phase (Wieland and Wieland 1972). There has never been any doubt that the late stages, characterized by fulminant liver failure, are primarily caused by the amatoxins. However, some studies suggested that the early symptoms, which occur 6–12 h after ingestion and include abdominal pain, vomiting, and diarrhea, are not due to amatoxins damaging the epithelial cells of the gastrointestinal tract. This uncertainty was resolved by Fiume et al. (1973), using dogs instead of rodents (i.e., mice and rats), which are not affected by oral amanitin. Dogs fed amanitin showed clear gastrointestinal distress (vomiting and diarrhea) as well as histological damage to the gut epithelium, and therefore both the early and the late symptoms are caused by the amatoxins.

The symptoms of *Galerina* and *Lepiota* poisoning are also consistent with amatoxins causing both the early and late phases. In many published reports, early (6–12 h) abdominal pain, nausea, vomiting, and diarrhea are listed as symptoms of consuming amatoxin-containing species of these two genera (Ben Khelil et al. 2010; Kaneko et al. 2001; Mottram et al. 2010; Varvenne et al. 2015). Since *Galerina* and *Lepiota* do not make phallotoxins (Sgambelluri et al. 2014), and are unlikely to make other secondary metabolites in common with *Amanita*, these reports support the conclusion that the early symptoms are caused by amatoxins. On the other hand, some species of *Amanita* outside sect. *Phalloideae* are known to make other toxic compounds, so not all symptoms should be assumed to be caused by amatoxins in all cases of *Amanita* mushroom poisoning (Chap. 2).

7.1.3 Advances in Treatment of Amatoxin Poisoning

Treatment of amatoxin-related mushroom poisonings has a long and checkered history. Dozens of potential antidotes and treatments have been proposed and tried on numerous theoretical and empirical grounds. The small sample sizes and the large number of modifying factors surrounding each case of mushroom poisoning render it difficult to reliably judge efficacy of any particular treatment. Unfamiliarity of most physicians with mushroom poisoning combined with the severity of the reaction encourages a "try anything" approach. Gastric lavage (stomach pumping) is useful only in the first few hours before all of the toxins are absorbed. Hydration to increase renal elimination is critically important, because only ~60% of the toxins are absorbed by the liver at each pass through the enterohepatic circulatory system, and toxins in the blood can eventually be cleared by the kidneys as long as good urine flow is maintained (Jaeger et al. 1993; Zuliani et al. 2016). Transplantation is the last resort for acute liver failure and has been performed many times as a "cure" for amatoxin poisoning (e.g., Grabhorn et al. 2013).

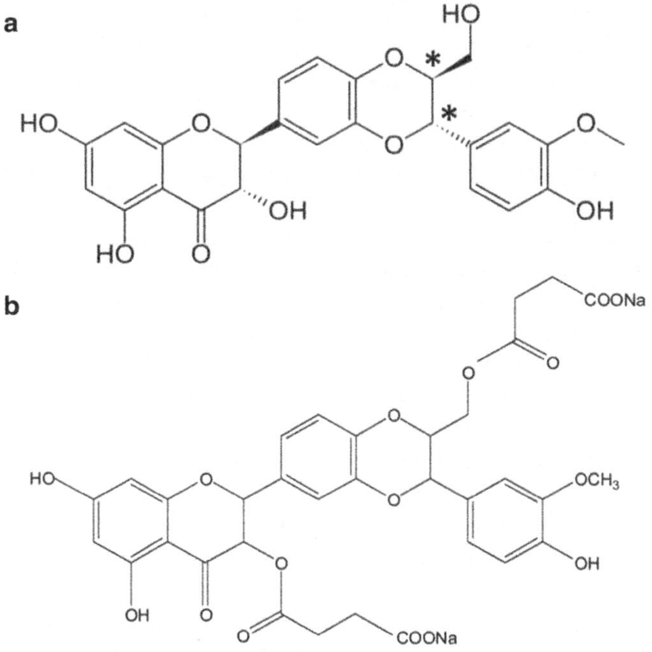

Fig. 7.1 (**a**) Silibinin. Silibinin A and B differ at the stereochemistry of the two bonds indicated with an asterisk (*) (Davis-Searles et al. 2005). (**b**) Silibinin-C-2′,3-dihydrogen succinate (Mengs et al. 2012)

Many chemical antidotes have been tried in attempts to counteract amatoxin poisoning, with varying degrees of success. The rationale for some, such as silibinin, rifampicin, cyclosporin, penicillin G, and paclitaxel, is supported by studies showing that they competitively inhibit uptake of amatoxins into hepatocytes (Letschert et al. 2006). Other drugs that have been tried include thioctic acid, acetylcysteine, and steroids (Enjalbert et al. 2002).

Silibinin, also known as *silybin*, a compound extracted from the milk thistle plant *Silybum marianum,* is currently considered to be a promising antidote (Beug 2016; Lacombe and St-Onge 2016; Mengs et al. 2012). Silibinin is a mixture of two flavonolignan diastereoisomers, silibinin A and B (Fig. 7.1) (Davis-Searles et al. 2005). A standardized milk thistle extract, containing silibinin and other flavonolignans, is also known by the name *silymarin*. The commercial formulation of silibinin suitable for injection, called Legalon® SIL, is the disodium succinate salt (Fig. 7.1). The efficacy of Legalon® SIL alone (monotherapy) or in combination with penicillin G is substantiated by numerous clinical reports, although solid statistical support is still lacking due to the small number of cases (Lacombe and St-Onge 2016; Mengs et al. 2012).

Legalon® SIL has been approved for amanitin poisoning in Germany since 1982 and is also registered in other European countries. Its use in the United States has been pioneered by Dr. Todd Mitchell and colleagues at Dominican Hospital in Santa

Cruz, California (Mengs et al. 2012; Mitchell and Olson 2008). *A. phalloides* is invasively prolific in the region around Santa Cruz, resulting in an exceptionally high density of amatoxin poisonings in the last 10 years (Padojino 2016; Vo et al. 2017). Dr. Mitchell is currently leading an NIH-sponsored clinical study of Legalon® SIL (https://clinicaltrials.gov/ct2/show/NCT00915681). For information, the Legalon® SIL hotline is 866-520-4412. Extracts of milk thistle are sold in pill form as a nutritional supplement under the name silymarin, but the active ingredients have poor oral availability, i.e., they are not absorbed well from the gastrointestinal tract. Therefore, oral silibinin should not be trusted to provide an effective treatment for amatoxin poisoning. Some formulations of silymarin (e.g., mixed with liposomes, microparticles, or cyclodextrin conjugates) show enhanced oral bioavailability, but the disuccinate salt is preferred for injection (Javed et al. 2011; Mengs et al. 2012; Wang et al. 2015). Silymarin has also been successfully tested in dogs (Floersheim et al. 1978; Vogel et al. 1984).

The probable basis of silibinin's protective action is competitive inhibition of α-amanitin uptake into hepatocytes through the OATP1B3 membrane transporter (Letschert et al. 2006; Wieland 1986; Chap. 5). Inhibition of biliary secretion of the amatoxins and other secondary effects might also contribute to the efficacy of silibinin (Mengs et al. 2012).

There is continued exploration of alternative surgical and pharmaceutical treatments for amatoxin poisoning. Bile duct drainage to remove amatoxins from the enterohepatic system has been performed at least once in humans and also in dogs (www.petsreferralcenter.com) (Dog Heirs Team 2012; Frank and Cummins 1987). The overall negative consensus of several medical doctors on biliary drainage has been reviewed by Beug (2016). Another recent novel approach is the development of short single-strand DNA molecules, called aptamers, that bind amanitin. Although very preliminary, such ligands might be able to scavenge amanitin from blood and the intestinal tract (Muszyńska et al. 2017).

7.2 Advances in Detection Methods

There have been significant developments in toxin extraction and detection methodologies in the past 30 years. Encompassing advances in speed, detection limits, and specificity, they have been exploited for a better understanding of the ecological functions of the *Amanita* cyclic peptide toxins, delimiting their taxonomic and tissue distribution, and to support clinical diagnoses.

HPLC and mass spectrometry (MS), alone and in combination (LC/MS), are now firmly established as the standard methods for separation and quantitation of the whole range of amatoxins, phallotoxins, and other cycloamanides (Chap. 2). Recent papers on extraction methods and toxin analyses from mushrooms, organs, and bodily fluids such as serum, bile, and urine include Filigenzi et al. (2007), Gicquel et al. (2014), Robinson-Fuentes et al. (2008), Tomková et al. (2015), and Zuliani et al. (2016). ^{15}N-Labeled α-amanitin suitable for use as an internal MS

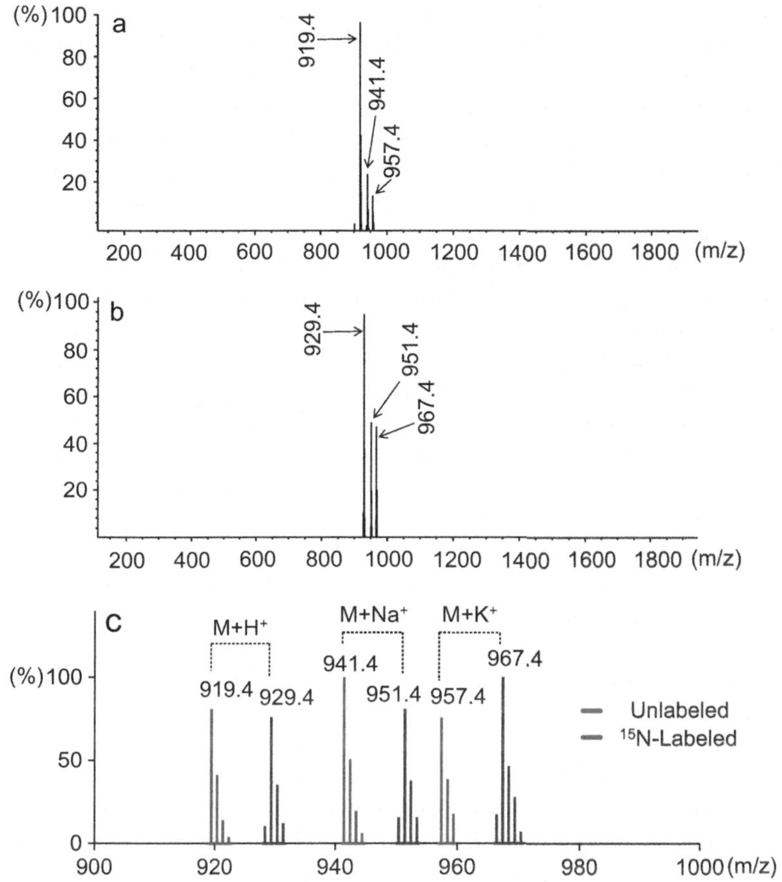

Fig. 7.2 Mass spectrometric analysis of ^{15}N-labeled α-amanitin synthesized in vivo by *Galerina marginata*. (**a**) Native ^{14}N-amanitin. [M + H$^+$] = 919.4, [M + Na$^+$] = 941.4, and [M + K$^+$] = 957.4. (**b**) ^{15}N-labeled amanitin. [M + H$^+$] = 929.4, [M + Na$^+$] = 951.4, and [M + K$^+$] = 967.4. (**c**) Overlaid spectra of ^{14}N-amanitin (blue) and ^{15}N-amanitin (red), shown in expanded scale (Reprinted from Luo et al. 2015 with permission of Elsevier)

standard for analyzing toxins in complex matrices such as bile, urine, and blood has been synthesized in vivo using *Galerina marginata* (Luo et al. 2015) (Fig. 7.2).

One important conclusion that has been substantiated by all of the newer analytical studies is that amatoxins are apparently not metabolized after ingestion. There is no experimental evidence that they are chemically modified by any mechanism in any organism, including known detoxification pathways such as oxidation by CYP450's or FMO's, or conjugation to glucuronic acid or glutathione, even though some gene expression studies suggested that chemical modification might be a mechanism of natural toxin resistance in insects (Chap. 6). Only amatoxins in their native forms have been found in the blood and urine.

With the development of PCR and other molecular genetic methods, it is now possible to identify poisonous mushrooms by their DNA. PCR-based approaches can offer high specificity and sensitivity in a reasonably short time (<12 h from sample preparation to results) and can identify mushrooms of the genus *Amanita* in general and toxin-producing strains in particular. Epis et al. (2010) designed real-time PCR primers specific for *A. phalloides*, *Lepiota cristata*, *L. brunneoincarnata*, and *Inocybe asterospora*. The first three species are frequent causes of human poisonings in Italy and make amatoxins, whereas *I. asterospora* does not make amatoxins but does make other as-yet unidentified toxins. The PCR primers were based on the ITS (intervening transcribed sequence) regions of the 5S ribosomal genes (rDNA), which are widely used for fungal taxonomic research (Chap. 1). The method was validated against 31 species of poisonous and nonpoisonous mushrooms from northern Italy for mushroom DNA extracted from clinical samples (gastric aspirates) and from cooked mushrooms. Analysis time was ~1 h after the DNA was extracted. While useful for the four mushrooms studied, this method will require further development and validation to be useful in other geographical areas where amatoxin-producing mushrooms other than *A. phalloides* grow, such as North America and Asia. Gausterer et al. (2014), Harper et al. (2011), and Maeta et al. (2008) have also developed DNA-based methods for specific detection of poisonous and other mushrooms in a variety of matrices including cooked foods, stomach contents, feces, and vomit.

Based on the sequences of the amatoxin-encoding genes (Hallen et al. 2007; Luo et al. 2012), PCR primers can be designed that are specific for the toxin genes themselves. Primers based on *AMA1* from *A. bisporigera* and *A. phalloides* have been used to amplify homologous genes from other species of *Amanita* (Hallen et al. 2007; Li et al. 2014; Wołoszyn and Kotłowski 2017). Such PCR primers should be able to distinguish poisonous from even closely related nonpoisonous species. With appropriate controls, the presence of a PCR product would indicate that a mushroom had the genetic potential to make amatoxins, and a negative result would indicate that the mushroom was toxin-free. A critical control in all PCR experiments is the inclusion of generic primers that will give a positive amplification signal with all fungal DNA. This is necessary to guarantee the quality of the DNA template (Wołoszyn and Kotłowski 2017).

There are two obstacles to the success of the PCR approach to identify amatoxin-producing fungi. First, the extent of natural variability of the amatoxin genes across *Amanita* is still not fully resolved. There are >50 toxic species and some species have multiple amatoxin genes. A single nucleotide polymorphism in the PCR primer binding region might not affect production of the amatoxins (i.e., a silent mutation) but could prevent successful PCR amplification. Second, the amino acid and encoding DNA sequences of the genes for α-amanitin (*AMA1*) in species of *Amanita* species are quite divergent from those of *Galerina* and *Lepiota* (Fig. 4.4) such that it is unlikely that any single pair of PCR primers could amplify the amatoxin genes from all three genera.

In conclusion, there are now multiple, highly specific chemical and biological tests for identifying and quantitating the cyclic peptide toxins and the mushrooms that produce them. However, it remains uncertain whether any of these tests will become relevant to emergency clinical situations where technical resources and especially time are critically limiting. They will mainly find utility in the confirmation of what astute physicians have hopefully already correctly diagnosed.

7.3 Pharmaceutical Applications of the *Amanita* Toxins

Because of their unique and specific modes of action, the amatoxins and phallotoxins have been exploited as reagents for biochemistry and cell biology research (Chap. 5). For example, as a specific inhibitor of eukaryotic DNA-dependent RNA polymerase II, α-amanitin revealed that trypanosome glycoproteins are transcribed by pol I and not pol II, and, in contrast, viroid replication utilizes pol II (Chung et al. 1993; Kooter and Borst 1984; Mühlbach and Sänger 1979). As a specific marker of F-actin, fluorescent phallacidin was used to discriminate between the involvement of F-actin and G-actin in chromosome migration during mitosis (Barak et al. 1981).

The utility of the *Amanita* cyclic peptides goes beyond their use as reagents for basic research. Of particular significance, amanitin can be harnessed as a "warhead" in antibody-drug conjugates (ADCs) in order to selectively kill undesirable eukaryotic targets such as cancer cells (Sievers and Senter 2013; Teicher 2014). The principle of ADCs is to attach a toxic compound to a cancer cell-specific antibody, which is then taken up by the cancer cells through endocytosis. The toxic compound is released by the action of lysosomal enzymes, such as esterases or proteases, and the toxic compound migrates to its site of action and kills the cells. Amanitin has several advantages over other toxic warheads, such as ricin, diphtheria toxin, maytansinoids, taxol, or duocarmycins. For example, its higher hydrophilicity compared to most anticancer drugs enhances the antibody coupling reaction, reduces the possibly of drug aggregation, promotes urinary excretion of any amanitin that is accidentally released into the bloodstream, and makes it a poorer substrate for multidrug resistance (MDR) transporters (Anderl et al. 2013).

Investigation of amanitin as a selective killing agent, particularly against cancer cells, has a long history based on early observations that amanitin-protein conjugates are more toxic than native amanitin (Wieland 1986). Being relatively large and polar, native amanitin is not taken up well by most cells, which lack the specific OATP1B3 (also known as SLCO1B3) transporter of hepatocytes (Chap. 5). The higher toxicity of amanitin-protein conjugates is probably due to their enhanced uptake through endocytosis (Barbanti-Brodano et al. 1974; Faulstich and Fiume 1985). As discussed in Chap. 5, the strong expression of OATP1B3 in some cancer cells has raised the possibility of developing anticancer therapies based on another cyclic peptide, microcystin (Niedermeyer et al. 2014).

Amanitin has been coupled by several methods to different proteins, including antibodies. Coupling can occur through ester linkages to the carboxyl group of β-amanitin (Asp #7), the hydroxyl group of Ile #1, or through the 6-hydroxyl group of Trp #2. Davis and Preston (1981) coupled amanitin to an antibody against a specific T-cell surface protein (Thy_1.2) and showed its selective toxicity against murine T lymphoma S49.1 cells. The antibody conjugate was 19- to 46-fold more toxic than free amanitin. Moldenhauer et al. (2012) conjugated amanitin to an antibody against human epithelial cell adhesion molecule (EpCAM), a protein overexpressed in many cancers. The amanitin-anti-EpCAM conjugates were inhibitory at 5–250 pM against several cancer cell lines, including human pancreatic, colorectal, breast, and bile duct, and also showed efficacy in immunocompromised mice bearing xenograft tumors.

Early amanitin-antibody conjugates were insufficiently stable, allowing the release of excessive free amanitin (Parmley 2014). Liu et al. (2015a, b) created an ADC with α-amanitin cross-linked to an anti-EpCAM antibody (HEA125) that was highly specific for adenocarcinomas including colorectal cancer (Bradner 2015). The amanitin was conjugated to lysine residues of HEA125 through a plasma-stable urea structure that would release free amanitin when degraded in the lysosome. The precise details of how the conjugate was made, including the location of the linkage within the amanitin molecule, were not provided, but based on other work from the same laboratory (Heidelberg Pharma AG) probably involved a protease-sensitive linker attached to the 6-hydroxy group of Trp #2 (Hechler et al. 2014).

There are several challenges to any ADC approach, including those based on amanitin as the warhead. First, the antibody must be highly specific. Many antigens are expressed in multiple cell types, and therefore untoward side effects and reduced tolerance are possible. For example, Thy_1 proteins are also made by normal T cells, some stem cells, and neurons, and EpCAM is a normal constituent of healthy epithelial cells. Second, the linker must be carefully designed. If too labile, excessive amounts of the toxic payload will be released and potentially damage nontarget cells. On the other hand, the linker must be readily cleaved inside the target cells. Third, the released compound, which might be slightly modified depending on how the linker is cleaved, must retain the high biological activity (toxicity) of the native compound.

The unique rationale for using α-amanitin in the study by Liu et al. (2015a) was that many cancers involve loss or mutation of the *TP53* gene, estimated at more than half in total, including 53% of colorectal cancers, 62% of breast cancers, 75% of ovarian cancers, and 41% of pancreatic cancers. TP53 (also known as tumor protein p53) is a tumor suppressor protein whose loss predisposes cells to undergo uncontrolled division (Aubrey et al. 2016). Fortuitously, the *POLR2A* gene encoding the large subunit of RNA polymerase II (pol II) is closely linked to *TP53*, such that many human and mouse *TP53* deletion mutants are also hemizygous for *POLR2A* (Fig. 7.3a). (Loss of both copies of *POLR2A* is lethal). As discussed in Chap. 5, α-amanitin is a potent and specific inhibitor of pol II, and cells that are hemizygous for *POLR2A* have enhanced sensitivity to α-amanitin and to an amanitin-antibody conjugate (Fig. 7.3b, c). In a mice orthotopic tumor model using cancer cells hemizygous for *POLR2A*, the amanitin-HEA125 ADC strongly inhibited tumor growth at 3 μg/kg, including complete tumor regression in ten out of ten mice treated with 90 μg/kg and six out of ten mice treated with 10 μg/kg. There were no apparent side effects (Liu et al. 2015a).

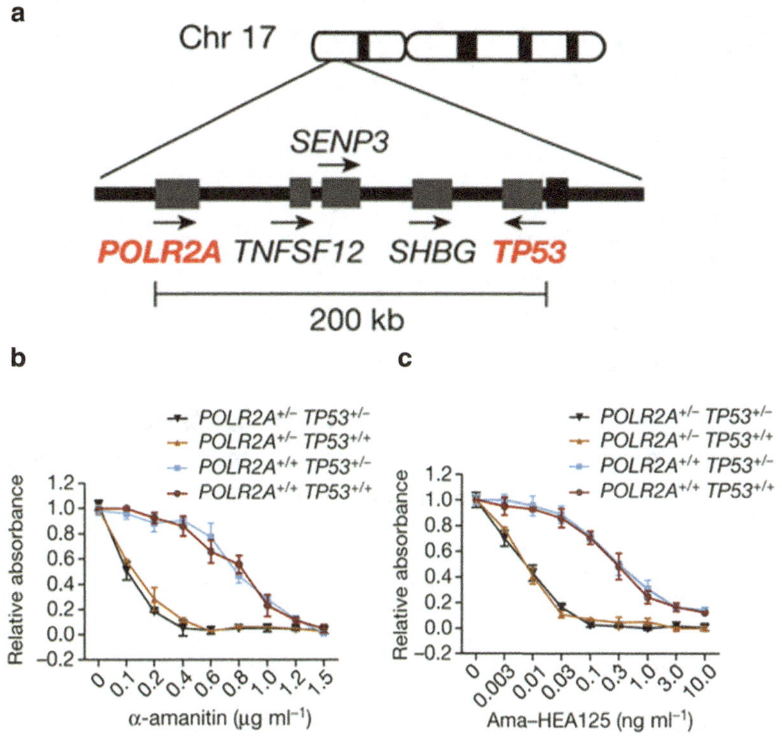

Fig. 7.3 (**a**) Map of region of human chromosome 17 showing tight linkage between POLR2A (encoding subunit 1 of pol II) and TP53 (encoding the tumor suppressor protein p53). (**b**, **c**) Enhanced sensitivity of mammalian cells (xhCRC – xenografted human primary colorectal cancer cells) that are hemizygous for pol II (POLR2A) to **b** amanitin or **c** amanitin-HEA125 antibody conjugate. +/− indicates cells are hemizygous for the indicated gene; +/+ indicates they are homozygous for that gene. (From Liu et al. 2015a, reprinted by permission from Macmillan Publishers Ltd)

In principle, amanitin-ADCs should also be effective against normal cells with two functional copies of *POLR2A*, since it is an essential gene. However, Liu et al. (2015a) found that the amanitin-HEA125 ADC gave significant tumor growth inhibition against cancer cells with two copies of *POLR2A* only at 90 μg/kg, the highest dose tested.

Several other amanitin ADCs against other cancer cell types are under development. Amanitin has been coupled to antibodies directed against prostate-specific membrane antigen (PSMA, also known as FOLH1 or GCPII) for use against prostate cancer (Hechler et al. 2014) and against B-cell maturation antigen (BCMA, also known as CD269) for treatment of multiple myeloma (HDP-101-ATAC) (Pálfi et al. 2016). Other amanitin ADCs are being developed by Wilex AG (Munich) and its subsidiary Heidelberg Pharma AG against hematological tumors and metastatic pancreatic cancer (www.wilex.de and heidelberg-pharma.com).

Amatoxins and phallotoxins can also be conjugated to lipophilic or polycationic moieties (e.g., oleic acid or polylysine) to promote cellular uptake and toxicity

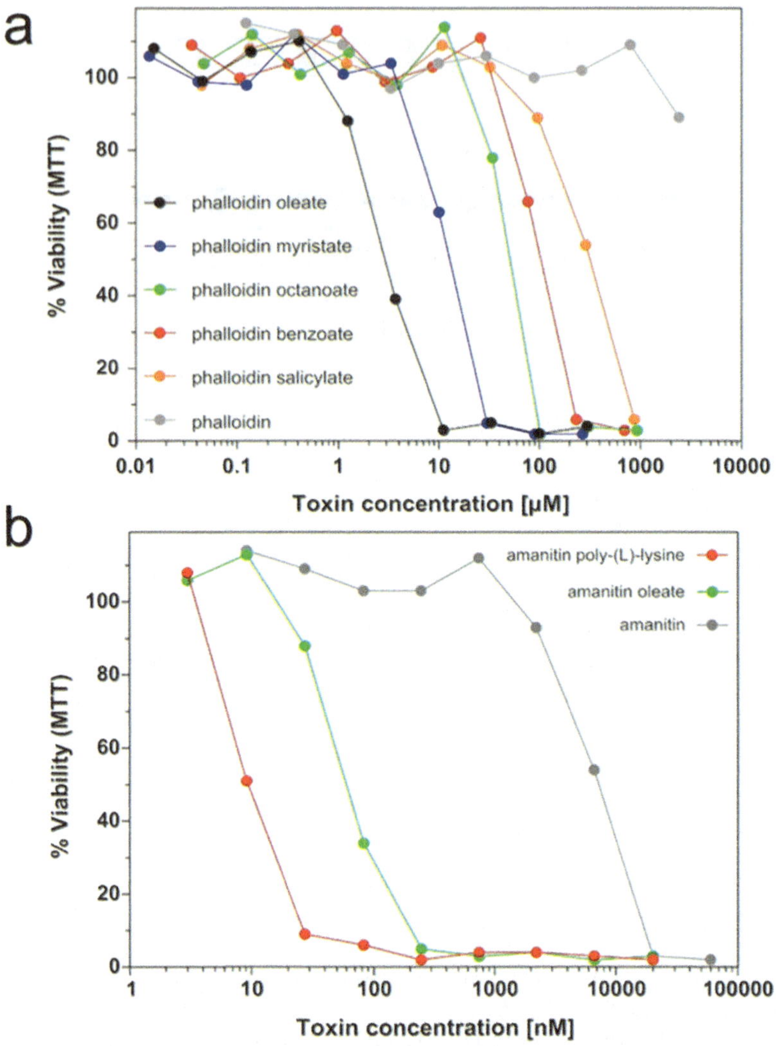

Fig. 7.4 (**a**) Enhanced toxicity of phalloidin derivatives against 3T3 mouse fibroblasts. (**b**) Enhanced toxicity of amanitin derivatives. (From Anderl et al. 2012, Creative Commons Attribution license)

against mammalian cells (Anderl et al. 2012) (Fig. 7.4). For the phallotoxins, conjugation occurred through the δ-hydroxyl group of Leu #3, and for amanitin the attachment site was the δ-hydroxyl group of Ile #1 or the 6-hydroxyl group of Trp #2. Like protein conjugates, the small molecule conjugates are taken up by endocytosis.

Species of *Amanita* that make amatoxins and phallotoxins also make a family of unmodified homodetic cyclic peptides known as the cycloamanides (Chap. 2). About six cycloamanides have been chemically characterized to date (Wieland

1986; Pulman et al. 2016), but *Amanita* and *Lepiota* species have the genetic potential to make many additional cycloamanides (Chap. 4). Extrapolating from the current genomic results (Pulman et al. 2016), the genus *Amanita* as a whole is predicted to have the genetic capacity to make hundreds of uncharacterized cycloamanides. Some of them have pharmaceutically interesting effects such as immunosuppression (Bolewska-Pedyczak et al. 1993; Wieczorek et al. 1993), but nothing is known about the biological activities of most cycloamanides.

7.4 Amatoxin Production Methods

The potential therapeutic utility of amanitin, dramatically illustrated by the work of Liu et al. (2015a), could create a large demand for this molecule. Currently, the only source of amanitin is by extraction from basidiocarps (mushrooms) collected in the wild. Since mushroom fruiting fluctuates widely with season and weather, the supply is unreliable. This situation has spurred attempts to obtain a more reliable source of amanitin. One approach is complete chemical synthesis; although this has recently been accomplished (Matinkhoo et al. 2018), it might remain too expensive to be practical (Chap. 2). Another approach is cultivation (fermentation) of toxin-producing mushrooms. Limited success has been obtained culturing *A. exitialis* and *G. marginata* (Benedict et al. 1966; Benedict and Brady 1967; Luo et al. 2015; Zhang et al. 2005), but this is also impractical due to the frustratingly slow growth of all known amatoxin-producing fungi and the low levels that they produce in culture (Fig. 7.5). Furthermore, the toxins are not secreted into the medium and thus must be extracted from the mycelium. Someday it may be possible to harness the genes and biochemical machinery to make amatoxins in vitro or in a heterologous host (Luo et al. 2014). However, to date only the biosynthetic steps as far as cyclization have been elucidated; the essential later steps (i.e., hydroxylations, tryptathionine cross-bridging, and sulfoxidation) are unknown as of this writing (Chap. 4).

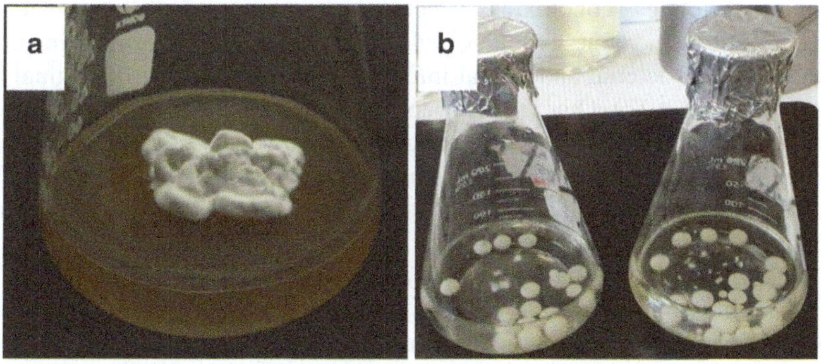

Fig. 7.5 (**a**) *Amanita exitialis* growing on solid medium for 40 days (Reprinted from Zhang et al. 2005 by permission of Oxford University Press). (**b**) *G. marginata* growing in shake culture. (Reprinted from Luo et al. 2015 with permission from Elsevier)

7.5 Biotechnological Applications of the *Amanita* Cyclic Peptide Genes

Some of the properties of the *Amanita* cyclic peptides, such as high oral bioavailability, stability, cell permeability, and target specificity, are desirable or necessary in drugs. The amatoxins have what is known in the pharmacological sciences as a good ADME profile, where ADME stands for absorption, distribution, metabolism, and excretion (wikipedia.org/wiki/ADME). Although fulminating liver necrosis is obviously not a preferred outcome for a pharmaceutical compound, it might be possible to synthesize molecules based on amanitin that retain its desirable pharmacokinetic properties but have a different molecular target.

The knowledge gleaned from studies of the cyclic peptide toxin biosynthetic pathway opens potential applications above and beyond applications of the native toxins themselves. Toxin-producing species of *Amanita* have the genetic capacity to make a large number of cyclic peptides, known collectively as the cycloamanides, utilizing the same biochemical pathway that they use to make the amatoxins and phallotoxins (Hallen et al. 2007; Pulman et al. 2016) (Chap. 4). In particular, prolyl oligopeptidase B (POPB) has been shown to be a versatile peptide macrocyclase (Sgambelluri et al. 2018). Therefore, the *Amanita* toxin biosynthetic machinery could possibly be harnessed to make novel cyclic peptides that retain the desirable ADME profile of the amatoxins.

7.5.1 *Fundamental Properties of Cyclic Peptides*

The majority of drugs used in human medicine are native or modified natural products, and cyclic peptides are one of the most important classes (Bhat et al. 2015; Göransson et al. 2012; Smith et al. 2013; White and Craik 2016). Biologically significant cyclic peptides include cyclosporin, pristinamycin, daptomycin, bacitracin, HC-toxin, cyclotides, conotoxins, vancomycin, valinomycin, gramicidin, polymyxin B, and nisin. Nine cyclic peptide drugs have been approved in the past 10 years against bacterial and fungal infections, cancer, and gastrointestinal disorders (Zorzi et al. 2017). Other cyclic peptides under development as pharmaceuticals include an inhibitor of the RAS oncogene (Upadhyaya et al. 2015); the modified griselimycins, which have promise against multidrug-resistant tuberculosis (Kling et al. 2015); and a cyclotide that activates the p53 tumor suppressor pathway (Ji et al. 2013). A human commensal bacterium produces a cyclic hexapeptide called lugdunin, which blocks colonization of the human nasal passages by pathogenic bacteria such as *Staphylococcus aureus* (Zipperer et al. 2016). A cyclic peptide inhibitor of influenza hemagglutinin was designed based on known neutralizing antibodies (Kadam et al. 2017). Synthetic bicyclic peptides have shown high serum stability with excellent binding to Grb2 (growth factor receptor-bound protein 2) (Quartararo et al. 2017).

Why are cyclic peptides so pharmaceutically advantaged? Their many desirable attributes include:

- **Chemical diversity.** The side groups of the 20 proteinogenic amino acids possess a wide variety of different chemistries, and this diversity is multiplied exponentially when amino acids are assembled into peptides; for example, there are 2.56×10^{10} possible octapeptides. Furthermore, all amino acids except glycine have at least one chiral center, providing additional three-dimensional complexity not found in smaller, planar molecules.
- **Stability.** Linear peptides are susceptible to the action of peptidases and proteases, rendering them unstable in most biological contexts. Cyclic peptides, on the other hand, are resistant to proteases because their peptide bonds are not accessible to the active sites of either exo- or endoproteases (Bockus et al. 2013; Craik et al. 2010). In addition, cyclic peptides are more resistant to degradation or denaturation by extremes of temperature and pH.
- **Defined shape.** Cyclization of a linear peptide reduces its number of possible conformations. Being more rigid enhances a molecule's ability to bind to target sites. This is believed to be due to reduced entropic penalty, i.e., because the molecule is already constrained in its free state, there is less of a decrease in entropy when the peptide is further constrained by binding to its target site. Therefore, more constrained molecules have a more favorable change in free energy (ΔG) upon binding. For example, a cyclic peptide mimic of a human antihemagglutinin antibody showed stronger binding affinity (as well as stability) compared to its linear counterpart (Kadam et al., 2017). The entropic benefit of cyclization is possibly an oversimplification because cyclization also changes other potentially critical molecular properties such as solvation (Udugamasooriya and Spaller 2008). Monocyclic peptides can be made even more rigid and stable by the introduction of internal cross-bridges such as disulfide linkages or tryptathionine to make them bicyclic, as in the case of the cyclotides, amatoxins, and other natural and synthetic compounds (Bionda and Fasan 2017; Rhodes and Pei 2017; Weidmann and Craik 2016).
- **Oral bioavailability and cell permeability.** Successful pharmaceuticals delivered orally must be absorbed by the digestive system and be able to cross multiple cell membranes. A number of chemical properties affect bioavailability, including lipophilicity, molecular weight, number of hydrogen bond donors and acceptors, and charge (Beck et al. 2012; Veber et al. 2002; Wang et al. 2015). Poor cellular delivery of therapeutic molecules is a common bottleneck of modern pharmacology. Cell-permeable peptides (CPPs) are able to cross the plasma membrane despite being relatively large and polar (Cascales et al. 2011; Hewitt et al. 2015; Wang and Craik 2016). The *Amanita* cyclic peptides are natural cell-permeable peptides; they are taken up by cells more readily than their size and polarity would predict. Although native amatoxins and phallotoxins enter hepatocytes through the OATP1B3 and OATP1B1 transporters, they also readily penetrate other cell types including plants (Letschert et al. 2006;

Meier-Abt et al. 2004; Theologis et al. 1985). Amatoxins conjugated to small molecules or proteins are efficiently endocytosed by all mammalian cell types (Anderl et al. 2012; Faulstich and Fiume 1985).

- **High surface area.** Interference with protein-protein interactions, which modulate many cellular processes, is an area of strong pharmaceutical interest. However, protein-protein interactions are difficult to drug because they involve large, flat surfaces without major binding pockets (Dougherty et al. 2017). Larger ligands provide more possibilities for binding between them and their targets. Small cyclic peptides (ranging from 1–3 kDa), which naturally have many *sp3* bonds and chiral centers, have intrinsically higher surface area than smaller molecules (<500 Da) with a higher degree of *sp2* bond character. They are thus more likely to stick to flat protein surfaces (Cardote and Ciulli 2016). Examples of successful macrocycles affecting protein-protein interactions include inhibitors of the CD2:CD58 interaction (Sable et al. 2016), HF-1 heterodimerization (Miranda et al. 2013), and antagonists of tumor necrosis factor-α (Lian et al. 2013).

In conclusion, the *Amanita* cyclic peptides possess many of the desirable characteristics shared generally by cyclic peptides. α-Amanitin is resistant to heat and the mammalian digestive tract, it is readily absorbed into the bloodstream and across the plasma membrane, and it is a highly specific inhibitor of an essential enzyme. The generalized amanitin backbone might therefore provide the starting point from which to design *beneficial* bioactive molecules.

7.5.2 *Novel Cyclic Peptides Based on the* Amanita *Cyclic Peptides*

Because of the valuable properties of cyclic peptides discussed above, new and better methods to synthesize them are being continually developed. Cyclic peptides can be synthesized chemically and also by combinations of in vivo and in vitro biological methods (Bhat et al. 2015; Bionda and Fasan 2017; Clark et al. 2005; Jagadish et al. 2015; Kritzer et al. 2009; Mootz 2017; Tavassoli 2017).

Natural cyclic peptides are biosynthesized either by nonribosomal peptide synthetases (NRPSs) or on ribosomes (RiPPs). The RiPP pathway offers several advantages as a platform for biosynthesizing cyclic peptides. Due to their large size, NRPSs are difficult to genetically manipulate. The basis of amino acid preference in NRPS modules is poorly defined, and many NRPS modules are promiscuous, leading to mixtures of products (Kries 2016). RiPPs are more precise, more amenable to directed evolution approaches, and their cognate macrocyclases have high substrate versatility. Hence they are well suited for synthetic biology and genome mining approaches (Hetrick and van der Donk 2017).

7.5.3 POPB and Other Peptide Macrocyclases

POPB, the peptide macrocyclase involved in cycloamanide biosynthesis (Chap. 4), can be used to make a range of monocyclic peptides. Systematic testing showed that it tolerates different amino acids in its core region, including β-amino acids, N-methylated amino acids, and D-amino acids, and can create cyclic peptides with ring sizes up to 16 amino acids (Sgambelluri et al. 2018). It cannot cyclize a phalloidin precursor peptide containing all L amino acids, but it can partially cyclize a phalloidin peptide containing the naturally occurring D-Thr at position #5. POPB can cyclize peptides en masse, as shown for a pool of ten precursor peptides (Fig. 7.6). A scheme to make larger libraries (>30,000) is shown in Fig. 7.7 (Sgambelluri et al. 2018).

Several other peptide macrocyclases from plants, bacteria, and fungi have been characterized (Truman 2016). Of these, butelase 1 from the cyclotide-producing plant *Clitoria ternatea* (family Fabaceae), POPB from *Galerina marginata*, and maybe PCY1 from the orbitide-producing plant *Vaccaria hispanica* (Caryophyllaceae) seem to be the only ones known to date with sufficient kinetic efficiency and versatility to be practical for making libraries of cyclic peptides. Published k_{cat} values for PCY1 range from ~1 to 216 h^{-1} and for POPB from 35 to 342 min^{-1}. Butelase 1 has a k_{cat} of ~140 min^{-1}. Being highly dependent on substrate, assay conditions, and data quality, all of these values must be considered to be rough approximations (Barber et al. 2013; Chekan et al. 2017; Czekster and Naismith 2017; Luo et al. 2014; Nguyen et al. 2014).

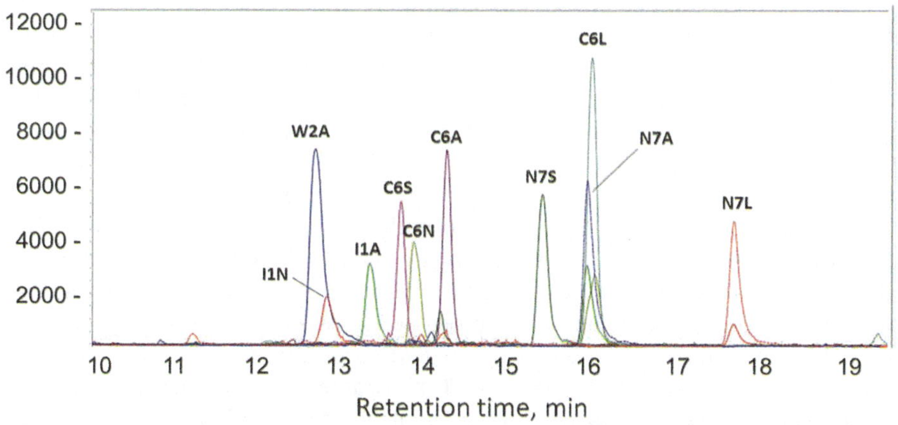

Fig. 7.6 Production of novel cyclic octapeptides (cycloamanides) with POPB. Ten genes encoding 35-amino acid precursor peptides with variant core regions were expressed in *E. coli* fused to maltose-binding protein (MBP). The bacterial extracts were then combined, purified on an amylose affinity column, treated with Factor X protease to remove the MBP, and cyclized with POPB. The resulting mixture of ten cyclic peptide products was analyzed by LC/MS. The peptides differed as indicated from α-amanitin (amino acid sequence IWGIGCNP) at the corresponding amino acid. y-Axis: Relative MS response. (From Sgambelluri et al. 2018, Creative Commons Attribution license)

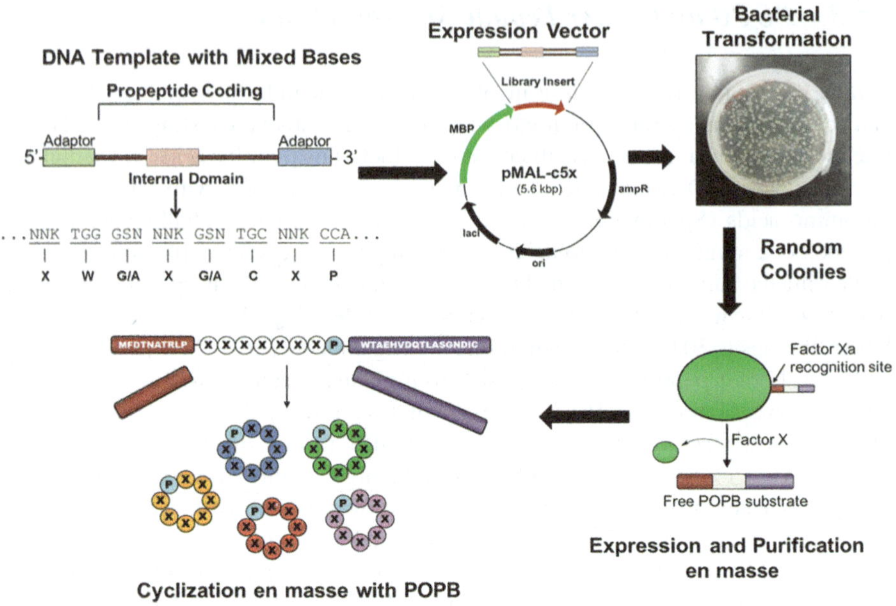

Fig. 7.7 Scheme for generating mixed cyclic peptide libraries based on the cycloamanide biosynthetic pathway. From Sgambelluri et al. (2018), Creative Commons Attribution license

Butelase 1 requires only a three amino acid recognition motif (Asn-His-Val) (Fig. 7.8a). It can cyclize peptides as small as 12 amino acids and as large as 240 amino acids if the amino and carboxy termini are in proximity, and it can also catalyze intermolecular ligation between peptides with the appropriate terminal amino acids (i.e., Asn-His-Val) (Nguyen et al. 2014, 2015, 2016). Butelase 1 can cyclize peptides containing D amino acids, but its reactivity with peptide substrates containing other unusual amino acids is unknown (Nguyen et al. 2016). As of this writing, the sole source of butelase 1 is extraction from the plant source.

PCY1 from *S. vaccaria*, involved in biosynthesis of segetalin A (Chap. 4), is, like POPB, a member of the prolyl oligopeptidase (POP) subfamily of serine proteases (Barber et al. 2013; Chekan et al. 2017). Segetalin A is processed from a ribosomally biosynthesized 32mer that is processed in two steps by two distinct proteases. PCY1 catalyzes the second cleavage/transamidation (Fig. 7.8b). Little is known about the substrate versatility of PCY1 (Barber et al. 2013), but based on its crystal structure, it is predicted to be tolerant of amino acid changes in the core peptide region (Chekan et al. 2017).

POPB, the peptide macrocyclase involved in biosynthesis of the cycloamanides (Chap. 4), converts a 35-amino acid peptide into a cyclic octapeptide in two nonprocessive steps (Fig. 7.8c). Like butelase 1, one essential invariant amino acid remains in the final product, i.e., Asn in the case of butelase 1 and Pro in the case of POPB. POPB cyclizes the 25mer produced by the first step as efficiently as the 35mer, and therefore the 10-amino acid leader can be omitted from synthetic

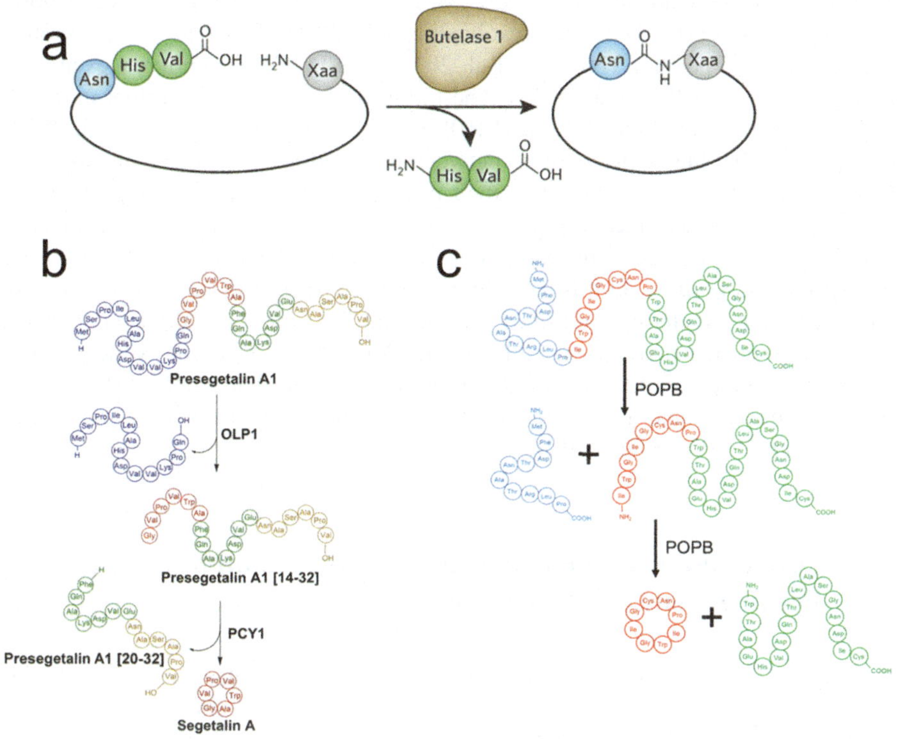

Fig. 7.8 Comparison of the reactions catalyzed by three peptide macrocyclases. (**a**) Reaction catalyzed by butelase 1. (From Heinis 2014, reprinted by permission from Macmillan Publishers Ltd). (**b**) Reaction catalyzed by PCY1. The first hydrolysis step is catalyzed by another enzyme, OLP1. (From Chekan et al. 2017, reprinted with permission from the National Academy of Sciences). (**c**) Reactions catalyzed by POPB (Luo et al. 2014). POPB catalyzes two nonprocessive steps

substrates (Luo et al. 2014). POPB has been shown in vitro to cyclize peptides from 7 to 16 amino acids; the upper limit is not yet known but is probably not much larger due to restricted access to the active site of POPB, which is gated by its β-propeller domain (Sgambelluri et al. 2018). For comparison, the natural cycloamanides of *Amanita* range from six to ten amino acids (Chaps. 2 and 4). POPB can be produced at high titers in either yeast or *E. coli* (Czekster and Naismith 2017; Luo et al. 2014).

Luo et al. (2014) identified some of the features of the conserved leader and follower peptides necessary for POPB macrocyclase activity. Czekster et al. (2017) showed that at least some cyclic product can still be made when the length of the follower peptide is reduced to five or six amino acids. The 13mer (i.e., eight amino acids of the core region plus five amino acids of the C-terminal linker) had a lower K_m (higher affinity) but produced more linear product than the 14mer. Because shorter precursors are less expensive to *chemically* synthesize, this might reduce the cost of using GmPOPB to make cyclic peptides for biotechnology applications. However, it would provide no significant cost advantage when the precursor peptides are made in a microbial host cell (Sgambelluri et al. 2018).

The comparative advantages of butelase 1, PCY1, and POPB as general peptide macrocyclases are still to be fully resolved. Butelase 1 has some amino acid preferences at the P2" position (Nguyen et al. 2016), and POPB has preferences at sites P3 and P5 (Sgambelluri et al. 2018). The follower peptides are important in length and composition for both PCY1 and POPB (Luo et al. 2014; Barber et al. 2013) (Chap. 4). The low conservation between the follower peptides of *Amanita*, *Galerina*, and *Lepiota* indicates that their primary sequences are not important, but an α-helix secondary structure probably is (Figs. 4.4 and 4.8). Based on in vitro work as well as analysis of the cycloamanide superfamily in *Galerina*, *Lepiota*, and two species of *Amanita*, POPB has a high tolerance for amino acid substitutions in the core regions, including multiple Pro, Phe, and Trp residues (Pulman et al. 2016; Sgambelluri et al. 2018). It is possible that the POPB from one of the cyclic peptide-producing species of *Amanita* might be more versatile than GmPOPB, because *Amanita* species make a wide range of cyclic peptides whereas *Galerina* makes only α-amanitin (Luo et al. 2012).

7.6 Other Biotechnological Aspects of the *Amanita* Cyclic Peptide Genes and Enzymes

Some of the amatoxins such as γ-amanitin contain 4-hydroxyisoleucine (Fig. 2.6). This amino acid by itself has antidiabetic properties (Jetté et al. 2009; Zafar and Gao 2016). 4-Hydroxyisoleucine is found naturally in fenugreek (*Trigonella foenum-graecum*), where the hydroxylation is catalyzed by a dioxygenase (Haefelé et al. 1997). The mechanism of hydroxyisoleucine biosynthesis in cyclic toxin-producing fungi is currently unknown (Chap. 4). Identification of the genes and enzymes that catalyze isoleucine hydroxylation in the agaric biosynthetic pathway might therefore provide a useful path to the production of 4-hydroxyisoleucine for pharmaceutical uses.

Kaya et al. (2014) studied the dermal (skin) toxicity of α-amanitin in mice and found no absorption into the blood. They concluded that α-amanitin might find use in treating ectoparasites and superficial infections (although it would be expected to be effective only against eukaryotes; see Chap. 5).

In the 20 years prior to March, 2018, there were 486 patents containing the keyword "amanitin" and 2111 with the keyword "phalloidin." There is clearly a continued high interest in commercial applications of the *Amanita* cyclic peptides.

7.7 Sources of Information about Mushroom Poisoning

For people and animals:

North American Mycological Association: http://www.namyco.org/poisonings.php
(San Francisco) Bay Area Mycological Society: http://www.bayareamushrooms.org/poisonings/index.html

Chemical Safety Information from Intergovernmental Organizations/International Programme on Chemical Safety: http://www.inchem.org/documents/pims/fungi/amatox.htm
Poison Control Centers online tool: https://www.poison.org

The Legalon® SIL hotline (see Sect. 7.1.3) is 866-520-4412.
For animals:

Focus on dog poisonings: http://mushrooms911.blogspot.com/
North American Mycological Assocation: http://www.namyco.org/mushroom_poisonings_in_dogs_an.php
PetMD: http://www.petmd.com/dog/conditions/digestive/c_dg_mushroom_poisoning

References

Ammirati JF, Traquair JA, Horgen PA (1985) Poisonous mushrooms of the Northern United States and Canada. University of Minnesota Press, Minneapolis, and Fitzhenry & Whiteside, Ltd., Markham

Anderl J, Echner H, Faulstich H (2012) Chemical modification allows phallotoxins and amatoxins to be used as tools in cell biology. Beilstein J Org Chem 8:2072–2084. https://doi.org/10.3762/bjoc.8.233

Anderl J, Faulstich H, Hechler T, Kulke M (2013) Antibody–drug conjugate payloads. Methods Mol Biol 1045:51–70. https://doi.org/10.1007/978-1-62703-541-5_4

Aubrey BJ, Strasser A, Kelly GL (2016) Tumor-suppressor functions of the TP53 pathway. Cold Spring Harb Perspect Med 6: pii: a026062. doi: https://doi.org/10.1101/cshperspect.a026062

Barak LS, Nothnagel EA, DeMarco EF, Webb WW (1981) Differential staining of actin in metaphase spindles with 7-nitrobenz-2-oxa-1,3-diazole-phallacidin and fluorescent DNase: is actin involved in chromosomal movement? Proc Natl Acad Sci U S A 78:3034–3038

Barbanti-Brodano G, Derenzini M, Fiume L (1974) Toxic action of a phalloidin-albumin conjugate on cells with a high protein uptake. Nature 248:63–64

Barber CJS, Pjuara PT, Reed DW, Chiwocha S, Zhang H, Covello PS (2013) The two-step biosynthesis of cyclic peptides from linear precursors in a member of the plant family Caryophyllaceae involves cyclization by a serine protease-like enzyme. J Biol Chem 288:12500–12510. https://doi.org/10.1074/jbc.M112.437947

Bazzicalupo A, Berch S, Callan B, Ceska O, Kroeger P, Li R, McIntyre L, Miao V, Stanwick R, Trudell S, Vellinga EC, Walton JD, Arevalo W, Asay A, Auxier B, Ceska A, Dee J, Gamiet S, Landry B, Le Renard L, Schwarz C, Super L, van der Meer B, Berbee M (2018) White paper on strategies to reduce risks and expand appreciation of foraged wild mushrooms. McIlvanea 26. http://www.namyco.org/docs/EdiblePoisonousReport20170914.pdf

Beck JG, Chatterjee J, Laufer B, Kiran MU, Frank AO, Neubauer S, Ovadia O, Greenberg S, Gilon C, Hoffman A, Kessler H (2012) Intestinal permeability of cyclic peptides: common key backbone motifs identified. J Am Chem Soc 134:12125–12133. https://doi.org/10.1021/ja303200d

Ben Khelil M, Zhioua M, Bakir O, Allouche M, Gloulou F, Banasr A, Haouet S, Hedhili A, Hamdoun M (2010) Intoxication mortelle par *Lepiota brunnioincarnata*: à propos de 4 cas. Ann Biol Clin (Paris) 68:561–567. https://doi.org/10.1684/abc.2010.0467

Benedict RG, Brady LR (1967) Further studies on fermentative production of toxic cyclopeptides by *Galerina marginata* (Fr Kühn). Lloydia 30:372–378

Benedict RG, Tyler VE Jr, Brady LR, Weber LJ (1966) Fermentative production of *Amanita* toxins by a strain of *Galerina marginata*. J Bacteriol 91:1380–1381

Benjamin DR (1995) Mushrooms: poisons and panaceas. A handbook for naturalists, mycologists, and physicians. WH Freeman, New York

Beug MW (2016)Amatoxinmushroom poisoning in North America 2015–2016. McIlvainea 25. Accessed Sept 2017, at www.namyco.org/mcilvainea.php

Bhat A, Roberts LR, Dwyer JJ (2015) Lead discovery and optimization strategies for peptide macrocycles. Eur J Med Chem 94:471–479. https://doi.org/10.1016/j.ejmech.2014.07.083

Bionda N, Fasan R (2017) Ribosomal synthesis of thioether-bridged bicyclic peptides. Methods Mol Biol 1495:57–76. https://doi.org/10.1007/978-1-4939-6451-2_5

Bockus AT, McEwen CM, Lokey RS (2013) Form and function in cyclic peptide natural products: a pharmacokinetic perspective. Curr Top Med Chem 13:821–836. https://doi.org/10.2174/1568026611313070005

Bolewska-Pedyczak E, Wieczorek Z, Zimecki M, Siemion IZ (1993) Immunosuppressive activity of some peptides related to cycloamanide A (CyA A) and to the active fragment of a peptide immunomodulator from ovine colostrum. Arch Immunol Ther Exp 41:297–301

Bradner JE (2015) An essential passenger with p53. Nature 520:626–627. https://doi.org/10.1038/nature14390

Bresinsky A, Besl H (1990) A colour atlas of poisonous fungi: a handbook for pharmacists, doctors, and biologists. Wolfe Publishing, London

Cardote TAF, Ciulli A (2016) Cyclic and macrocyclic peptides as chemical tools to recognise protein surfaces and probe protein-protein interactions. ChemMedChem 11:787–794. https://doi.org/10.1002/cmdc.201500450

Cascales L, Henriques ST, Kerr MC, Huang YH, Sweet MJ, Daly NL, Craik DJ (2011) Identification and characterization of a new family of cell-penetrating peptides: cyclic cell-penetrating peptides. J Biol Chem 286:36932–36943. https://doi.org/10.1074/jbc.M111.264424

Chekan JR, Estrada P, Covello PS, Nair SK (2017) Characterization of the macrocyclase involved in the biosynthesis of RiPP cyclic peptides in plants. Proc Natl Acad Sci U S A 114:6551–6556. https://doi.org/10.1073/pnas.1620499114

Chung HM, Lee MG, Dietrich P, Huang J, Van der Ploeg LH (1993) Disruption of largest subunit RNA polymerase II genes in *Trypanosoma brucei*. Mol Cell Biol 13:3734–3743. https://doi.org/10.1128/MCB.13.6.3734

Clark RJ, Fischer H, Dempster L, Daly NL, Rosengren KJ, Nevin ST, Meunier FA, Adams DJ, Craik DJ (2005) Engineering stable peptide toxins by means of backbone cyclization: stabilization of the α-conotoxin MII. Proc Natl Acad Sci U S A 102:13767–13772. https://doi.org/10.1073/pnas.0504613102

Craik DJ, Mylne JS, Daly NL (2010) Cyclotides: macrocyclic peptides with applications in drug design and agriculture. Cell Mol Life Sci 67:9–16. https://doi.org/10.1007/s00018-009-0159-3

Czekster CM, Naismith JH (2017) Kinetic landscape of a peptide bond-forming prolyl oligopeptidase. Biochemistry 56:2086–2095. https://doi.org/10.1021/acs.biochem.7b00012

Czekster CM, Ludewig H, McMahon SA, Naismith JH (2017) Characterization of a dual function macrocyclase enables design and use of efficient macrocyclization substrates. Nat Commun 8:1045. https://doi.org/10.1038/s41467-017-00862-4

Davis MT, Preston JF 3rd (1981) A conjugate of α-amanitin and monoclonal immunoglobulin G to Thy 1.2 antigen is selectively toxic to T lymphoma cells. Science 213:1385–1388. https://doi.org/10.1126/science.6115471

Davis-Searles PR, Nakanishi Y, Kim NC, Graf TN, Oberlies NH, Wani MC, Wall ME, Agarwal R, Kroll DJ (2005) Milk thistle and prostate cancer: differential effects of pure flavonolignans from *Silybum marianum* on antiproliferative end points in human prostate carcinoma cells. Cancer Res 65:4448–4457. https://doi.org/10.1158/0008-5472.CAN-04-4662

Dog Heirs Team (2012) http://www.dogheirs.com/dogheirs/posts/1541-new-procedure-saves-dog-from-mushroom-poisoning-may-save-future-lives

Dougherty PG, Qian Z, Pei D (2017) Macrocycles as protein-protein interaction inhibitors. Biochem J 474:1109–1125. https://doi.org/10.1042/BCJ20160619

Enjalbert F, Rapior S, Nouguier-Soulé J, Guillon S, Amouroux N, Cabot C (2002) Treatment of amatoxin poisoning: 20-year retrospective analysis. J Toxicol Clin Toxicol 40:715–757

Epis S, Matinato C, Gentili G, Varotto F, Bandi C, Sassera D (2010) Molecular detection of poisonous mushrooms in different matrices. Mycologia 102:747–754. https://doi.org/10.3852/09-124

Faulstich H, Fiume L (1985) Protein conjugates of fungal toxins. Methods Enzymol 112:225–237

Filigenzi MS, Poppenga RH, Tiwary AK, Puschner B (2007) Determination of α-amanitin in serum and liver by multistage linear ion trap mass spectrometry. J Agric Food Chem 55:2784–2790. https://doi.org/10.1021/jf063194w

Fiume L, Derenzini M, Marinozzi V, Petazzi F, Testoni A (1973) Pathogenesis of gastro-intestinal symptomatology during poisoning by *Amanita phalloides*. Experientia 29:1520–1521

Floersheim GL, Eberhard M, Tschumi P, Duckert F (1978) Effects of penicillin and silymarin on liver enzymes and blood clotting factors in dogs given a boiled preparation of *Amanita phalloides*. Toxicol Appl Pharmacol 46:455–462

Frank IC, Cummins L (1987) *Amanita* poisoning treated with endoscopic biliary diversion. J Emerg Nurs 13:132–136

Garcia J, Costa VM, Carvalho A, Baptista P, de Pinho PG, de Lourdes BM, Carvalho F (2015) *Amanita phalloides* poisoning: mechanisms of toxicity and treatment. Food Chem Toxicol 86:41–55. https://doi.org/10.1016/j.fct.2015.09.008

Gausterer C, Penker M, Krisai-Greilhuber I, Stein C, Stimpfl T (2014) Rapid genetic detection of ingested *Amanita phalloides*. Forensic Sci Int Genet 9:66–71. https://doi.org/10.1016/j.fsigen.2013.11.002

Gicquel T, Lepage S, Fradin M, Tribut O, Duretz B, Morel I (2014) Amatoxins (α- and β-amanitin) and phallotoxin (phalloidin) analyses in urines using high-resolution accurate mass LC-MS technology. J Anal Toxicol 38:335–340. https://doi.org/10.1093/jat/bku035

Göransson U, Burman R, Gunasekera S, Strömstedt AA, Rosengren KJ (2012) Circular proteins from plants and fungi. J Biol Chem 287:27001–27006. https://doi.org/10.1074/jbc.R111.300129

Grabhorn E, Nielsen D, Hillebrand G, Brinkert F, Herden U, Fischer L, Ganschow R (2013) Successful outcome of severe *Amanita phalloides* poisoning in children. Pediatr Transplant 17:550–555. https://doi.org/10.1111/petr.12108

Haefelé C, Bonfils C, Sauvaire Y (1997) Characterization of a dioxygenase from *Trigonella foenum-graecum* involved in 4-hydroxyisoleucine biosynthesis. Phytochemistry 44:563–566

Hallen HE, Luo H, Scott-Craig JS, Walton JD (2007) Gene family encoding the major toxins of lethal *Amanita* mushrooms. Proc Natl Acad Sci U S A 104:19097–19101. https://doi.org/10.1073/pnas.0707340104

Harper KA, Smart CD, Davis RM (2011) Development of a DNA-based macroarray for the detection and identification of *Amanita* species. J Forensic Sci 56:1003–1009. https://doi.org/10.1111/j.1556-4029.2011.01739.x

Hechler T, Kulke M, Müller C, Pahl A, Anderl J (2014) Amanitin-based antibody-drug conjugates targeting the prostate-specific membrane antigen. In: Proceedings of the 105th annual meeting of the American Association for Cancer Research. Cancer Res 74 (19 Suppl): Abstract 664. doi:https://doi.org/10.1158/1538-7445.AM2014-664

Heinis C (2014) Tools and rules for macrocycles. Nat Chem Biol 10:696–698. https://doi.org/10.1038/nchembio.1605

Hetrick KJ, van der Donk WA (2017) Ribosomally synthesized and post-translationally modified peptide natural product discovery in the genomic era. Curr Opin Chem Biol 38:36–44. https://doi.org/10.1016/j.cbpa.2017.02.005

Hewitt WM, Leung SS, Pye CR, Ponkey AR, Bednarek M, Jacobson MP, Lokey RS (2015) Cell-permeable cyclic peptides from synthetic libraries inspired by natural products. J Am Chem Soc 137:715–721. https://doi.org/10.1021/ja508766b

Jaeger A, Jehl F, Flesch F, Sauder P, Kopferschmitt J (1993) Kinetics of amatoxins in human poisoning: therapeutic implications. J Toxicol Clin Toxicol 31:63–80

Jagadish K, Gould A, Borra R, Majumder S, Mushtaq Z, Shekhtman A, Camarero JA (2015) Recombinant expression and phenotypic screening of a bioactive cyclotide against α-synuclein-induced cytotoxicity in baker's yeast. Angew Chem Int Ed Eng 54:8390–8394. https://doi.org/10.1002/anie.201501186

Javed S, Kohli K, Ali M (2011) Reassessing bioavailability of silymarin. Altern Med Rev 16:239–249

Jetté L, Harvey L, Eugeni K, Levens N (2009) 4-Hydroxyisoleucine: a plant-derived treatment for metabolic syndrome. Curr Opin Investig Drugs 10:353–358

Ji Y, Majumder S, Millard M, Borra R, Bi T, Elnagar AY, Neamati N, Shekhtman A, Camarero JA (2013) In vivo activation of the p53 tumor suppressor pathway by an engineered cyclotide. J Am Chem Soc 135:11623–11633. https://doi.org/10.1021/ja405108p

Kadam RU, Juraszek J, Brandenburg B, Buyck C, Schepens WBG, Kesteleyn B, Stoops B, Vreeken RJ, Vermond J, Goutier W, Tang C, Vogels R, Friesen RHE, Goudsmit J, van Dongen MJP, Wilson IA (2017) Potent peptidic fusion inhibitors of influenza virus. Science 358:496–502. https://doi.org/10.1126/science.aan0516

Kaneko H, Tomomasa T, Inoue Y, Kunimoto F, Fukusato T, Muraoka S, Gonmori K, Matsumoto T, Morikawa A (2001) Amatoxin poisoning from ingestion of Japanese *Galerina* mushrooms. J Toxicol Clin Toxicol 39:413–416

Kaya E, Surmen MG, Yaykasli KO, Karahan S, Oktay M, Turan H, Colakoglu S, Erdem H (2014) Dermal absorption and toxicity of alpha amanitin in mice. Cutan Ocul Toxicol 33:154–160. https://doi.org/10.3109/15569527.2013.802697

Kling A, Lukat P, Almeida DV, Bauer A, Fontaine E, Sordello S, Zaburannyi N, Herrmann J, Wenzel SC, König C, Ammerman NC, Barrio MB, Borchers K, Bordon-Pallier F, Brönstrup M, Courtemanche G, Gerlitz M, Geslin M, Hammann P, Heinz DW, Hoffmann H, Klieber S, Kohlmann M, Kurz M, Lair C, Matter H, Nuermberger E, Tyagi S, Fraisse L, Grosset JH, Lagrange S, Müller R (2015) Targeting DnaN for tuberculosis therapy using novel griselimycins. Science 348:1106–1112. https://doi.org/10.1126/science.aaa4690

Kooter JM, Borst P (1984) Alpha-amanitin-insensitive transcription of variant surface glycoprotein genes provides further evidence for discontinuous transcription in trypanosomes. Nucleic Acids Res 12:9457–9472

Kries H (2016) Biosynthetic engineering of nonribosomal peptide synthetases. J Pept Sci 22: 564–570. https://doi.org/10.1002/psc.2907

Kritzer JA, Hamamichi S, McCaffery JM, Santagata S, Naumann TA, Caldwell KA, Caldwell GA, Lindquist S (2009) Rapid selection of cyclic peptides that reduce α-synuclein toxicity in yeast and animal models. Nat Chem Biol 5:655–663. https://doi.org/10.1038/nchembio.193

Lacombe G, St-Onge M (2016) Towards evidence-based emergency medicine: best BETs from the Manchester royal infirmary. BET 1: silibinin in suspected amatoxin-containing mushroom poisoning. Emerg Med J 33:76–77. https://doi.org/10.1136/emermed-2015-205558.1

Letschert K, Faulstich H, Keller D, Keppler D (2006) Molecular characterization and inhibition of amanitin uptake into human hepatocytes. Toxicol Sci 91:140–149. https://doi.org/10.1093/toxsci/kfj141

Li P, Deng W, Li T (2014) The molecular diversity of toxin gene families in lethal *Amanita* mushrooms. Toxicon 83:59–68. https://doi.org/10.1016/j.toxicon.2014.02.020

Lian W, Upadhyaya P, Rhodes CA, Liu Y, Pei D (2013) Screening bicyclic peptide libraries for protein–protein interaction inhibitors: discovery of a tumor necrosis factor-α antagonist. J Am Chem Soc 135:11990–11995. https://doi.org/10.1021/ja405106u

Lincoff G, Mitchel DH (1977) Toxic and hallucinogenic mushroom poisoning: a handbook for physicians and mushroom hunters. Van Nostrand Reinhold, New York

Liu Y, Zhang X, Han C, Wan G, Huang X, Ivan C, Jiang D, Rodriguez-Aguayo C, Lopez-Berestein G, Rao PH, Maru DM, Pahl A, He X, Sood AK, Ellis LM, Anderl J, Lu X (2015a) TP53 loss creates therapeutic vulnerability in colorectal cancer. Nature 520:697–701. https://doi.org/10.1038/nature14418

Liu Y, Wang L, Lu X (2015b) A new way to target p53-defective colorectal cancer. Future Oncol 11:3101–3104. https://doi.org/10.2217/fon.15.223

Luo H, Hallen-Adams HE, Scott-Craig JS, Walton JD (2012) Ribosomal biosynthesis of α-amanitin in *Galerina marginata*. Fungal Genet Biol 49:123–129. https://doi.org/10.1016/j.fgb.2011.12.005

Luo H, Hong SY, Sgambelluri RM, Angelos E, Li X, Walton JD (2014) Peptide macrocyclization catalyzed by a prolyl oligopeptidase involved in α-amanitin biosynthesis. Chem Biol 21:1610–1617. https://doi.org/10.1016/j.chembiol.2014.10.015

Luo H, DuBois B, Sgambelluri RM, Angelos ER, Li X, Holmes D, Walton JD (2015) Production of ^{15}N-labeled α-amanitin in *Galerina marginata*. Toxicon 103:60–64. https://doi.org/10.1016/j.toxicon.2015.06.018

Maeta K, Ochi T, Tokimoto K, Shimomura N, Maekawa N, Kawaguchi N, Nakaya M, Kitamoto Y, Aimi T (2008) Rapid species identification of cooked poisonous mushrooms by using real time PCR. Appl Environ Microbiol 74:3306–3309. https://doi.org/10.1128/AEM.02082-07

Matinkhoo K, Pryyma A, Todorovic M, Patrick BO, Perrin DM (2018) Synthesis of the death cap mushroom toxin α-amanitin. J Am Chem Soc, in press. https://doi.org/10.1021/jacs.7b12698

Meier-Abt F, Faulstich H, Hagenbuch B (2004) Identification of phalloidin uptake systems of rat and human liver. Biochim Biophys Acta 1664:64–69. https://doi.org/10.1016/j.bbamem.2004.04.004

Mengs U, Pohl RT, Mitchell T (2012) Legalon® SIL: the antidote of choice in patients with acute hepatotoxicity from amatoxin poisoning. Curr Pharm Biotechnol 13:1964–1970. https://doi.org/10.2174/138920112802273353

Miranda E, Nordgren IK, Male AL, Lawrence CE, Hoakwie F, Cuda F, Court W, Fox KR, Townsend PA, Packham GK, Eccles SA, Tavassoli A (2013) A cyclic peptide inhibitor of HIF-1 heterodimerization that inhibits hypoxia signaling in cancer cells. J Am Chem Soc 135:10418–10425. https://doi.org/10.1021/ja402993u

Mitchell ST, Olson KR (2008) Intravenous silibinin in the management of amatoxin poisoning: first usage in an American cohort. Clin Toxicol 46:396. (Abstract)

Moldenhauer G, Salnikov AV, Lüttgau S, Herr I, Anderl J, Faulstich H (2012) Therapeutic potential of amanitin-conjugated anti-epithelial cell adhesion molecule monoclonal antibody against pancreatic carcinoma. J Natl Cancer Inst 104:622–634. https://doi.org/10.1093/jnci/djs140

Mootz HD (ed) (2017) Split inteins: methods and protocols, Meth Mol Biol, vol 1495. Springer, New York. https://doi.org/10.1007/978-1-4939-6451-2

Mottram AR, Lazio MP, Bryant SM (2010) *Lepiota subincarnata* J.E. Lange induced fulminant hepatic failure presenting with pancreatitis. J Med Toxicol 6:155–157. https://doi.org/10.1007/s13181-010-0062-1

Mühlbach HP, Sänger HL (1979) Viroid replication is inhibited by alpha-amanitin. Nature 278:185–188

Muszyńska K, Ostrowska D, Bartnicki F, Kowalska E, Bodaszewska-Lubaś M, Hermanowicz P, Faulstich H, Strzałka W (2017) Selection and analysis of a DNA aptamer binding α-amanitin from *Amanita phalloides*. Acta Biochim Pol 64:401–406. https://doi.org/10.18388/abp.2017_1615

Nguyen GK, Wang S, Qiu Y, Hemu X, Lian Y, Tam JP (2014) Butelase 1 is an Asx-specific ligase enabling peptide macrocyclization and synthesis. Nat Chem Biol 10:732–738. https://doi.org/10.1038/nchembio.1586

Nguyen GK, Kam A, Loo S, Jansson AE, Pan LX, Tam JP (2015) Butelase 1: a versatile ligase for peptide and protein macrocyclization. J Am Chem Soc 137:15398–15401. https://doi.org/10.1021/jacs.5b11014

Nguyen GK, Hemu X, Quek JP, Tam JP (2016) Butelase-mediated macrocyclization of D-amino-acid-containing peptides. Angew Chem Int Ed Eng 55:12802–12806. https://doi.org/10.1002/anie.201607188

Niedermeyer TH, Daily A, Swiatecka-Hagenbruch M, Moscow JA (2014) Selectivity and potency of microcystin congeners against OATP1B1 and OATP1B3 expressing cancer cells. PLoS One 10:e91476. https://doi.org/10.1371/journal.pone.0091476

Olson KR, Pond SM, Seward J, Healey K, Woo OF, Becker CE (1982) *Amanita phalloides*-type mushroom poisoning. West J Med 137:282–289

Padojino J (2016) https://sfbay.ca/2016/12/08/hospitals-see-uptick-in-wild-mushroom-poisoning/

Pálfi A, Hechler T, Müller C, Pahl A, Kulke M (2016) CD269 – a promising target for amanitin based ADCs. In: Proceedings of the 107th annual meeting of the American Association for Cancer Research. Cancer Res 76 (14 Suppl): Abstract 2973. doi: https://doi.org/10.1158/1538-7445.AM2016-2973

Parmley S (2014) Resurrecting a magic bullet. SciBX 7. https://doi.org/10.1038/scibx.2014.1397

Pulman JA, Childs KL, Sgambelluri RM, Walton JD (2016) Expansion and diversification of the MSDIN family of cyclic peptide genes in the poisonous agarics *Amanita phalloides* and *A. bisporigera*. BMC Genomics 17:1038. https://doi.org/10.1186/s12864-016-3378-7

Quartararo JS, Wu P, Kritzer JA (2017) Peptide bicycles that inhibit the Grb2 SH2 domain. Chembiochem 13:1490–1496. https://doi.org/10.1002/cbic.201200175

Rhodes CA, Pei D (2017) Bicyclic peptides as next-generation therapeutics. Chemistry 23:12690–12703. https://doi.org/10.1002/chem.201702117

Robinson-Fuentes VA, Jaime-Sánchez JL, García-Aguilar L, Gómez-Peralta M, Vázquez-Garcidueñas MS, Vázquez-Marrufo G (2008) Determination of α- and β-amanitin in clinical urine samples by capillary zone electrophoresis. J Pharm Biomed Anal 47:913–917. https://doi.org/10.1016/j.jpba.2008.03.032

Sable R, Durek T, Taneja V, Craik DJ, Pallerla S, Gauthier T, Jois S (2016) Constrained cyclic peptides as immunomodulatory inhibitors of the CD2:CD58 protein-protein interaction. ACS Chem Biol 11:2366–2374. https://doi.org/10.1021/acschembio.6b00486

Santi L, Maggioli C, Mastroroberto M, Tufoni M, Napoli L, Caraceni (2012) Acute liver failure caused by *Amanita phalloides* poisoning. Int J Hepatol 2012:487480. https://doi.org/10.1155/2012/487480

Sgambelluri RM, Epis S, Sassera D, Luo H, Angelos ER, Walton JD (2014) Profiling of amatoxins and phallotoxins in the genus *Lepiota* by liquid chromatography combined with UV absorbance and mass spectrometry. Toxins 6:2336–2347. doi: 10.3390/toxins6082336

Sgambelluri RM, Smith MO, Walton JD (2018) Versatility of prolyl oligopeptidase B in peptide macrocyclization. ACS Synth Biol 7:145–152. https://doi.org/10.1021/acssynbio.7b00264

Sievers EL, Senter PD (2013) Antibody-drug conjugates in cancer therapy. Annu Rev Med 64: 15–29. https://doi.org/10.1146/annurev-med-050311-201823

Smith MR, Davis RL (2016) Mycetismus: a review. Gastroenterol Rep 4:107–112. https://doi.org/10.1093/gastro/gov062

Smith JM, Frost JR, Fasan R (2013) Design, synthesis, and diversification of ribosomally derived peptide macrocycles. Curr Opin Struct Biol 23:571–580. https://doi.org/10.1016/j.sbi.2013.06.015

Spoerke DG, Rumack BH (eds) (1994) Handbook of mushroom poisoning: diagnosis and treatment. CRC Press, Boca Raton

Tang S, Zhou Q, He Z, Luo T, Zhang P, Cai Q, Yang Z, Chen J, Chen Z (2016) Cyclopeptide toxins of lethal amanitas: compositions, distribution and phylogenetic implication. Toxicon 120:78–88. https://doi.org/10.1016/j.toxicon.2016.07.018

Tavassoli A (2017) SICLOPPS cyclic peptide libraries in drug discovery. Curr Opin Chem Biol 38:30–35. https://doi.org/10.1016/j.cbpa.2017.02.016

Teicher BA (2014) Antibody drug conjugates. Curr Opin Oncol 26:476–483. https://doi.org/10.1097/CCO.0000000000000108

Theologis A, Huynh TV, Davis RW (1985) Rapid induction of specific mRNAs by auxin in pea epicotyl tissue. J Mol Biol 183:53–68

Tomková J, Ondra P, Válka I (2015) Simultaneous determination of mushroom toxins α-amanitin, β-amanitin and muscarine in human urine by solid-phase extraction and ultra-high-performance liquid chromatography coupled with ultra-high-resolution TOF mass spectrometry. Forensic Sci Int 251:209–213. https://doi.org/10.1016/j.forsciint.2015.04.007

Truman AW (2016) Cyclisation mechanisms in the biosynthesis of ribosomally synthesised and post-translationally modified peptides. Beilstein J Org Chem 12:1250–1268. https://doi.org/10.3762/bjoc.12.120

Udugamasooriya DG, Spaller MR (2008) Conformational constraint in protein ligand design and the inconsistency of binding entropy. Biopolymers 89:653–667. https://doi.org/10.1002/bip.20983

Upadhyaya P, Qian Z, Selner NG, Clippinger SR, Wu Z, Briesewitz R, Pei D (2015) Inhibition of Ras signaling by blocking Ras–effector interactions with cyclic peptides. Angew Chem Int Ed 54:7602–7606. https://doi.org/10.1002/anie.201502763

Varvenne D, Retornaz K, Metge P, De Haro L, Minodier P (2015) Amatoxin-containing mushroom (*Lepiota brunneoincarnata*) familial poisoning. Pediatr Emerg Care 31:277–288. https://doi.org/10.1097/PEC.0000000000000399

Veber DF, Johnson SR, Cheng HY, Smith BR, Ward KW, Kopple KD (2002) Molecular properties that influence the oral bioavailability of drug candidates. J Med Chem 45:265–2623. On-line analysis at www.molinspiration.com

Vo KT, Montgomery ME, Mitchell ST, Scheerlinck PH, Colby DK, Meier KH, Kim-Katz S, Anderson IB, Offerman SR, Olson KR, Smollin CG (2017) *Amanita phalloides* mushroom poisonings – Northern California, December 2016. MMWR Morb Mortal Wkly Rep 66:549–553. https://doi/org/10.15585/mmwr.mm6621a1

Vogel G, Tuchweber B, Trost W, Mengs U (1984) Protection by silibinin against *Amanita phalloides* intoxication in beagles. Toxicol Appl Pharmacol 73:355–362

Wang CK, Craik DJ (2016) Cyclic peptide oral bioavailability: lessons from the past. Biopolymers 106:901–909. https://doi.org/10.1002/bip.22878

Wang M, Xie T, Chang Z, Wang L, Xie X, Kou Y, Xu H, Gao X (2015) A new type of liquid silymarin proliposome containing bile salts: its preparation and improved hepatoprotective effects. PLoS One 10:e0143625. https://doi.org/10.1371/journal.pone.0143625

Ward J, Kapadia K, Brush E, Salhanick SD (2013) Amatoxin poisoning: case reports and review of current therapies. J Emerg Med 44:116–121. https://doi.org/10.1016/j.jemermed.2012.02.020

Weidmann J, Craik DJ (2016) Discovery, structure, function, and applications of cyclotides: circular proteins from plants. J Exp Bot 67:4801–4812. https://doi.org/10.1093/jxb/erw210

White AM, Craik DJ (2016) Discovery and optimization of peptide macrocycles. Expert Opin Drug Discovery 11:1151–1163. https://doi.org/10.1080/17460441.2016.1245720

Wieczorek Z, Siemion IZ, Zimecki M, Bolewska-Pedyczak E, Wieland T (1993) Immunosuppressive activity in the series of cycloamanide peptides from mushrooms. Peptides 14:1–5

Wieland T (1986) Peptides of poisonous *Amanita* mushrooms. Springer, New York

Wieland T, Wieland O (1972) The toxic peptides of *Amanita* species. In: Kadis S, Ciegler A, Ajl SJ (eds) Microbial toxins, vol 8. Academic, New York, pp 249–280

Wołoszyn A, Kotłowski R (2017) A universal method for the identification of genes encoding amatoxins and phallotoxins in poisonous mushrooms. Rocz Panstw Zakl Hig 68:247–251

Zafar MI, Gao F (2016) 4-Hydroxyisoleucine: a potential new treatment for type 2 diabetes mellitus. BioDrugs 30:255–262. https://doi.org/10.1007/s40259-016-0177-2

Zhang P, Chen Z, Hu J, Wei B, Zhang Z, Hu W (2005) Production and characterization of amanitin toxins from a pure culture of *Amanita exitialis*. FEMS Microbiol Lett 252:223–228. https://doi.org/10.1016/j.femsle.2005.08.049

Zipperer A, Konnerth MC, Laux C, Berscheid A, Janek D, Weidenmaier C, Burian M, Schilling NA, Slavetinsky C, Marschal M, Willmann M, Kalbacher H, Schittek B, Brötz-Oesterhelt H, Grond S, Peschel A, Krismer B (2016) Human commensals producing a novel antibiotic impair pathogen colonization. Nature 535:511–516. https://doi.org/10.1038/nature18634

Zorzi A, Deyle K, Heinis C (2017) Cyclic peptide therapeutics: past, present, and future. Curr Opin Chem Biol 38:24–29. http://dx.doi.org/10.1016/j.cbpa.2017.02.006

Zuliani AM, Kabar I, Mitchell T, Heinzow HS (2016) Akutes Leberversagen nach Knollenblätterpilz-Ingestion. Dtsch Med Wochenschr 141:940–942. https://doi.org/10.1055/s-0042-103987

Chapter 8
Future Outlook

In the Preface, a series of commonly asked questions regarding the biology and chemistry of the *Amanita* cyclic peptide toxins were posed, questions that were an important impetus for writing this book. However, despite an impressively major expansion of our knowledge in the past three decades, some of these key questions remain only partially or fully unanswered.

Which of these questions are the most important, and which are the most likely to be solved in the coming years? As Peter Medawar (1968) wrote, "Science is the art of the soluble," by which he didn't mean the challenge of finding the right solvent for your favorite cycloamanide, but rather that often the most profound questions in science are not answerable with the available techniques, and therefore we must settle for choosing questions that we *can* answer. Such is the march of scientific progress that we must often be satisfied to narrow the possible answers to our questions, for example, by proving some to be false, and thereby we converge over time on the most probable or least unlikely explanations.

The relative importance of questions in any research field is subjective to some extent, so the following list strongly reflects, without apology, the author's perspective. Other scientists would, with perfect legitimacy, give these questions different priorities or even pose a different set.

- **What adaptive advantages do the peptide toxins confer on the producing fungi?** This age-old question in amatoxicology remains unsatisfactorily answered. Furthermore, in light of recent progress on the genetics of cyclic peptide biosynthesis in *Amanita*, *Galerina*, and *Lepiota*, we have to ask this question not only of the amatoxins and phallotoxins but of the large cycloamanide superfamily as well. In the absence of being able to genetically manipulate the fungi and study the behavior of toxin mutants in the wild, we may never have a definitive answer. However, there are a number of interesting experiments that can be done that would provide relevant information. For example, it should be possible with modern molecular genetics and biochemistry to elucidate the mechanism or mechanisms of resistance to amatoxins among wild mycophagous fruit flies. Is it due to a general detoxification strategy or something more specific? Another way to approach the role of the toxins in mediating interactions with other organisms might be to observe the presence of insects and other potential mycovores on

J. Walton, *The Cyclic Peptide Toxins of Amanita and Other Poisonous Mushrooms*, https://doi.org/10.1007/978-3-319-76822-9_8

natural toxin-free specimens of otherwise poisonous mushrooms, such as *A. bisporigera*, compared to toxin-producing siblings growing nearby. Spontaneous loss of toxin biosynthesis is not rare in the *A. bisporigera* species complex, and individual mushrooms could be monitored in the field with the Wieland-Meixner assay. To address the question whether any mollusks are resistant to amatoxins, slugs could be collected from known toxin-producing mushrooms and kept for observation and/or controlled feeding studies in the laboratory.

Ecologically, it may be overly simplistic to think of the functions of the cycloamanides including amatoxins and phallotoxins in terms of binary interactions. It is now appreciated that plant and microbial secondary metabolites can mediate tritrophic and even higher levels of ecological interactions. For example, more consideration should be given to possible *Amanita*/*Drosophila*/nematode interactions, or possibly *Amanita*/mollusk/mollusk predators. In a tritrophic interaction, amanitin could be turned to the primary consumer's advantage in the context of its struggle with its own ecological antagonists. For example, perhaps flies or other mycophagous insects accumulate amanitin in their bodies as a mechanism of protection against parasites or predators, as suggested by Jaenike (1985); this possibility is easily testable with modern analytical methods. It might also be necessary to consider not only the possible functions of the toxins in fresh basidiocarps but also in living mycelium or in the soil.

It is not yet possible to formulate any but the most speculative hypotheses about the possible functions of the monocyclic cycloamanides in light of how little we know about their biological activities. However, considering that some species of *Amanita* have the genetic potential to make more than 30 cycloamanides and that there is very little overlap between species (Pulman et al. 2016), perhaps we should think beyond the contribution to mushroom survival of single compounds but rather of combinations of compounds. The capacity to generate rapidly evolving combinatorial mixtures of cycloamanides might have its own evolutionary rationale. To explore these possibilities, one need to first determine how many of the cycloamanides in the various species of *Amanita* and *Lepiota* are actually biosynthesized, which is a tractable question with current analytical methods.

- **What is the origin of the cycloamanide gene family, and what evolutionary forces are acting on it to create and maintain its diversity?** Most "normal" protein-encoding genes can be traced back evolutionarily to distantly related ancestors, but the cycloamanide gene family has no homologs in any cycloamanide nonproducing organism. How, when, and where did these genes apparently spring from nowhere?

  A particularly intriguing question is how amanitin biosynthesis arose in three unrelated subgeneric lineages of agarics. Was it the result of convergent evolution, horizontal gene transfer, loss from intervening taxa, or perhaps a combination of these? Nothing excludes the possibility that one or more of the amanitin-producing fungi were preadapted to accommodate the amanitin precursor peptides by having a promiscuous prolyl oligopeptidase (POP) with incipient macrocyclase potential. It is intriguing in this regard that a POP is also involved in the biosynthesis of omphalotin A, another agaric RiPP (Ramm et al. 2017). Identification of the accessory enzymes and their encoding genes that catalyze the hydroxylations, bicyclization, and sulfoxidation of the amatoxins could help decipher the evolutionary trajectory, especially if such studies were to indicate that these steps are not catalyzed by evolutionarily related (homologous) enzymes in the three agarics.

- **What are the phylogenetic and chemical relationships among the toxin-producing fungi, especially in the large and rapidly evolving genus *Amanita*?** A relevant sub-question is the breadth of the genetic diversity of the cycloamanide family in *Amanita*. More genome sequencing of additional species of *Amanita*, which is in progress in several labs as this book is being written, will help reveal the extent of genus-wide diversity. Another project underway is to sequence multiple (even thousands) of specimens of the same toxin-producing species, which will reveal the degree of variation in the cycloamanide gene family within a species over short evolutionary times. Analysis of the degree of clustering of the cycloamanide genes in different species will help understand how the gene family is expanding and diversifying. More, and more sophisticated, chemical analyses would indicate how many of the cycloamanide genes are expressed in any particular species of *Amanita* or *Lepiota*. Although it has not been observed to date, it is possible that some of the predicted cycloamanides undergo posttranslational modifications such as α-carbon epimerization or hydroxylation, as occurs in the amatoxins and phallotoxins.

  In order to make sense of cycloamanide gene family diversity and expansion, we need definitive resolution of the number of species of *Amanita* sect. *Phalloideae*, their phylogeny, and their geographic distribution. It would be especially informative to resolve the *A. bisporigera* species complex in North American. With participation of amateur mycology clubs, voucher specimen certification, ITS sequencing, and chemical fingerprinting, this is a feasible goal.

- **How are the remaining steps in cycloamanide biosynthesis catalyzed?** The cycloamanide families in *Amanita*, *Galerina*, and *Lepiota* are all probably cyclized by cognate prolyl oligopeptidases. For the classic cycloamanides, the biosynthetic pathway ends with monocyclization, but the amatoxins and phallotoxins undergo several additional modifications for which we do not yet know the genes or enzymes. Comparative genomics of the three amanitin-producing genera of agarics might help identify the involved genes. The function(s) of candidate genes could be tested by the time-proven but laborious processes of heterologous expression combined with in vitro enzyme assays. Targeted gene disruption to test the involvement of specific genes in the pathway will remain problematic until a toxin-producing fungus that grows faster than *Galerina* is found, which seems unlikely. It is important to keep an open mind about what the genes and enzymes might look like, e.g., epimerization and hydroxylation can be catalyzed by multiple classes of enzymes. Furthermore, the other steps are not necessarily catalyzed by homologous enzymes in the three agarics, because cycloamanide biosynthesis may have evolved by convergent evolution.

A particularly intriguing gap in our knowledge concerns the stereochemistry of hydroxyPro (Hyp) in the amatoxins and phallotoxins. The amatoxins have the common *trans* isomer of Hyp, whereas the phallotoxins have the unusual *cis* isomer. Both hydroxylations seem to be catalyzed by stereospecific enzymes, so could the same enzyme or class of enzyme (e.g., P450, dioxygenase) catalyze both reactions? Unhydroxylated and *trans*-hydroxylated phallotoxins are inactive, but it is apparently not known if *cis*-Hyp-amanitin is active or not. Furthermore, it seems that we do not know with certainty that the Hyp hydroxylations in the amatoxins from *Lepiota* and *Galerina* are *trans* like they are in *Amanita*.

- **Why are some DNA-dependent RNA polymerases resistant to amatoxins?** Even yeast pol II, for which we know the structure complexed with amanitin, is for unknown reasons tenfold less sensitive than mammalian pol II. The basis of pol II resistance in protists is also not fully understood, nor is the basis of resistance of mammalian pol I and pol III. Is the strong resistance of pol II from *Aspergillus nidulans* and the partial resistance of *Agaricus bisporus* pol II characteristic of other ascomycetes and basidiomycetes? It is theoretically possible to answer this question, albeit technically challenging, insofar as it would require stringent purification of pol II from nonmodel organisms. Immuno-affinity chromatography based on antibodies against the highly conserved C-terminal domain (CTD) found in all pol II's might provide a path to the rapid and specific separation of pol II from pol I and pol III (Zaborowska et al. 2016). [^{15}N]-Amanitin could be helpful in studying the interaction between amanitin and the different RNA polymerases by NMR (Luo et al. 2015).
- **What is the basis of self-protection?** That is, why don't amatoxin- and phallotoxin-producing fungi kill themselves? A plausible explanation, albeit one supported less by experimental evidence than by elimination of alternatives, is sequestration into specialized intracellular compartments (Luo et al. 2010). This hypothesis could be addressed in greater detail with modern microscopic and cell fractionation methods. Sequestration could be tied to biosynthesis, i.e., some or all of the enzymes might be in specialized compartments, as has been found for trichothecene mycotoxins in *Fusarium graminearum* (Boenisch et al. 2017; Kistler and Broz 2015). Sequestration could thus provide a dual function: self-protection as well as more efficient biosynthesis through metabolic channeling.
- **Is there a single superior approach to the treatment of amatoxin poisoning?** Recent results with Legalon SIL®, when combined with treatments that promote general liver and kidney function, suggest that we are, in fact, reaching that point.
- **Will our knowledge of the cyclic peptide toxins and their biosynthetic pathway lead to practical applications?** Within the next few years, we should know if the amatoxins can be harnessed as anticancer payloads in antibody-drug conjugates

and if the enzymes of the biosynthetic pathway can be exploited to produce libraries of monocyclic and bicyclic peptides for a variety of potential uses. Should amanitin prove useful as an ADC payload, there will be a need for a source that is more reliable than extraction from wild mushrooms. Recent advances in synthesis (Matinkhoo et al. 2018) and biosynthesis (Sgambelluri et al. 2018), or a combination of the two, hold promise for alleviating this bottleneck.

The study of the cyclic peptide toxins of *Amanita* and other poisonous mushrooms is still very much open for future scientists to make significant contributions.

References

Boenisch MJ, Broz KL, Purvine SO, Chrisler WB, Nicora CD, Connolly LR, Freitag M, Baker SE, Kistler HC (2017) Structural reorganization of the fungal endoplasmic reticulum upon induction of mycotoxin biosynthesis. Sci Rep 13:44296. https://doi.org/10.1038/srep44296

Jaenike J (1985) Parasite pressure and the evolution of amanitin tolerance in *Drosophila*. Evolution 39:1295–1301. https://doi.org/10.1111/j.1558-5646.1985.tb05695.x

Kistler HC, Broz K (2015) Cellular compartmentalization of secondary metabolism. Front Microbiol 6:68. https://doi.org/10.3389/fmicb.2015.00068

Luo H, Hallen-Adams HE, Scott-Craig JS, Walton JD (2010) Co-localization of amanitin and a candidate toxin-processing prolyl oligopeptidase in *Amanita* basidiocarps. Eukaryot Cell 9:1891–1900. https://doi.org/10.1128/EC.00161-10

Luo H, DuBois B, Sgambelluri RM, Angelos ER, Li X, Holmes D, Walton JD (2015) Production of ^{15}N-labeled α-amanitin in *Galerina marginata*. Toxicon 103:60–64. https://doi.org/10.1016/j.toxicon.2015.06.018

Matinkhoo K, Pryyma A, Todorovic M, Patrick BO, Perrin DM (2018) Synthesis of the death cap mushroom toxin α-amanitin. J Am Chem Soc, in press. https://doi.org/10.1021/jacs.7b12698

Medawar PB (1968) The art of the soluble. Methuen, London

Pulman JA, Childs KL, Sgambelluri RM, Walton JD (2016) Expansion and diversification of the MSDIN family of cyclic peptide genes in the poisonous agarics *Amanita phalloides* and *A. bisporigera*. BMC Genomics 17:1038. https://doi.org/10.1186/s12864-016-3378-7

Ramm S, Krawczyk B, Mühlenweg A, Poch A, Mösker E, Süssmuth RD (2017) A self-sacrificing N-methyltransferase is the precursor of the fungal natural product omphalotin. Angew Chem Int Ed Eng 56:9994–9997. https://doi.org/10.1002/anie.201703488

Sgambelluri RM, Smith MO, Walton JD (2018) Versatility of prolyl oligopeptidase B in peptide macrocyclization. ACS Synth Biol 7:145–152. doi:10.1021/acssynbio.7b00264

Zaborowska J, Egloff S, Murphy S (2016) The pol II CTD: new twists in the tail. Nat Struct Mol Biol 23:771–777. https://doi.org/10.1038/nsmb.3285

Index

J. Walton, *The Cyclic Peptide Toxins of Amanita and Other Poisonous Mushrooms*, https://doi.org/10.1007/978-3-319-76822-9

P

R

The manufacturer's authorised representative in the EU is Springer Nature Customer Service Centre GmbH, Europaplatz 3, 69115 Heidelberg, Germany. If you have any concerns regarding our products, please contact ProductSafety@springernature.com

Printed and bound by CPI Group (UK) Ltd, Croydon, CR0 4YY
30/09/2026
02225658-0006